Planning and Executing Credible Experiments

Planning and Executing Credible Experiments

A Guidebook for Engineering, Science, Industrial Processes, Agriculture, and Business

Robert J. Moffat
Stanford University, USA

Roy W. Henk, Ph.D., P.E.
Kyoto University, Japan (retired)

This edition first published 2021

Registered Offices
John Wiley & Sons, Inc., 111 River Street, Hoboken, NJ 07030, USA
John Wiley & Sons Ltd, The Atrium, Southern Gate, Chichester, West Sussex, PO19 8SQ, UK

Editorial Office
The Atrium, Southern Gate, Chichester, West Sussex, PO19 8SQ, UK

For details of our global editorial offices, customer services, and more information about Wiley products visit us at www.wiley.com.

Wiley also publishes its books in a variety of electronic formats and by print-on-demand. Some content that appears in standard print versions of this book may not be available in other formats.

Library of Congress Cataloging-in-Publication data applied for
ISBN HB: 9781119532873

Cover Design: Wiley
Cover Image: © NASA

Set in 10/12pt Warnock by SPi Global, Pondicherry, India
Printed and bound by CPI Group (UK) Ltd, Croydon, CR0 4YY

10 9 8 7 6 5 4 3 2 1

In memory of Lou London
RJM

For Cherrine and Neal and those they love
RWH

Robert J. Moffat
Stanford University, USA

Roy W. Henk
Kyoto University, Japan

Contents

Detailed Contents

About the Authors

Dr. Robert J. Moffat is Professor Emeritus at Stanford University and former President of Moffat Thermosciences, Inc. Professor Moffat started his professional career at the General Motors Research Laboratories in the Gas Turbine Laboratory. He and a small group of engineers designed, built, and tested a high efficiency, two-spool regenerative gas turbine that, starting from 30 below zero Fahrenheit, could deliver 350 HP in less than 2 minutes. He graduated from the University of Michigan and received his Master of Science at Wayne State University. Enrolling at Stanford University, he earned degrees in Mechanical Engineering as Master of Science, Engineer, and Ph.D. He became a Stanford professor and served as chairman of the Thermosciences Division for 13 years.

Professor Moffat's research efforts have involved three areas: convective heat transfer in engineering systems, experimental methods in heat transfer and fluid mechanics, and biomedical thermal issues. His largest body of work concerns convective heat transfer. One program focused on gas turbine blade heat transfer. A second program aimed at convective cooling of electronic components covering forced, free and mixed convection. Several contributions arose including invariant descriptors, a new heat transfer coefficient for electronics cooling, and a simple correlation based on turbulence intensity.

His second area of research concerned experimental methods in the thermosciences namely full-field imaging techniques for temperature, heat flux, and heat transfer coefficient measurement using thermochromic liquid crystals. He contributed regularly to the theory of uncertainty analysis. Dr. Moffat was an invited lecturer for 40 consecutive years in the Measurement Engineering Series, for more than 20 years in the Instrument Society of America Test Measurements Division and, for ten years in the ASME Professional Development program.

Dr. Moffat worked on biomedical engineering problems, in particular the thermal protection of newborn infants. He jointly developed a self-contained, portable incubator which provided a neutral thermal environment for the infant while allowing free access by the attending physicians. Used on almost every continent where cold-weather transport is needed, it received the ASME Holley Medal Award, 1987. He founded Moffat Thermosciences as a vehicle for consulting, research, and teaching in Heat Transfer and Experimental Methods. He delivered short courses in Electronics Cooling, Experimental Methods, and Uncertainty Analysis.

Dr. Roy W. Henk, professor in the Graduate School of Energy Science at Kyoto University, Japan, earned his bachelors degree at Virginia Tech and his masters and doctorate at Stanford University. Professor Henk taught courses within natural and experimental philosophy, currently known as classical physics and mechanical engineering.

Professor Henk's experience includes industry, government labs, and academia. His work included wind tunnel tests at Virginia Tech and Japan's Mach 5 tunnel at NAL, and water tunnels at the U.S. Naval Research Lab and at Stanford. He worked internationally in the aerospace industry, designing and testing advanced engine components with IHI (石川島播磨重工業株式会社). He spearheaded improvements to the turbine engine design process. A registered Professional Engineer, Dr. Henk is keenly interested in appropriate energy.

Dr. Henk's recent work focused on experimental methods. He has done diverse experiments, from field work on environmental flows, to tests in pristine laboratories, to materials tests for structures. He founded Royal HanMi 로열韓美 to promote international energy science and design of experiments.

Professor Henk has taught at universities, public and private, internationally as well as in the USA. Two of Dr. Henk's courses, Experiment Design and Statistical Modeling, drew students from the medical school, business, engineering and environmental schools. Researchers learned advanced strategies to select data and how to draw strong defensible conclusions from data.

Together with colleagues at Kyoto University (京都大學), Science University of Tokyo, IHI, and Tokyo University they helped the nation brave the impact of the Fukushima Tsunami. Together with colleagues at LeTourneau, Virginia Military Institute, Handong (韓東大學校, 한동대학교) University, he trained our future. Dr. Henk has served within the Commonwealth of Virginia's Governor's Schools and STEM Academies.

Dr. Henk's youth science book, UnLock Rocks, explores how rocks and crystals reveal the age of the Earth. Simple experiments, using kitchen ingredients, make concepts tangible and tasty. Two books were published in South Korea.

Titles
UnLock Rocks
Physics 1 Lab with Experiment Design (RoK)
Physics 2 Lab with Experiment Design (RoK)

with R.J. Moffat
Planning and Executing Credible Experiments

Preface

Laboratories (and businesses) need people who can answer difficult questions experimentally. If data are needed for decision-making, this book is for you.

Science, medicine, the environment, agriculture, and engineering depend heavily on experiments. Business does as well, when surveying the added complexity of human taste and markets. In any of these fields, an experiment must be credible or it wastes time and resources. This book offers a tested guide, developed and used for decades at Stanford University, to equip novice researchers and experienced technicians to plan and execute credible experiments.

This book prepares the reader to anticipate the choices one will face from the launch of your project to the final report. We come alongside the reader, emphasizing the strategies of experiment planning, execution, and reporting.

The foundation of this book originated in "Planning Experimental Programs," a set of class notes by Robert Moffat, our first author. Our second author, Roy Henk, is one of thousands whose own experiments benefited from Moffat's notes. This book blends our approaches to developing and planning experimental programs.

Bob Moffat: Our goal for this book was to collect, organize, summarize, and present what we think are important ideas about developing, running, and reporting reproducible, provably accurate experiments – with minimum wasted effort.

My contribution has to do with the process of developing experimental programs that can be trusted to produce trustworthy data. I used the word "developing" because "new" experiments generally require quite a few changes to eliminate unforeseen (but finally recognized) errors before they qualify as "good" experiments. This is called "the shakedown phase."

This raises a question, however: how can you spot an error in a "new" experiment? The only way I know is to run a test for which you already know the answer. We have the three conservation laws we can trust (conservation of mass, momentum, and energy), but they are a bit hard to work with. I think it generally requires less work to simply stretch the operable domain of the proposed experiment until it includes some conditions that have already yielded data that you trust. That data become the "qualification test" for your experiment.

I've been designing and running experiments for more than 60 years, and my experience has been that the development of a good experiment is always iterative. It takes a lot of work during the shakedown period to eliminate all the sources of error. Surprisingly, this aspect of experiment planning isn't generally "taught." That's really what started this book! The suggestions presented here reflect things I learned "on the job" during my first 10 years in industry and then refined during the rest of my career at Stanford.

When I got my BSME from the University of Michigan, I was offered a job in the General Motors (GM) Research Laboratories (GMR). I stayed with GMR almost exactly 10 years, rising from working on little problems like calibrating instruments to developing a high-precision, high-temperature wind tunnel for testing aircraft engine thermocouple probes up to 1600 °F, finding what caused the bearings to fail on an experimental engine, and developing high-effectiveness heat exchangers for gas turbine engines. In the last two years of my stay, I was part of a small group of engineers who designed, built, and tested a high-efficiency, two-spool regenerative gas turbine that, starting from 30° below zero Fahrenheit, could deliver 350 HP in less than two minutes. That was developed for GM trucks and buses, but a descendant of that engine powered the US Army's M1A1 battle tank – the one designed for cold weather.

The increasingly interesting problems I encountered in the last three or four years made it clear to me that I needed to go back to school – there was too much that I didn't know.

I applied to Stanford with references from GMR and was offered a scholarship, which lead to a PhD, a Stanford professorship, 15 years as head of the mechanical engineering department, and 25 years of research and teaching on engineering problems. In my spare time, I worked with Dr. Alvin Hackel, a Stanford pediatrician, to develop the Stanford Transport Incubator. Dr. Hackel provided the medical insights and I did the engineering. His patients, usually newborn and sometimes premature, often needed intensive medical care during transport between hospitals. That incubator won the American Society of Mechanical Engineers' 1987 Holley Medal for Service for the Benefit of Mankind and is displayed on the cover.

I retired and began to put this book together.

Now a word from my colleague, Roy Henk.

Roy Henk: With the aim to continually improve, experiments can fill one's personal daily life. My experimenting began on a small farm with vegetables and livestock (plus honeybees). Later experiments involved concrete mixing, internal combustion engines, molten metal, and measuring ore. At Virginia Tech, I joined low-noise precision wind-tunnel tests on flow transition; I have used wind tunnels at Stanford and in Japan at the National Aerospace Laboratory of Japan (NAL), even joining tests in a Mach 5 tunnel. NAL together with JAXA is Japan's equivalent to NASA. My research used water tunnels at the US Naval Research Lab and at Stanford, and Bob's notes. Ten years in industry had me designing advanced components of commercial aerospace engines. Compared to air flow around an airplane, thermo-fluid flow inside an aerospace engine is about the most challenging area of classical physics. Those engines have flown on aircraft for years without incident.

I have thrilled with experimental success, yet failures too provided valuable data and lessons. For example, as a young researcher I did not realize the advantage of randomizing the order of data collection. How did this matter? In the middle of one particularly costly data run, a valve failed staying open. Had I randomized, I would have had two valuable datasets at lower resolution. Instead that data had little use. Design of Experiments spared me from much learning the hard way, like ancient wisdom.

No longer a bother, Uncertainty Analysis now informs every stage, fortifying results, averting disaster. To postpone uncertainty analysis starves one's experiment.

While teaching Design of Experiments, I asked Bob Moffat about his book. Bob graciously allowed his notes to supplement my course. Over 18 years, I had developed notes on certain powerful, open-source software for experiment design and analysis. Bob at Stanford and I, now as professor of energy science at Kyoto University, began exploring how to merge our individual works into a single book. This text is the fruit of our efforts.

This book serves as a generalist guide to experiment planning, execution, and analysis. The text also introduces two powerful, free, open-source software tools: Gosset to optimize experiment design, and R for statistical computing. This book addresses renewed demands by the public and science community that science be credible. Recently the purported reliability of many landmark medical studies has been undermined due to poor experiment design, execution, or analysis. Separate studies could not reproduce their results. Furthermore, a rash of scientific papers have been retracted due to fraud. Credible science needs a solid new footing.

Audience

We encourage readers to use our text while planning and executing an actual experiment. If the reader already has a problem that must be answered by experiment, that provides the best motivation for each chapter. Each chapter proposes questions that an experimenter will need to ask and answer during each stage of planning and execution.

Major portions of this text have been class-tested at Stanford in graduate and undergraduate mechanical engineering courses for decades (since the late 1970s). This text has also been class-tested internationally, serving engineering, physics, chemistry, agriculture, industrial processes, medical, and business students. Drafts of this book have been used for a continuing education course by researchers at the National Renewal Energy Laboratory, among other national laboratories and industrial laboratories.

Our book (hereafter referred to as M&H) is designed to be a time-proven, single-source guide for an experimentalist, from initial conception of a need, through execution of the experiment, to final report. Our book will stand alone in the lab, yet introduces researchers into specialist texts.

Accompanying Material

The open-source software referenced in this text is free on the internet. The software, along with example data from the text, is additionally provided on an online website for this text.

Recommended Companion Texts

Our text forms a close companion with Hugh W. Coleman and W. Glenn Steele, *Experimentation, Validation, and Uncertainty Analysis for Engineers*, 4th ed., published by Wiley (hereafter referred to as C&S) (2018). Coleman was a student of Moffat's. C&S section 1.2, "Experimental Approach," outlines our book in one page.

Statistics for Experimenters, 2nd ed., by George E. P. Box, J. Stuart Hunter, and William G. Hunter (hereafter referred to as BH&H), is a Wiley classic (2005).

Response Surface Methodology, 4th ed., by Raymond H. Myers, Douglas C. Montgomery, and Christine M. Anderson-Cook (hereafter referred to as MM&AC), is also published by Wiley (2016).

How Is This Book Used for Teaching?

Our book has been used in Experiment Design courses as well as in diverse laboratory courses. Students in a variety of fields, including physics, engineering, chemistry, genetics, economics, medicine, and environmental studies, have used this book. Our text has also found a home directly in the lab (independent of classroom instruction) as a guidebook. It is written for self-study and continuing education; as such, it has been the text for short courses at national labs.

Acknowledgments

We thank N.J.A. Sloane and R.H. Hardin for developing the Gosset Program for Design of Experiments. We deeply appreciate their releasing Gosset to the public domain in 2018. We expect that researchers will benefit from the capabilities of Gosset for decades to come.

We are grateful to the thousands upon thousands of users of the R statistical language, contributors and researchers worldwide who have evolved R into a powerful tool for statistical analysis, modeling, and plotting. Thank you all for promoting this open-source, free software.

Robert thanks and dedicates this book to Lou London of Stanford University who taught me, inspired me, mentored me, and changed my life, opening doors across the world.

Roy especially thanks Helen L. Reed, William S. Saric, and William C. Reynolds for launching me into this most fascinating field, experimental thermal-fluid physics. R.J. Hansen and R. Rollins at the U.S. Naval Research Lab introduced me to advanced instruments and signal processing. Stanford colleagues and students ensured that all angles were considered and defensible.

Colleagues at Kyoto University (京都大學), IHI (石川島播磨重工業株式会社), Science University of Tokyo, NAL, and Tokyo University provided a convivial environment while I worked in the aerospace engine industry and energy science. Together we braved the national impact of the Fukushima tsunami. Colleagues and students at Handong (韓東大學校, 한동대학교) and LeTourneau provided valued support and feedback.

D.T. Kaplan and his books are outstanding for training statistical modeling.

We thank science editor A. Hunt of Wiley who paved the way for us as authors. S. Benjamin assembled documents. S. Brown ensured protocol. Ashwani Veejai Raj and her team converted diverse formats into typeset proofs. Eagle eyes of Cherrine R., Elaine H. and Lucy K. spotted stray prey. L. Poplawski corralled our project into final form. We thank you all.

Bob Moffat and his wife, Karina, have been constant encouragement as this work progressed. Whether on the other side of the world or in their dining area, they have been most generous.

My family and their enduring support are treasure beyond measure.

About the Companion Website

This book is accompanied by a companion website:

www.wiley.com/go/moffat/Planning

Scan this QR code to visit the companion website.

The website includes:

- Problems
- Sample data sets
- Archived sources to install Gosset and R
- Solution Manuals

1

Choosing Credibility

No one believes a theory except its originator.
Everyone believes an experiment except the experimenter.[1]

The decision to design a credible experiment sets you on a path to research with impact. Along this path you will make many decisions. This book prepares you to anticipate the choices you will face to plan and achieve an experiment that you the experimentalist also can believe.

There are two kinds of material to consider with respect to experimental methods: the mechanics of measurement and the strategy of experimentation. This book emphasizes the strategy and tactics of experiment planning.

The fact is that laboratories need people who can answer difficult questions experimentally. We offer this text to answer this need, to promote balanced competence in the field of experimental work. Abraham Lincoln is attributed with the saying "Give me six hours to chop down a tree and I will spend the first four sharpening the axe."[2] If there is anything like a "law of compound interest" in experimental work, then the effort spent to improve an experiment plan will return a bigger payoff than the same effort applied to fine-tuning transducers.

It is relatively easy to deal with sensors, calibrations, and corrections: those are concrete, factual bits of technology. The basic mechanisms are well known and tested. Could this be why there are so many technical references to transducers and so few to experiment planning?

Strategy is not concrete, however; it contains large elements of opinion, experience, and personal preference. One cannot prove that a particular strategy is best! What seems to be a "clever insight" to one person may seem "dull and pedestrian" to another.

1 Variations of the quote are attributed to Albert Einstein and to William Ian Beardmore Beveridge. In *The Art of Scientific Investigation* (1950), p. 65, "A theory is something nobody believes, except the person who made it. An experiment is something everybody believes, except the person who made it."
http://en.wikiquote.org/wiki/William_Ian_Beardmore_Beveridge.
2 https://quoteinvestigator.com/2014/03/29/sharp-axe asserts no evidence of Lincoln writing this. Having grown up on a small farm with one chore of clearing hundreds of pines out of our pasture, I (RH) can attest to its advice. Would rail-splitter Lincoln not agree?

Planning and Executing Credible Experiments: A Guidebook for Engineering, Science, Industrial Processes, Agriculture, and Business, First Edition. Robert J. Moffat and Roy W. Henk.

Companion website: www.wiley.com/go/moffat/planning

The ideas presented in this text were developed over 60 years of teaching and consulting on experimental methods. They provide an outline of a systematic way of designing experiments of provable accuracy. Not the only way, certainly, but at least one way. Each reader who has much experimental experience will have techniques to add to the list – we ask and welcome your feedback.

Novices might be inclined to take this text too literally, as though experiment planning were a quantitative discipline with rules that always worked. That would be a mistake. One must be flexible in the laboratory – following a sound basic philosophy – but taking advantage of the specific opportunities each experiment offers.

1.1 The Responsibility of an Experimentalist

People have always been impressed by data, as though it could never be wrong. As a young experimentalist, the saying that leads this chapter impressed me greatly. At first, I thought it was a clever play on words. Then I began to take it more seriously. It is true. People do seem to put more credence to experimental results than in analysis. I cannot tell you how many times I have heard someone say, with a tone of absolute finality, "Well, you can't argue with the data!" Fact is, you can and should.

This places a heavy responsibility on the experimenter. One may respond to such responsibility by taking great pains to establish the credibility of the data before actually taking data "for the record." Another response is to simply crash ahead and take data on a plausible but not proven experiment, because you can't think of anything you have done wrong. After all, these latter folks seem to think, "If I don't like the data, I will do it again, differently."

That latter view undermines credibility. If you feel free to ignore data because you don't like it or don't understand it, then you haven't run an experiment, you have just been playing around in the lab.

Many of the suggestions offered in this work are related to establishing credibility before the production data are taken: calibration of instruments, running baseline tests, etc. In this respect, an experiment is like an iceberg – 90% of the effort is "unseen," whereas only about 10% of the effort is invested in taking the production data.

Practically all of the experimentalists we know find the saying as a unique privilege and honor and as exhortation to produce the very best science. It gives us pause that people may base their design choices, their engineering, or even their lifestyle choices on their trust of our reported results and conclusions. This is a weighty responsibility worthy of doing our very best.

1.2 Losses of Credibility

Sadly, not all reported science is credible. In 2005, the *Journal of the American Medical Association* published "Contradicted and Initially Stronger Effects in Highly Cited Clinical Research," by J.P.A. Ioannidis (now at Stanford University) (Ioannidis 2005). Among his findings, "Five of six highly-cited nonrandomized studies had been contradicted or had found stronger effects" than other, better-designed studies. In other words, only one in six (17%) of these highly influential medical studies was credible.

Some scientific studies lost credibility due to poorly designed experiments. We offer this text to our readers to help them plan and execute experiments that are credible. There is no excuse for poorly designed experiments.

Retractions of scientific studies are no longer rare. The *New York Times* reported in 2011, that a "well-known psychologist...whose work has been published widely in professional journals falsified data and made up entire experiments" (Carey 2011). In his case, more than 50 articles were retracted. A site devoted to retracted science, Retraction Watch, can be found at http://retractionwatch.com.

A whole industry of fact-checkers has come into existence, purportedly to expose the false and reveal the true. We know, however, even fact-checkers must be checked. Let this motivate us.

1.3 Recovering Credibility

Science and engineering recently found a champion in Ioannidis.

By careful planning and execution of our own experiments, we too become champions of credibility.

As experimentalists, we must be our own front-line fact-checker, tackling errors as they arise. We put effort into uncovering every reason to not believe our method, equipment, results, and especially our modeling. We diligently report the uncertainty of our measurements.

When you have completed an experiment, you must have assembled so much evidence of credibility that, like it or not, you have to believe the data. The experimenter must be on guard all the time, looking for anomalies, looking for ways to challenge the credibility of the experiment. If something unexpected happens, the diligent experimenter will find a way to challenge it and either confirm it or refute it. Let "except the experimenter" spur you to create experiments you can believe and defend.

1.4 Starting with a Sharp Axe

In this text, we reintroduce one of the sharpest tools for designing an experiment, the computer program Gosset. Gosset was developed at AT&T Bell Laboratories in the 1980s. Although a few companies and researchers have used Gosset extensively, experimentalists across the world can benefit from its powerful features. The developers of Gosset released it for free to the public domain in 2018. We demonstrate Gosset in Chapter 9. With skill using Gosset, your axe will be sharper.

We also encourage use of one of the sharpest tools for data analysis, the R language. The R language is open-source software, available for free. In addition to the basic package, thousands of researchers from around the world have contributed specialized enhancements to R, also available for free. We demonstrate how to use R for data analysis, statistics, model design, and uncertainty analysis in Chapters 7–10 and 13. If you use an alternative commercial package for statistical analysis, you can test and compare.

1.5 A Systems View of Experimental Work

An experiment is a system designed to make a measurement. The system consists of the hardware (test rig and specimens), the instruments (including sensors, amplifiers, extension wires, etc.), and the interpretive software (calibration routines, data reduction programs, etc.). The whole system is an instrument designed to make a particular kind of measurement. As such, the system must be designed so that it can be calibrated. The system must be calibrated before it is used to generate new data. It must be possible, using diagnostic tests on the system, to confirm the accuracy of the measurements made with the system.

This view of an experiment is illustrated in Figure 1.1, which also shows some of the necessary features of the system design.

Perhaps the most important feature of this view of experimental work is the important role given to uncertainty analysis. There are uncertainties in every measurement and, therefore, in every parameter calculated using experimental data. When the results of an experiment scatter (i.e. are different on repeated trials), the question always arises, "Is this scatter due to the uncertainties in the input data or is something changing in the experiment?"

Uncertainty analysis provides the proven credible way to answer that question. By quantifying and reporting the uncertainty of each value, we allow our client confidence and credence in our results. Figure 1.1 shows the uncertainty analysis as a key part of the data reduction program, although it is too often neglected. Using either

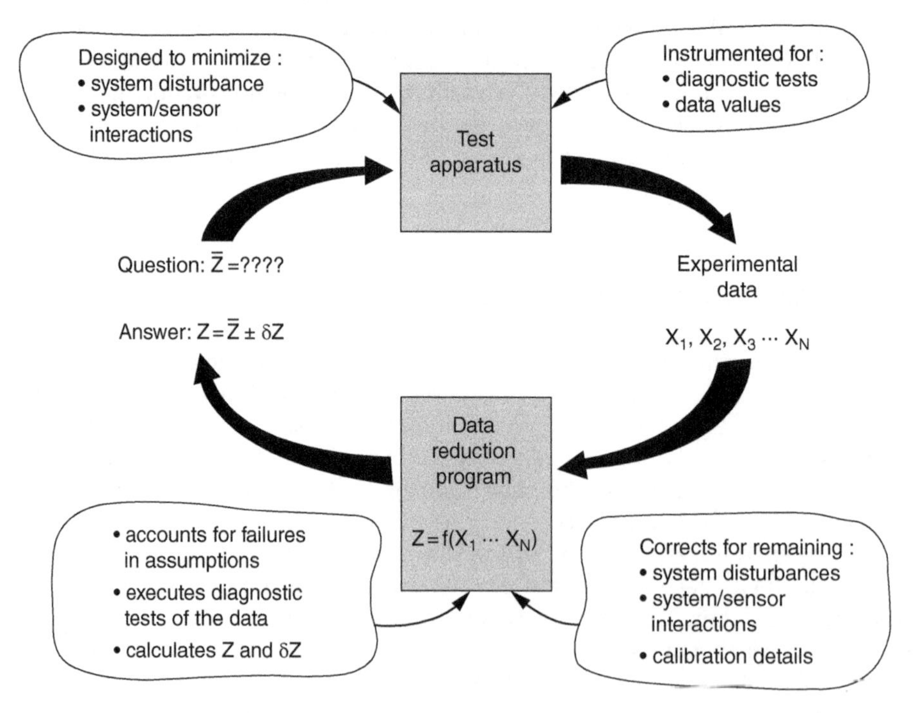

Figure 1.1 The experiment viewed as an instrument. Adjust the instrument by analyzing Uncertainty in each Bubble.

Root-Sum-Squared estimation or Monte Carlo simulation, the uncertainty in experimental results can be calculated with little additional effort on the part of the experimenter.

1.6 In Defense of Being a Generalist

The last point we wish to make before sending you off into this body of work has to do with the level of expertise one needs to run good experiments.

We think it is more important for an experimentalist to have a working knowledge of many areas than to be a specialist in any one. The lab is a *real* place; Mother Nature never forgets to apply her own laws. If you are unaware of the Coanda effect, you will wonder why the water runs under the counter instead of falling off the edge. If you haven't heard of Joule–Thomson cooling, you will have a tough time figuring out why you get frost on the valve of a CO_2 system.

Accordingly, if you aren't aware of the limitations of statistics, then using a statistical software package may lead you to indefensible conclusions.

It is not necessary to be the world's top authority on any of the mechanisms you encounter in the lab. You simply have to know enough to spot anomalies, to recognize that something unexpected or interesting is happening, and to know where to go for detailed help.

As an experimentalist, always beware of assumptions and presuppositions. See Figure 1.2 and "The Bundt Cake Story" (Panel 1.1). Step forward and predict. Then be ready and humble to course correct.

The lab is a great place for an observant generalist. The things that happen in the lab are real and reflect real phenomena. When something unexpected happens in the lab, if you are alert, you may learn something! As Pasteur said, "Chance favors only the prepared mind" (Pasteur 1854).

Let's now launch toward planning and executing credible experiments.

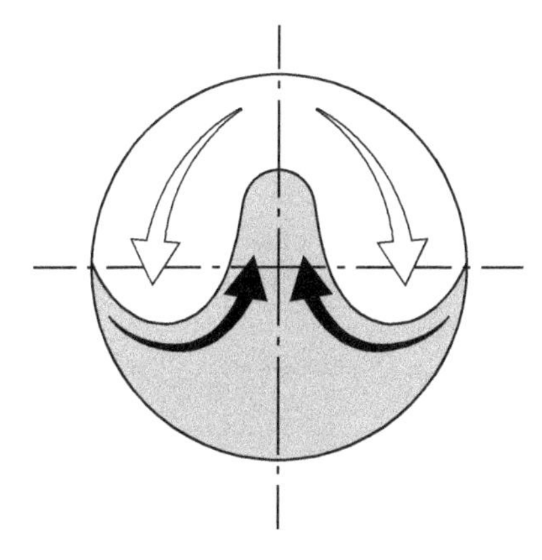

Figure 1.2 The Bundt cake as delivered. A high heat-transfer coefficient lifts the fluid batter like a hot air balloon. But which stagnation point is up, and which is down?

Panel 1.1 The Bundt Cake Story

One night, years ago, my wife baked a Bundt cake (chocolate and vanilla batter layered in a toroidal pan). When she presented me with a slice of that cake for dessert, I was impressed. But, also, I noticed something interesting about the pattern the batter had made as it cooked.

I recognized that the flow pattern, as drawn in figure 1.2, was related to the heat-transfer coefficient distribution around the baking pan.

I tried to impress my wife with my knowledge of heat transfer by explaining to her what I thought I saw. "Look," I said, "see how the batter rose up in the center, and came down on the sides. That means that the batter got hot in the center sooner than it did on the edges. That means that the heat-transfer coefficient is highest at the bottom center stagnation point for a cylinder in free convection with a negative Grashof number."

My wife was silent for a minute, then gently corrected me "I baked the cake upside down."

Of course, as soon as I learned that, I was able to say with confidence that "The heat-transfer coefficient is lowest at the bottom center stagnation point and high on the sides, for a cylinder in free convection with a negative Grashof number."

The Moral of This Story?

It is critically important that you can trust your data before you try to interpret it. Beware! Once we accept our results as valid, how can we avoid constructing or searching for an explanation? Does not the scientific method and our human nature spur us to do so?

References

Carey, B. (2011). Fraud case seen as a red flag for psychology research. http://www.nytimes.com/2011/11/03/health/research/noted-dutch-psychologist-stapel-accused-of-research-fraud.html.

Ioannidis, J.P.A. (13 July 2005). *Contradicted and initially stronger effects in highly cited clinical research. JAMA* 294 (2): 218–228. https://doi.org/10.1001/jama.294.2.218. PMID 16014596.

Pasteur L. (Dec. 7, 1854). "*Dans les champs de l'observation le hasard ne favorise que les esprits préparés,*" translated as "In the fields of observation chance favours only the prepared mind." Lecture, University of Lille. http://en.wikiquote.org/wiki/Louis_Pasteur.

Homework

1.1 Following the guide in Appendix D, Section D.1, download and install the statistical language R, which is open source and free. Please consider this software tool essential.

1.2 Following the guide in Appendix D, Section D.2, download and install LibreOffice, open source and free.

LibreOffice is compatible with msOffice documents. LibreOffice can even read and write antiquated *.doc and *.xls files of obsolete versions of msOffice better than ms does. The interface is more accessible and less bloated than msOffice. Consider this optional, but highly recommended and free.

1.3 Following the guide in Appendix D, Section D.4, consider R-Studio. Please consider this software tool optional.

2

The Nature of Experimental Work

It doesn't matter how beautiful your theory is, it doesn't matter how smart you are. If it doesn't agree with experiment, *it's wrong.*

Richard Feynman[1]

Science, medicine, and engineering depend heavily on experiments. Business does too, where product design and marketing experiments must account for the added complexity of human nature and taste. In every case in any field, an experiment must be credible or it wastes time and resources.

Engineering problems are solved using three tools: insight, analysis, and experiment. Usually all three are brought to bear on any given problem; each complementing the others. A sudden insight or inspiration may be enough to suggest what must be done next, a new analysis or a new experiment, but to actually get an answer takes work.

2.1 Tested Guide of Strategy and Tactics

This book is about the strategy and tactics of experimental work – the techniques whereby one plans and executes an experiment which insight or inspiration has suggested. Our strategies and tactics are a superset which includes the concepts of design of experiments (DoE) and extends beyond.

Have you heard how designed experiments improve results while reducing the amount of data needed? In our experience, few engineering students have learned DoE in their undergraduate labs, so we are including DoE concepts. Many DoE techniques were pioneered by Ronald Fisher (1890–1962) for agricultural experiments in the early 1900s. As the advantages of DoE become known, we all benefit.

This book offers a tested guide so that you can design effective experiments. We teach the techniques by which your insight can result in a high-impact experiment. You know of a need that requires a credible solution. We hope to launch from your creativity and reinforce it. Could something with absolutely no creativity (e.g. a computer) be successful in either analysis or experiment? It seems not. As humans, our needs often inspire our inventiveness and creativity. Creativity is boosted by seeing successful and failed

---

1 This quotation is from a Feynman lecture at Cornell in 1964. A select portion of the lecture can be viewed at https://youtu.be/OL6-x0modwY. A lengthier quote is given in Pomeroy (2012).

Planning and Executing Credible Experiments: A Guidebook for Engineering, Science, Industrial Processes, Agriculture, and Business, First Edition. Robert J. Moffat and Roy W. Henk.

Companion website: www.wiley.com/go/moffat/planning

solutions. The art and science of engineering advances through decisions – decisions based on experience gained through experimental results.

The purpose of an experiment is to get provably accurate, relevant, and credible data – data that are reliable enough to serve as the basis for answering questions and making decisions. Most experiments arise out of questions which must be answered, such as "Does this device behave the way it is supposed to?" or "How much cooling do we need?" or "What is the relationship between X and Y?" or "Which of these designs performs best?" In many cases, such questions lead to experiments, and the data from those experiments lead to decisions.

The three key words are "accurate," "relevant," and "credible." One point to remember is that when the work is all done, it will be your signature on the report – and that report will be around for a long time! It is not enough for you to be personally convinced that your results are accurate. You must be able to establish the credibility of the work "beyond a reasonable doubt" or at least well enough so that a prudent engineer would be willing to accept your results as valid when you are no longer around to answer his or her questions.

The process of establishing credibility begins with the experiment plan and only ends when the results have been presented in such form that they can easily be understood. The experiment plan must make provision for the appropriate checks and balances: baseline checking, repeatability tests, and the other diagnostics which guard against error. Showing agreement with a baseline dataset is one of the most convincing pieces of evidence that can be offered to support the credibility of an experiment. The data presentation must include a quantitative description of the residual uncertainty in the results.

In large measure, that is what this book is all about: designing and executing experiments for credibility. Before we get to the main issue, however, there are some key points to consider about experiments in general.

2.2 What Can Be Measured and What Cannot?

A quantitative property can be measured. A categorical property can be recorded but not measured.

The act of measurement is an ordering in a scalar system involving a "less than, equal to, or greater than" test. We assign a value to the measurand by comparing it with a standard interval and counting the number of intervals equal to the measurand. The only attributes of any system that can be measured are those which can be put into one-to-one correspondence with points on the real number line. Since only the real number system has the order property, only real numbers (scalars) can be measured.

2.2.1 Examples Not Measurable

Only the simplest attributes of systems can be directly measured – the rest are inferred.

For example, tensors, complex numbers,[2] and vectors cannot be ordered and, therefore, cannot be "measured." These can be described by ordered sets of scalars (dyads,

2 Rather than divide a complex number into real and imaginary parts, shall we describe it as revealed and concealed parts? Our reasoning is that both parts of a complex number have reality in our universe. For example, in dynamic systems, the revealed part corresponds to position and the concealed part corresponds to velocity. Is velocity just as real as location? Of course. We, you and us, can see the location of a car relative to its surroundings. The speed (scalar) is revealed to us via the speedometer. The change of velocity is revealed to us via our inner ear. For deeper insight, Roger Penrose has provided a most compelling motivation for complex numbers in his book *Road to Reality* (2005, chapters 1 through 14).

triads, and two or three-dimensional arrays of scalars), but those are simply the components of the entity, not the entity itself.

Many times people seem to recognize this problem but don't know how to describe their malaise. They want information but find themselves talking about measurements. For example, a former governor of Mississippi has been quoted as saying (and I paraphrase), "When putting money in education, everyone wants to see some measurable return for the money. Yet it is the one area that has the greatest degree of intangibles." There is no measurable attribute of education except a purely artificial one: "test scores." Teachers know that test scores don't measure educational achievement. But when people insist on measurements, they will get measurements.

A great deal of the information we use in daily decision-making is nonscalar and, therefore, intrinsically not measurable. For example, we cannot measure the appearance of a face, the sound of a voice, or the taste of tomato soup, and yet with no difficulty at all, we greet our friends, recognize their voices, and enjoy our dinners. The information transfer by sight, hearing, and through taste represents very complex information handling using arrays of scalars and correlations between pairs of scalars (temporal and spatial). No instrumentation system can do as sophisticated a job of pattern recognition as the human eye/mind combination, or of frequency analysis/correlation as the ear/mind combination, or of chemical analysis as the taste-bud/mind combination. We are not denying the improving capabilities of neural networks, wavelet transformations, or AI deep learning – we are just marveling.

2.2.2 Shapes

Shape cannot be measured – not even simple shapes, such as circles. Simple shapes can be described by names that we all understand by experience, but they cannot be measured. For example, a "circle" is defined as the locus of points lying in a plane and at the same distance from a common point, called "the center." Given that definition and a value for the radius, you can draw a circle and look at it, and you know exactly what was meant – but that does not constitute measuring the shape of the circle. The shape information was conveyed using the reserved word "circle"; only the size was described by the radius and location by the center.

Shape can have a delicious impact, when tested by experiment! For example, the design of rice cookers has made it possible to make chef-quality rice at home. A few years ago, product design engineers in Japan focused on the cooking profile of rice (not the shape of the device). By experiment, they found the shape of the temperature versus time curve, $T = f(t)$, as shown in Figure 2.1, which optimizes rice taste when the ingredient is washed white, short-grain rice. They also found a different shape that is best for unwashed rice and yet another shape for brown rice. Although aspects of the shape can be measured, such as the time at each temperature, it is the shape of the cooking profile that enhances the taste of the rice. By offering more delicious rice, manufacturers gained market share over older cookers that merely boiled rice.[3]

Another example comes from the physics of fluid flows. In boundary layer studies, the shape of the velocity profile is of considerable interest and often needs to be

3 The superiority of the cooking profile trajectory is such that I (RH) witnessed a 50-person block of Asian tourists who each traveled to Japan primarily to buy a rice cooker to bring home to their country. There was insufficient room in the overhead compartments for all the rice cookers, so the flight departure was delayed 40 minutes as all the cookers were relocated to the checked-baggage compartment.

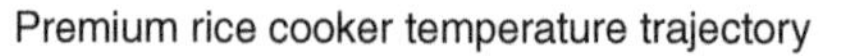

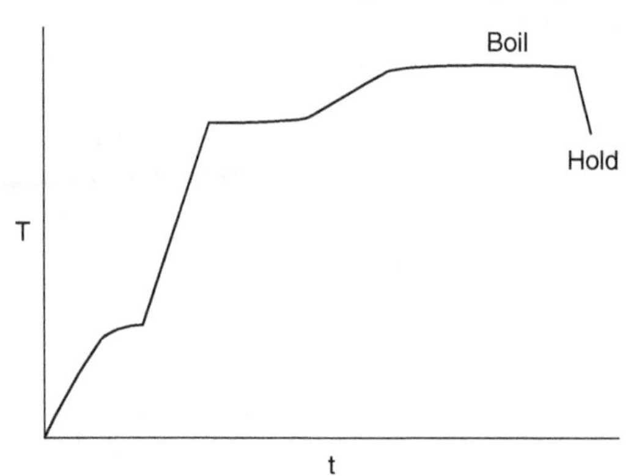

Figure 2.1 Rice cooker design trajectory.

recorded. One way to deal with this is to present $u = f(y)$, a set of ordered pairs (u, y). This allows the viewer to draw the shape and look at it. Another way, conveying less information but sometimes enough, is to present the shape factor: the ratio of two integral measures of the boundary layer thickness, the displacement thickness divided by the momentum thickness. Turbulent boundary layers have larger values of the shape factor than laminar boundary layers. Researchers who know approximately what the velocity profile looks like (i.e. what family of shapes to which it belongs) can communicate quite a bit of information to one another by quoting shape factor values. Yet the value of the shape factor itself is not a measure of shape. Only if the boundary layer is known to be laminar or turbulent or somewhere in between does the shape factor convey information. If the family of possible shapes is not specified either explicitly or implicitly, then it takes a very large number of scalar pairs to describe shape – enough data points to plot the shape so it can be looked at. Presenting $u(y)$ throughout the boundary layer allows the viewer to see the shape, but that does not constitute a measurement of shape any more than a photograph of a face is a measurement of the shape of that face. The derived scalar "measures" of shape, such as displacement thickness, momentum thickness, and shape factor, can convey significantly wrong impressions of the shape of a boundary layer velocity distribution when they are reported for a "pathological boundary layer," i.e. one whose velocity distribution is significantly different than usual. They convey the right information only when they are applied to boundary layers with generally typical velocity distributions.

2.2.3 Measurable by the Human Sensory System

How amazing is the human sensory system and what it enables us to observe! Trying to extract as much information using scalar measuring instruments is quite a challenge. The human sensory system is very complex, and its receptors are very well tuned to our environment. If our eyes were just a few decibels more sensitive, we could see single photons; if our ears were just a few decibels more sensitive, we could hear the Brownian motion of individual air molecules as they bounced off our eardrums.

Consider our sense of touch. Our machinists claim to easily detect surface roughness of 50 mils (1 μm) with work-calloused hands. Can our body measure force? Can it measure temperature? We sense not temperature directly but heat-transfer rate; if you've ever dipped a cold toe into a warm bath, did the water feel boiling hot even if was just warm? (As in an Onsen hot spring in Japan.) We don't feel force directly but pressure and shear force. In contrast, in the lab we have simple instruments for measuring force and temperature but need sophisticated techniques to measure heat-transfer rate. Even touch is a marvel. Prepublication we learned: in the journal *Nature*, it has been "experimentally established that humans have the capacity to perceive single photons of light" (Tinsley et al. 2016). Furthermore "human tactile discrimination extends to the nanoscale ... within billionths of a meter" (Skedung et al. 2013). In part this explains how polished steel tactilely differs from smooth rubber.

2.2.4 Identifying and Selecting Measurable Factors

One of the first problems to be faced in exploratory research, and in development work, is identifying which scalars are significant to the issue at hand. Sometimes a single scalar is sufficient, such as a temperature, pressure, or velocity. Sometimes a compound scalar measure can be put together from a set of simple scalars. One example would be the Reynolds number, used for characterizing the state of the flow in a channel. Sometimes two or more scalars can be combined into one measure to reflect value judgments or "trade-offs" in desirability. For example, consider the problem of selecting the optimum heat exchanger for an engine application. In even the simplest situation, ignoring such considerations as cost, size, and durability, a heat exchanger for engine service has at least two important scalar descriptors: its effectiveness and its pressure drop. Typically, high effectiveness is "good," while high-pressure drop is "bad." Also typically, pressure drop goes up when effectiveness goes up. Neither by itself is a measure of goodness for the heat exchanger. A weighted sum of the two can be used as a "composite scalar" by finding two weight functions, one for effectiveness and one for pressure drop, which accounts for the effects of both on some single, important parameter – brake-specific fuel consumption is another example. Such "goodness factors" can be used to account for many factors at a time and to convert a nonmeasurable situation to a measurable one.

The choice of what to measure sets the course of the entire experiment, and that choice should be made with considerable care.

2.2.5 Intrusive Measurements

Always remember as you plan: making a measurement intrudes on the system measured. Ask: How much does the measurement intrusion alter the system? How does measurement affect the accuracy and precision of its results? Furthermore, the intrusion affects the uncertainty of the measurement. A prime motivator for inventing new probes and experimental techniques is the aim to intrude less when taking a measurement. Classic measurement is the focus of this text.

In certain quantum physics experiments, however, mere observation alters the system. In these experiments, if an observation is made the system has one result; if no observation is made the system shows contrary results. These tests have been reproduced worldwide. If you are interested, we recommend Richard Muller's book (Muller 2016). Or search "theory of measurement in physics."

2.3 Beware Measuring Without Understanding: Warnings from History

There is an unfortunate tendency, among engineers particularly, simply to measure everything which can be measured, report the results, and hope that someone, someday, will find the results useful. This is a deplorable state of affairs, even though it has a long and honored past. In many respects, it follows the scientific tradition which emerged from Europe in the nineteenth century, when the art of measurement expanded so rapidly in Western civilization.

William Thomson, Lord Kelvin, famously proclaimed: "When you can measure what you are talking about and express it in numbers then you have the beginning of knowledge." That is still true today but with some limitations (Thomson 1883).

Thomson's enthusiasm for measurements should be interpreted in terms of the times in which he lived. The European scientists of that period were infatuated with measurement. Every new measurement technique developed was applied to every situation for which it seemed to fit. There was no storehouse of knowledge about the physical world. Each new series of measurements revealed order in another part of the physical world, and it appeared that every measurement answered some question, and every question could be answered if only enough measurements were made. That was true partly because there were so many unanswered questions and partly because most of the questions which were being asked at that time could be answered by scalars.

Since that era, experimental work has become more expensive and more complex. The questions that lead us into the lab today usually involve the behavior of systems with many components or processes with several simultaneous mechanisms. It is not easy to translate a "need to know" into an experiment under such complex conditions. The first problem is deciding what scalars (i.e. what measurable items) are important to the phenomenon being investigated. This step often takes place so fast and so early in a test program that its significance is overlooked. When you choose what to measure, you implicitly determine the relevance of the results.

The early years of the automobile industry provide at least one good example of the consequences of "leaping in" to measurements. As more and more vehicles took to the road, it became apparent that some lubricating oils were "better" than others, meaning that automobile engines ran longer or performed better when lubricated with those oils. No one knew which attributes of the oils were important and which were not, and so all the easily measured properties of the "good" oils were measured and tabulated. The result was a "profile of a good oil." The oil companies then began trying to develop improved oils by tailoring their properties to match those of the "good oil profile." The result was a large number of oils that had all the desired properties of a good oil except one: they didn't run well in engines![4]

---

4 Dr. Loyd Withrow, GM Research Labs, ca. 1953. Personal communication.

2.4 How Does Experimental Work Differ from Theory and Analysis?

The techniques of experimentation differ considerably from the techniques of analysis, by the nature of the two approaches. It is worthwhile to examine some of these differences.

2.4.1 Logical Mode

Analysis is deductive and deals with the manipulation of a model function. The typical problem is: Given a set of postulates, what is the outcome? Experiment is inductive and deals with the construction of models. The typical problem is: given a set of input data and the corresponding output data, what is the form of the model function that connects the output to the input? The analytical techniques are manipulative, whereas the experimental techniques are those of measurement and inference.

2.4.2 Persistence

An analysis is a persistent thing. It continues to exist on paper long after the analyst lays down his or her pencil. That sheet (or that computer program) can be given to a colleague for review: "Do you see any error in this?"

An experiment is a sequence of states that exist momentarily and then are gone forever. The experimenter is a spectator, watching the event – the only trace of the experiment is the data that have been recorded. If those data don't accurately reflect what happened, you are out of luck.

An experiment can never be repeated – you can only repeat what you think you did. Taking data from an experiment is like taking notes from a speech. If you didn't get it when it was said, you just don't have it. That means that you can never relax in the lab. Any moment when your attention wanders is likely to be the moment when the results are "unusual." Then, you will wonder, "Did I really see that?" [Please see "Positive Consequences of the Reproducibility Crisis" (Panel 2.1). The crisis, by way of the Ioannidis article, was mentioned in Chapter 1.]

The clock never stops ticking, and an instant in time can never be repeated. The only record of your experiment is in the data you recorded. If the results are hard to believe, you may well wish you had taken more detailed data. It is smart to analyze the data in real time, so you can see the results as they emerge. Then, when something strange happens in the experiment, you can immediately repeat the test point that gave you the strange result. One of the worst things you can do is to take data all day, shut down the rig, and then reduce the data. Generally, there is no way to tell whether unusual data should be believed or not, unless you spot the anomaly immediately and can repeat the set point before the peripheral conditions change.

2.4.3 Resolution

The experimental approach requires gathering enough input–output datasets so that the form of the model function can be determined with acceptable uncertainty. This is, at best, an approximate process, as can be seen by a simple example. Consider the differences between the analytical and the experimental approaches to the function $y = \sin(x)$. Analytically, given that function and an input set of values of x, the corresponding values of y can be determined to within any desired accuracy, by using the known behavior of the function $y = \sin(x)$. Consider now a "black box" which, when fed values of x, produces values of y. With what certainty can we claim that the model function (inside the box) is really $y = \sin(x)$? Obviously, the certainty is limited by the accuracy of the input and the output. What uncertainty must we acknowledge when we claim that the model function (inside the box) is $y = \sin(x)$? That depends on the accuracy of the input and the output data points and the number and spacing of the points. With a set of data having some specified number of significant figures in the input and the output, we can say only that the model function, "evaluated at these data points, does not differ from $y = \sin(x)$ by more than …," or alternatively, "$y = \sin(x)$ within the accuracy of this experiment, at the points measured."

That is about all we can be sure of because our understanding of the model function can be affected by the choice of the input values. Suppose that we were unfortunate enough to have chosen a sampling rate that caused our input data points (the test rig set points) to exactly match values of $n\pi$ with n being an integer. Then all of the outputs would be zero, and we could not distinguish between the "aliased" model function $y = 0$ and the true model function $y = \sin(x)$.

In general, with randomly selected values of x, the "resolution" of the experiment is limited by the accuracy of the input and output data. Consider Figure 2.2. In this case, $\sin(x)$ may be indistinguishable from $\{\sin(x) + 0.1 \sin(10x)\}$ if there is significant scatter in the data. In many cases, the scatter in data is, in reality, the trace of an unrecognized component of the model function that could be included. One of an experimenter's most challenging tasks is to interpret correctly small changes in the data: is this just "scatter," or is the process trying to show me something?

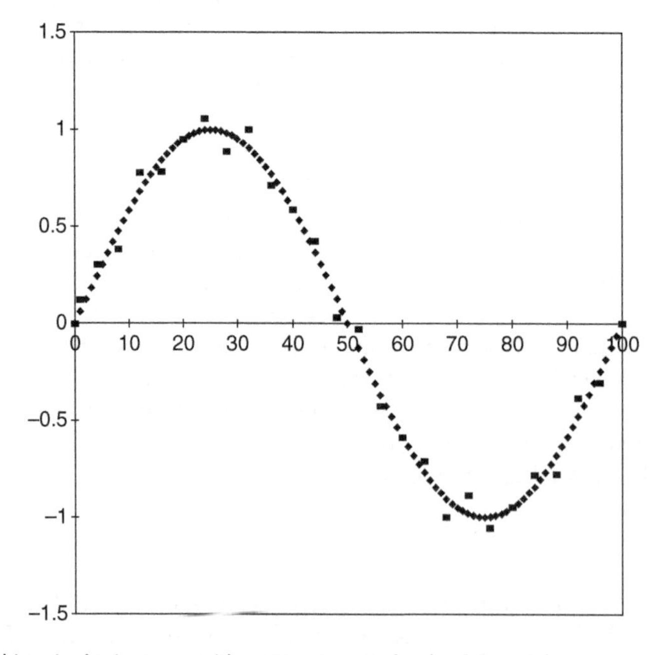

Figure 2.2 Is this a single sine wave with some scatter in the data? Does it have a superposed signal or both signal and scatter?

2.4.4 Dimensionality

Another difference between experiment and analysis that is important is their description in terms of "dimensionality." A necessary initial step in an analysis is to set the dimensional domain of relevant factors; e.g. $y = f(x_1, x_2, x_3, \ldots, x_N)$. Once an analyst has declared the domain, then the analyst may proceed with certainty by applying the rules appropriate for functions of N variables.

Experimentalists cannot make their results insensitive to "other factors" simply by declaration, as the analyst can. The test program must be designed to reveal and measure the sensitivity of the results to changes in the secondary variables.

Experiments are always conducted in a space of unknown dimensionality. Any variable which affects the outcome of the experiment is a "dimension" of that experiment. Whether or not we recognize the effect of a variable on the result may depend on the precision of the measurements. Thus, the number of significant dimensions for most experiments depends on the precision of the measurements of input and output and the density of data points as well as on the process being studied and the factors which are being considered.

For example, one might ask: "Does the width dimension of the test channel affect the measured value of the heat-transfer coefficient h on a specimen placed in the tunnel?" This is a question about the dimensionality of the problem. The answer will depend on how accurately the heat-transfer coefficient is being measured. If the scatter in the h-data is ±25%, then only when the blockage is high will the tunnel dimensions be important. If the scatter in h is ±1%, then the tunnel width may affect the measured value even if the blockage is as low as 2 or 3%.

One common approach to limit dimensionality of an experiment is to carefully describe the apparatus, so it could be duplicated if necessary, and then run the tests by holding constant as many variables as possible while changing the independent variables, one at a time. This is not the wisest approach. A one-at-a-time experiment measures the partial derivative of the outcome with respect to each of the independent variables, holding constant the values of the secondary variables. Although this seems to limit dimensionality, it does not. Running only a partial derivative experiment begs the question of sensitivity to peripheral factors: that is, holding the interaction effect constant does not make it go away, it simply makes it more difficult to find.

Wiser approaches are discussed in Chapters 8 and 9, whereby systematic investigation of the sensitivity of the results to the details of the technique and the equipment helps the dimensionality of the experiment to be known. Whatever remains unknown contributes to experimental uncertainty.

Be ready to defend your factors. As an experimentalist, there will be times when you work with a client, or with a theoretician, who fails to understand what cannot be measured. Or she may need guidance to tolerate uncertainty in measurements.

2.4.5 Similarity and Dimensional Analysis

Nature does not know how big an inch is (unless you experiment on inchworms), nor how big a centimeter is. The laws of physics are independent of the length, mass, and time scales familiar to us. For this reason, similarity analysis and the Buckingham Pi Π method are tools to cast the physics into nondimensional parameters which are independent of scale.

Perhaps the most familiar nondimensional parameter is the Mach number, the ratio of speed to the speed of sound. A fighter flying at Mach 2 is traveling at twice the speed of sound.

In thermo-fluid physics, we use various nondimensional parameters, including Reynolds number (Re), Strouhal number (St), Froude number (Fr), Prandtl number (Pr), Mach, etc. We will see these again in later chapters. The Reynolds number is a ratio relating size, speed, fluid density, and viscosity. When NASA tests a model plane in a wind tunnel, it matches the Re of the model to that of the full-size plane. Experimental results for the model plane and full-size plane relate even better by simultaneously matching Mach number. And so forth as more parameters match.

As an expert in your field, you know which nondimensional parameters pertain to your experiment. The applicability of your measurements expands via nondimensional parameters. In your experiment, plan to ensure that you record all the factors, including environmental factors, so that all pertinent nondimensional parameters can be reported.

Upon reflection, the value percent (%) is likely the most familiar nondimensional number.

2.4.6 Listening to Our Theoretician Compatriots

Experimentalists and theoreticians need each other.

Richard Feynman, whose quote leads this chapter, was an experimentalist as well as a theoretician.

Einstein, whose paraphrased quote lead off Chapter 1, received his Nobel Prize for explaining experiments on the photoelectric effect. Einstein's theory of Brownian motion showed that prior experiments provided indirect evidence that molecules and atoms exist.

Yet just as Feynman stated, Einstein's theory of general relativity was "just a theory" until Arthur Eddington gave it experimental verification during a total solar eclipse in 1919.

NASA provides a good example of the interdependence of theory and experiment. The National Advisory Council on Aeronautics (NACA) was the precursor of NASA; "Aeronautics" is the first A of NASA. As airplane designs rapidly advanced during the 1900s, NASA deliberately adopted a four-pronged approach: theory, scale-model testing (wind-tunnel experiments), full-scale testing (in-flight experiments), and numerical simulation (computational models verified by experiment). Each of the first three prongs have always been essential (Baals and Corliss 1981). Since the 1980s, numerical simulation has aided theory. Theory and experiment need each other. Since our numerical colleagues often refer to their "numerical experiments," we do advocate an appropriate way to report the uncertainties of their results, just as we experimentalists do.

The science of fluid flow remains important, as another quote (from a personal letter) from Feynman makes clear:

> Turbulence is the most important unsolved problem of classical physics.

Feynman spoke of basic turbulence. Turbulence can be further complicated by heat transfer; yet more complicated by mass transfer; yet more by chemical reactions or combustion; yet more complicated by electromagnetic interactions. Turbulence is key for weather, for breath and blood, for life, for flight, for circulation within celestial stars and their evolution. Turbulence remains unsolved to this day.

To consider more viewpoints, we include three panels:

- Panel 2.1, "Positive Consequences of the Reproducibility Crisis"
- Panel 2.2, "Invitations to Experimental Research, Insights from Theoreticians"
- Panel 2.3, "Prepublishing Your Experiment Plan"

This text focuses on experimental strategies, planning, techniques of analysis, and execution. That is our expertise, in addition to thermo-fluid physics. We have taught experimental planning to students in many fields using draft notes of this text for more than 60 years.

Panel 2.1 Positive Consequences of the Reproducibility Crisis

As researchers and instructors, we have been promoting experimental repeatability and uncertainty analysis for more than 60 years. When the work of Dr. J.P.A. Ioannidis brought the Reproducibility Crisis in the medical field to public awareness, we welcomed the positive impact it produced.

Two papers by Dr. Ioannidis in 2005 brought the Reproducibility Crisis to the fore. One was the *Journal of the American Medical Association* (*JAMA*) article mentioned in Chapter 1, "Contradicted and Initially Stronger Effects in Highly Cited Clinical Research" (Ioannidis 2005a). The second was "Why Most Published Research Findings Are False" (Ioannidis 2005b).

The two 2005 articles by Dr. Ioannidis appear to be a watershed moment for science. In various scientific disciplines, researchers have produced guidelines adopted by major publishers.

Going deeper into the 2005 *JAMA* article, Dr. Ioannidis chose a notably high criteria for the publications he evaluated. He considered only:

- "All original clinical research studies published in 3 major general clinical journals or high-impact-factor specialty journals
- in 1990–2003 and
- cited more than 1000 times in the literature…"

Dr. Ioannidis then compared these "results of highly cited articles … against subsequent studies of comparable or larger sample size and similar or better controlled designs. The same analysis was also performed comparatively for matched studies that were not so highly cited."

Although part of the same article, this collection of research studies fared better than those mentioned in our Chapter 1. "Of 49 highly cited original clinical research studies, 45 claimed that the intervention was effective. Of these, 7 (16%) were contradicted by subsequent studies, 7 others (16%) had found effects that were stronger than those of subsequent studies, 20 (44%) were replicated, and 11 (24%) remained largely unchallenged."[5]

In the same year, Dr. Ioannidis published "Why Most Published Research Findings Are False," a provocative title. Although the wording appears to encompass all fields, the

5 Thank you to Dr. Ioannidis for permission to include the lists from his articles.

examples in the article were medical experiments. In order to make his evaluations, he adopted a key metric called the "Positive Predictive Value" (PPV). From this research, Dr. Ioannidis deduced the following "corollaries about the probability that a research finding is indeed true":

Corollary 1: The smaller the studies conducted in a scientific field, the less likely the research findings are to be true.
Corollary 2: The smaller the effect sizes in a scientific field, the less likely the research findings are to be true.
Corollary 3: The greater the number and the lesser the selection of tested relationships in a scientific field, the less likely the research findings are to be true.
Corollary 4: The greater the flexibility in designs, definitions, outcomes, and analytical models in a scientific field, the less likely the research findings are to be true.
Corollary 5: The greater the financial and other interests and prejudices in a scientific field, the less likely the research findings are to be true.
Corollary 6: The hotter a scientific field (with more scientific teams involved), the less likely the research findings are to be true.

Be aware. Beware small studies (i) and small effects (ii). Beware explaining indiscriminately (iii). Beware lack of standards (iv). Beware political favors and conflicts of interest (v). Beware fashionable science (vi). The corollaries provide fair warning for all research, and do affect the Nature of Experimental Work.

Feynman's words suggest that experiments help science to be self-correcting. However, Ioannidis (2012) gave the warning "Why Science Is Not Necessarily Self-Correcting" in a more recent paper concerning psychological science.

Dr. J.P.A. Ioannidis has done more for us than exposing flaws in medical science. He invites better experiments in the biological and medical sciences. He sharpens experimental discernment.

Dr. Ioannidis is not alone. For example, Simera et al. (2010) report in the *European Journal of Clinical Investigation* how researchers collaborated and top publishers have agreed to guidelines regarding health research studies. There are many similar groups now.

Furthermore, there is incentive to resist bad and fraudulent research. RetractionWatch.com keeps track of research that has been retracted by author, publisher, or sponsor.

Overall, the concerted effort to assure the credibility of experiments keeps us in good company.

Panel 2.2 Selected Invitations to Experimental Research, Insights from Theoreticians

The desire to better understand how our world works invites us to credible experiments. Beyond measuring, we predict what we expect to measure. Our interest may be our company's factory floor, economics and marketing, agriculture and environment, medicine, information, engineering and technology, climate, or the sciences biology, chemistry, and physics.

Richard Feynman in his lecture "There's Plenty of Room at the Bottom" (1959) invited scientists into a new field which we now call "nanoscience." The launch of nanoscience has led to innovative products that affect our daily lives. Nanoscience, as a research field, reaches from measuring individual atoms within molecules, to manipulating atoms, to fluid nano-arrays for medical and pharmaceutical tests and beyond.

Better Invitation than a Nobel Prize

My (RH) first experimental mentor in aerodynamics was W.S. Saric. He told me the advantage our area had over other physics fields: in thermo-fluid physics, researchers shared their techniques. His reasoning: since Ludwig Prandtl, an early pioneer in fluids, had not won a Nobel Prize, no one in our area of classical physics would expect to win. Classic thermo-fluid physics was crucial to so many areas of science. The camaraderie of joint effort was inviting.

Essentially every Nobel Prize in Chemistry recognizes experimental work. Likewise, essentially every Nobel Prize in Physiology or Medicine recognizes experimental tests. How about for Physics Nobel Prizes? An accounting (Quantum Coffee 2014) of the Nobel Prizes in Physics up to 2014 divided as such: theory, 30.75 prizes (28.7%); experiment, 76.25 prizes (71.3%). Nobel Prizes in Physics for technical innovations (reckoned as experiment) were 22.2%, almost as much as theory.

The Higgs mechanism, theoretically predicted in 1962, eventually culminated in the announcement of the experimental discovery of the Higgs boson in 2012 at the Large Hadron Collider (LHC). The LHC employs thousands of scientists and engineers. F. Englert and P.W. Higgs received the 2013 Nobel Prize in Physics for their theoretical work. The Higgs Nobel Prize was a rare instance where theory preceded experiment. Einstein, like most theorists to win, won the prize for explaining an experiment.

Einstein's Theory Always Invites Tests

Einstein's theory of relativity is arguably of the most precisely tested theory in science, with experimental agreement to better than the 12th decimal place. The measurements allowing such fine precision and accuracy involved a binary pulsar (Antoniadis et al. 2013). Although it is so well tested, several times a decade we read about an experiment claiming to violate or refute Einstein's theory. The experimental results which appear to refute are expertly considered and critiqued. Invariably a flaw in technique or instrumentation is discovered, further confirming Einstein's theory rather than refuting it. Confidence in Einstein's theory increases with each test. In 2018 a test beyond our galaxy was reported and confirmed.

We trust Einstein's theory as far as it has been experimentally tested, not due to its popularity.

Observations of a Popular Theoretical Physics Field

In the early 1980s, String Theory became a popular physics field. Our particular interest for this text is twofold: (i) its fashionability invites comparison with Ioannidis corollaries 5 and 6; (ii) Ioannidis evaluated medical research based on falsifiable predictions, called "PPV," as discussed in Panel 2.1.

During the peak years of String Theory popularity, its math techniques flourished. It garnered the majority of physics funding; its proponents placed the majority of professorships. We avidly read about it.

Since String Theory is one of several competing theories, and outside our specialty, we continue to watch with interest all sides in the dispute.

String Theory notably depended on multiple spatial dimensions. E.A. Abbott's book *Flatland: A Romance of Many Dimensions* had already introduced us to imagining extra spatial dimensions. E. Witten was a top advocate of String Theory. A principal concern, stated by advocates and critics alike, was that String Theory lacked predictions that could be tested experimentally.

Beginning in 2006, the warnings in Ioannidis corollaries 5 and 6 compared with String Theory. Theoretical physicist L. Smolin raised an alert in his book *The Trouble with Physics: The Rise of String Theory, the Fall of a Science and What Comes Next* (2006). Noting that physics was rich in alternative "promising new directions," Smolin wrote to promote other areas of theoretical and experimental physics. Others noticed as well. P. Woit wrote his critique *Not Even Wrong: The Failure of String Theory and the Continuing Challenge to Unify the Laws of Physics* (2006).

Roger Penrose[6] took on three popular areas of physics, expressing similar concerns. Penrose's recent book (2016) is *Fashion, Faith, and Fantasy in the New Physics of the Universe*. The chapter entitled "Fashion" covers String Theory, adding his perspective. The chapter covering quantum mechanics is entitled "Faith." Techniques of quantum electrodynamics show success, the evidence being experimental validation to accuracies rivaling relativity. Penrose added an experimental test of macro-quantum superposition. The chapter entitled "Fantasy" deals with cosmologies beyond the big bang. Penrose highlighted "The Phenomenal Precision in the Big Bang"; of particular interest to us in thermofluids was the necessarily low entropy.

Regardless of a theory's popularity, political or otherwise, we urge experimental tests.

Sometimes an experiment planned for another purpose provides the answer. A prime example was the search for residual thermal evidence of the big bang. While a Princeton physics group was proposing to test the theory, a couple of astronomers at AT&T Bell Labs, Arno Penzias and Robert Wilson, were trying to eliminate noise which was contaminating the signal in their antenna. They even removed pigeon and bat residue. Failing to eliminate the noise, they spoke with Princeton professor Robert Dicke. They published "A Measurement of Excess Antenna Temperature at 4080 Mc/s" in the *Astrophysical Journal* (Penzias and Wilson 1965). In 1978, Penzias and Wilson won the Nobel Prize.

Another Invitation from Feynman

Richard Feynman was an experimentalist as well as a theoretician. CalTech once assigned him to teach introductory physics. His notes are immortalized in *The Feynman Lectures on Physics, Volumes 1–3* (1963). He remarked about fluid physics in volume 1, chapter 3 of his lectures:

> There is a physical problem that is common to many fields, that is very old, and that has not been solved. It is not the problem of finding new fundamental particles, but something left over from a long time ago – over a hundred years. Nobody in physics has really been able to analyze it mathematically satisfactorily in spite of its importance to the sister sciences. It is the analysis of circulating or turbulent fluids.

In volume 2, Feynman went into more depth in chapter 41 entitled "The Flow of Wet Water."

In a personal letter, Feynman admitted "The theory of turbulence (I have spent several years on it without success)." Turbulence was an area of classical physics beyond even

6 Penrose trained Stephen Hawking; together they wrote a number of landmark articles on the nature of the universe. Penrose's book *Road to Reality* (2005) highlights the math that underlies various areas of physics. His presentation of the reality and necessity of complex numbers is the best I've (RH) read.

Feynman's ability to solve. To this day, basic turbulence remains unsolved. As an added incentive, one of the seven Millennial Problems awaits solution of the Navier–Stokes equations which govern fluid flows.

Consider Feynman's challenge as an invitation to thermo-fluids, experimental or theoretical. It is the most challenging area of classical physics. Turbulence becomes further complicated by heat transfer; yet more complicated by mass transfer; yet more by chemical reactions or combustion; yet more complicated by electromagnetic interactions. It is important for flight, for weather, for breath and blood, for life, for engines, for circulation within celestial stars. Flows of liquids, gases, and plasmas are found at the microscopic scale within living cells to the astronomic scales between galaxies.

One of our colleagues, Professor Adrian Bejan, overlaps with us in the same field and the same publisher. Bejan's Constructal Theory has brought a fresh theoretical approach to thermo-fluid systems, to urban planning, and to appreciating design in living creatures as well as other fields.

Extra Invitations to Experiments

At the time of the first edition, two popular TV shows featured experimental scientists. *NCIS* featured Abby the forensic scientist and Dr. Mallard (Ducky) the coroner; not only are they scientists, they are the most well-rounded characters. *The Big Bang Theory* featured five experimentalists: Bernadette (biology), Leonard (physics), Amy (neuroscience), Raj (astrophysics), and Howard (engineer and astronaut); and one theoretician, Sheldon. The one nonscientist, Penny, was the most well rounded in reality and society.

Finally, a fun yet serious invitation. Physicists dispute a "theory of measurement" concerning mysterious quantum behavior. In certain experiments, if a measurement is made the experiment behaves one way; if no measurement is made the experimental results are opposite. The contrary results are repeatable. Include in the dispute: Is a human required as observer? Can a cat be the observer? If so, would Schrödinger's cat endanger itself? An internet search of this topic results in millions of hits, so we recommend Richard Muller's *Now: The Physics of Time* (2016).

Since the idea of this book is to equip you, dear Reader, to plan and execute taking measurements in your experiment, this dispute is moot.

Panel 2.3 Prepublishing Your Experiment Plan

We advise publishing and presenting your experiment plan to your client before collecting data. In the same vein, Button et al. in *Nature Reviews Neuroscience* (2013) made the following recommendations for researchers:

- Perform an a priori power calculation. [Our note: This helps estimate sample size.]
- Disclose methods and findings transparently.
- Preregister your study protocol and analysis plan.
- Make study materials and data available.
- Work collaboratively to increase power and replicate findings.

These recommendations combined with our strategies and guidelines help to achieve a credible experiment.

2.4.7 Surveys and Polls

If your type of experiment is a survey or poll, this text is also for you. The same strategies for sampling, statistics, and modeling apply. For a poll to promise any sort of accuracy, the sampling must be random and guided by good statistical practice. The principal challenges are designing nonleading questions and obtaining a representative sample. Otherwise a poll will mislead. Who has not witnessed and pained over a misleading poll?

Surveys of selective scientific audiences have a mixed history. One survey is particularly notable to us as experimentalists. In 2006, a survey of the International Astronomical Union (IAU) in Prague became memorable when the IAU demoted the planet Pluto. Pluto had been discovered and recognized as a planet in 1930. For decades teachers and textbooks across the world taught the nine planets in the solar system. Then the press in 2006 widely reported that the IAU declared that Pluto was no longer a planet. From our viewpoint, as IAU outsiders, it was a decree with huge impact on classrooms and the textbook industry.

As experimentalists, we are interested in the sampling, statistics, accuracy, and credibility of the Pluto survey. These all have relevance to us as we consider "The Nature of Experimental Work" (this chapter). We read that within the IAU there is dispute and controversy. How well did the vote represent the entire population?

Let's review details: The total population of IAU members is about 11,000. In 2012, the IAU reported 10,894 individual members from 93 countries worldwide. At the 2006 Prague conference, 2,400 members had registered. On the last day of the conference, just 424 astronomers remained and were polled on several issues. On some issues, exact numbers of the vote were reported (IAU 2006). However, the definitions that resulted in Pluto's demotion was reportedly decided by voice vote, approved by a majority.

How representative of the IAU population was the poll? For a sample size of 424 randomized members out of 11,000, a typical confidence interval could be the claimed 5%. The confidence interval surrounds an estimate, however a voice vote denies one. With a savvy, engaged population of size 11,000, why not take a vote of the full IAU population? Why rely on a poll to redefine international standards?

The IAU clearly has the right to set its own standards and definitions. Via the 2006 Pluto poll however, the IAU leaders decided for discriminating communication within its own community. How well did the IAU poll represent all science teachers when it demoted Pluto? By its decision, one consequence was to make existing text books obsolete, affecting the budgets of school systems worldwide.

Did the IAU leadership poorly serve its members, other scientists, and students across the world by relying on a poll instead of a full vote? From history, can we find examples where polls with small margins of error failed to predict the vote?

Another survey of note, recently conducted by the journal *Nature*, hearkened back to the Reproducibility Crisis announced by Ioannidis. *Nature* published (2016) a survey of researchers across the living and hard sciences asking "Is There a Reproducibility Crisis?" The author, Monya Baker, highlights a contradiction within the survey (Baker 2016): although researchers expected high confidence in their own fields, they admitted an inability to replicate results a majority of the time.

On a related note we may ask: what place have fashion and popularity in science? Fashion does have a notable impact on availability of funding. We defer to Panel 2.2, "Selected Invitations to Experimental Research, Insights from Theoreticians," because funding encompasses all research, not just experiments.

2.5 Uncertainty

Physical measurements are subject to errors from many sources. Try as we may, we can never entirely correct for all the possible error sources. Sometimes we may simply decide it is too expensive to correct for a small possible error. Sometimes, we don't know an error is present. In the best of situations, we must admit that even our corrected value is probably not exact. To deal with this situation we need a quantitative way to estimate the possible residual error and to describe it to potential users of the data. This problem was recognized by Airy (1879), who used the term "uncertainty," which he defined as "The possible value that the residual error may have."

Most engineering experiments involve several measurements, and the result of the experiment is derived from the measured values using a set of equations called the "data interpretation program." The challenge is to estimate the uncertainty in the derived result as a consequence of the recognized uncertainties in the measurements.

2.6 Uncertainty Analysis

Low uncertainty, by itself, is no assurance of accuracy in our results. The mathematical process by which we estimate the uncertainty in the result is referred to as "uncertainty analysis."

Uncertainty analysis is a powerful tool that, properly used, can help an experimenter develop a credible experiment. For example, we can use uncertainty analysis during the planning phase in order to select the approach offering the least uncertainty. We can use it to choose the most appropriate instruments. In the "shakedown and debugging phase" of an experiment, it helps to attribute which residual errors cause differences between our result and the expected result.

Uncertainty analysis was introduced to the American technical literature in a paper by Kline and McClintock (1953). Uncertainty analysis focuses on estimating how much uncertainty there is in the derived result as a consequence of the acknowledged uncertainties in the measurements. It follows a well-defined set of operations involving the data interpretation equations and the input data. The basic mathematics have not changed since that first description, but the techniques have been elaborated and extended considerably. See references at the end of this chapter and chapters 10, 11.

Uncertainty analysis is an essential element in experiment planning, before and during execution. Uncertainty analysis is a tool by which we can identify the significance of small changes in the output. If the observed difference is larger than the derived uncertainty, this is evidence that the process being studied is not well modeled by the equations used – in other words, something is going on that we don't know about. Chapters 10 and 11 are devoted to uncertainty analysis.

Experimentally, we never know we are "right." An essential element to a credible experiment is acknowledging that the best we can ever say is, "We have no evidence or suspicion that we are wrong."

Reporting the uncertainty of every reported value is essential for others to believe our science.

References

Airy, S.G.B. (1879). *Theory of Errors of Observation.* London, UK: Macmillan and Company.

Baals, D.D. and Corliss, W.R. (1981). Wind Tunnels of NASA. NASA SP-440.

Baker, M. (2016). Is there a reproducibility crisis? *Nature* 533: 452–454.

Button, K.S., Ioannidis, J.P.A., Mokrysz, C. et al. (2013). Power failure: why small sample size undermines the reliability of neuroscience.

Feynman, R. (1963). *The Feynman Lectures on Physics, Volumes 1–3.* MA: Addison-Wesley.

International Astronomical Union (2006). IAU 2006 General Assembly: Result of the IAU Resolution votes. https://www.iau.org/news/pressreleases/detail/iau0603.

Ioannidis, J.A. (2005a). *Contradicted and initially stronger effects in highly cited clinical research. JAMA* 294 (2): 218–228. https://doi.org/10.1001/jama.294.2.218. PMID 16014596.

Ioannidis, J.P.A. (2005b). Why most published research findings are false. *PLoS Medicine* 2 (8): e124.

Ioannidis, J.P.A. (2012). Why science is not necessarily self-correcting. *Perspectives on Psychological Science* 7 (6): 645–654.

Kline, S.J. and McClintock, F.A. (1953). Describing the uncertainties in single-sample experiments. *Mechanical Engineering* 75: 3–8.

Muller, R. (2016). *Now: The Physics of Time.* NY: W.W. Norton & Company Ltd. ISBN-13: 978-0393285239.

Penrose, R. (2005). *The Road to Reality: A Complete Guide to the laws of the Universe.* Knopf. ISBN-13: 978-0679454434.

Penrose, R. (2016). *Fashion, Faith, and Fantasy in the New Physics of the Universe.* NJ: Princeton University Press. ISBN-13: 978-0-691-11979-3.

Penzias, A.A. and Wilson, R.W. (1965). A Measurement of Excess Antenna Temperature at 4080 Mc/s. *Astrophysical Journal* 142: 419–421.

Pomeroy, S.R. (2012). The key to science (and life) is being wrong. *Scientific American.* https://blogs.scientificamerican.com/guest-blog/the-key-to-science-and-life-is-being-wrong/.

Quantum Coffee (2014). Nobel Prizes in physics: Theorists vs. experimentalists. https://quantumcoffee.wordpress.com/2014/06/08/nobel-prizes-in-physics-theorists-vs-experimentalists

Simera, I., Moher, D., Hoey, J. et al. (2010). A catalogue of reporting guidelines for health research. *European Journal of Clinical Investigation* 40 (1): 35–53.

Skedung, L., Arvidsson, M., Chung, J.Y. et al. (2013). Feeling Small: Exploring the Tactile Perception Limits. *Scientific Reports* 3: 2617.

Smolin, L. (2006). *The Trouble with Physics: The Rise of String Theory, the Fall of a Science, and What Comes Next.* Houghton Mifflin Harcourt. ISBN-13: 978-0618551057.

Thomson, W. (1883). Wikiquote. https://en.wikiquote.org/wiki/William_Thomson
Tinsley, J., Molodtsov, M., Prevedel, R. et al. (2016). Direct detection of a single photon by humans. *Nature Communications* 7: 12172.
Woit, P. (2006). *Not Even Wrong: The Failure of String Theory and the Continuing Challenge to Unify the Laws of Physics*. Basic Books. ISBN-13: 978-0465092765.

Homework

2.1 Prepare a lab computer for dual boot: Windows and the Linux operating system. Or ...

2.2 Prepare a dedicated lab computer for the Linux operating system. Or Exercise 2.1.

2.3 Following the guide in Appendix D3, download and install Gosset, public domain, open source, and free. Please consider this software tool essential.

3

An Overview of Experiment Planning

My background is mainly in research and development experiments in heat transfer and fluid physics. When I think of planning an experiment, I think about wind tunnels, heat exchangers, temperature, and flow control. I have tried to generalize my experience, but my background certainly colors my outlook.

Experimental work is expensive. Although costs vary for different fields and situations, the time costs are all similar: all laboratory work runs in real time, an hour for an hour, and there are no short cuts. It is important that experiments be well conceived, well executed, and well documented.

The purpose of experiments is to produce provably accurate data that answer an agreed-upon question about the behavior of a system. This is the kind of experiment I wish to discuss. Some of the important ideas will be transferable to other types of experiment, but the main thrust here is to deal with R&D experiments with tangible, numerical objectives.

The planning of such an experiment is necessarily iterative. The process starts with a tentative plan: a first impression of the goal, a plausible experimental approach, a possible suite of instruments, a tentative set of tests to run. This plan must then be challenged: will it produce the desired information with acceptable accuracy? Then the plan is refined, sometimes by improving the statement of objectives, sometimes by selecting a better approach, sometimes by improving the instrumentation.

3.1 Steps in an Experimental Plan

The general steps in an experimental program can be summarized in the following outline:

I) Assignment.

<u>The Iterative Loop</u>

II) Determine the objectives.
III) Select the experimental approach.
IV) Parametrically design an apparatus.

Planning and Executing Credible Experiments: A Guidebook for Engineering, Science, Industrial Processes, Agriculture, and Business, First Edition. Robert J. Moffat and Roy W. Henk.

Companion website: www.wiley.com/go/moffat/planning

V) Design apparatus hardware.
VI) Construct and install apparatus.
VII) Design analysis software; debug with fabricated sample data.
VIII) Perform shakedown, debugging, and qualification runs.

The Execution

IX) Collect data.
X) Reduce data and analyze.
XI) Report.

Seldom is a successful research experiment designed on a once-through basis. This is not surprising when you think about the amount of scratch paper usually generated in trying to develop an original analysis. Alternative experimental approaches must be investigated and trade-offs made between accuracy and convenience, range and speed, etc.

3.2 Iteration and Refinement

Research echoes sailing into uncharted territory. Sailboats cannot sail directly into the wind. The skipper must tack into the wind, repeatedly aiming right then left, iteratively correcting course.

This book takes an iterative approach to experiment planning. For example, it may not be clear how to execute some of the steps until later, when their background material has been developed. Some of the steps themselves may not even make sense until the background material has been developed. And some of the background material we ask for won't make sense until the need for it has been established. This circularity is typical of large-scale projects with interactions: They cannot be studied sequentially, they have to be approached integrating "all at once" and iterating.

We emphasize the iterative nature of experiments. What seems plausible at first may not prove acceptable later. Sometimes a preliminary "exploratory" experiment is a good investment – to test an approach. It may show that the original concept of the experiment is not a good one. You likely will iterate the experiment itself, as in Figure 1.1. Details of the planning steps and their objectives are dealt with in subsequent chapters.

An additional issue of risk assessment warrants early treatment because it affects every decision in the planning process.

3.3 Risk Assessment/Risk Abatement

Consider the following risks:

- The data may not answer the motivating question.
- The data may not be provably accurate.
- The schedule may not be met.
- The budget may be exceeded.

Before the experiment is authorized to proceed, it must be established to the satisfaction of the reviewers that the motivating question can be answered with the desired accuracy, within the time allowed, and within the allowable budget.

3.4 Questions to Guide Planning of an Experiment

Tables 3.1–3.3 elaborate the contents for the rest of this book – the topics that we feel should be addressed during the planning of an experiment.

Before some of the above topics can be addressed directly, some background material must be developed on the topology of experiments and the handling of experimental uncertainties.

Table 3.1 Overview of a research experiment plan.

Set up the experiment log.	*Keep a detailed log of your decisions.*
Identify:	
i. The motivating question.	*What question are you trying to answer?*
ii. The form of an acceptable answer.	*What should the answer look like?*
iii. The allowable uncertainty.	*What accuracy do you need?*
Design the data interpretation program (DIP).	*What equations provide the answer?*
Specify the data you need.	*Output data, peripheral data, and control values.*
Establish the allowable uncertainties.	*How accurately must variables be measured in order to get useful results?*
Select the instruments.	*Cross-check with required uncertainties.*
Specify the operable domain.	*What range of conditions must be covered?*
Estimate the shape of the response surface.	*What will be the likely outcome?*
Select the data trajectories and data-density distribution.	*How should the data points be distributed over the operating surface?*
Design the hardware.	*The apparatus must create the desired domain.*

Table 3.2 Review the program plan. Do risk assessment and plan risk abatement. If satisfactory, go ahead. If not, go back.

Build apparatus.	*Watch schedule and cost. Track critical path and critical person lines.*
Write the DIP.	*Convert equations and measurable inputs to selected response variables. Debug program.*
Shake down apparatus.	*First, make it repeatable. Then, make it work. Finally, make it work well.*
Execute qualification runs and document credibility.	*Calibrate. Document uniformity and stability. Certify baseline data.*

Table 3.3 Assess the credibility of the program. Do risk assessment and plan risk abatement. If satisfactory, go ahead. If not, go back.

Take production data.	*These are the required results.*
Interpret the results.	*What do the results mean to the client or target audience?*
Document the experiment.	*Present the results and interpret them. Record the data that support these conclusions and establish credibility of the experiment.*

Homework

3.1 From your review of prior work and research, what risks might your experiment encounter?

3.2 What risks will your client accept?

3.3 What risks will your client face by not pursuing this experiment?

3.4 How much will your client stake on successful completion of this experiment?

3.5 What is the cost of a null answer?

3.6 How much risk will your client assume?

3.7 Do the Occupational Safety and Health Administration (OSHA) or state laws or national codes limit the direction and extent that your proposed experiment can pursue?

3.8 Rinse and repeat from Exercise 3.1.

4

Identifying the Motivating Question

The motivating question is the question that, if answered, justifies the entire cost of running the experiment.

4.1 The Prime Need

I strongly believe in organizing every research-type experiment around a question, for several reasons.

i) When your goal is to answer a question, you know you can quit when you have an acceptable answer!
ii) If your experimental objective was "to study...," or "to investigate...," or "to document...," then you may never know when to quit. There is no end to "studying" or "investigating." You will quit only when the money runs out or you get bored or reassigned, but you will never be finished. If, on the other hand, you have a specific question to answer, you can quit when you have the answer or when you can prove that you can't get the answer by this kind of experiment.
iii) Knowing the motivating question helps in making the trade-off decisions during planning and debugging.

A research program generally begins with a "need to know," an urge on someone's part to solve a problem. Unfortunately, the person who first feels the urge may have in mind some steps toward what he/she thinks is the solution and may present an experiment plan that is a path to that particular "solution" rather than a path to solving the general problem.

To some extent, this can be avoided by specifically addressing the issue of "What question are we trying to answer?" instead of "What are we going to do?"

I (RM) came to this approach to experiment planning after many years of dealing with talented graduate students, each eager to get on with their programs but frequently stalling out when obstacles arose in the lab. I would find them at my door wondering what to do about the latest nuisance. Trying the simple approach of answering their questions, I found that the pace of research was then dictated by the number of hours I spent dealing with their problems. It was like pushing a rope! Every obstacle would bring the program to a halt. This sort of experience must be more common than I had thought,

Planning and Executing Credible Experiments: A Guidebook for Engineering, Science, Industrial Processes, Agriculture, and Business, First Edition. Robert J. Moffat and Roy W. Henk.

Companion website: www.wiley.com/go/moffat/planning

because it is the focus of an American folk song, "There's a Hole in My Bucket," which is presented below for your amusement (Panel 4.1).

With the help of (and some resistance from) my students, I discovered the problem: I was not communicating well enough. The students and I did not always share the same

Panel 4.1 There's a Hole in My Bucket

(American folk song, origin unknown)
A dialog between Henry and Liza over a chore with never-ending quandaries!

Henry	Liza
There's a hole in my bucket! Dear Liza, Dear Liza.	Well, mend it! Dear Henry, Dear Henry, Dear Henry.
There's a hole in my bucket, Dear Liza,	Well, mend it! Dear Henry, Dear Henry,
A hole!	Mend it!
With what shall I mend it? Dear Liza, Dear Liza.	With a straw! Dear Henry, Dear Henry, Dear Henry.
With what shall I mend it? Dear Liza,	With a straw! Dear Henry, Dear Henry,
With what?	A straw!
But the straw is too long! Dear Liza, Dear Liza.	Then cut it! Dear Henry, Dear Henry, Dear Henry.
But the straw is too long! Dear Liza,	Then cut it! Dear Henry, Dear Henry,
Too long!	Cut it!
With what shall I cut it? Dear Liza, Dear Liza.	With a knife! Dear Henry, Dear Henry, Dear Henry.
With what shall I cut it? Dear Liza,	With a knife! Dear Henry, Dear Henry,
Cut it?	A knife!
But the knife is too dull! Dear Liza, Dear Liza.	Then whet it! Dear Henry, Dear Henry, Dear Henry.
But the knife is too dull! Dear Liza,	Then whet it! Dear Henry, Dear Henry,
Too dull!	Whet it! (*Whet* means "*to sharpen.*")
With what shall I whet it? Dear Liza, Dear Liza.	With a stone! Dear Henry, Dear Henry, Dear Henry.
With what shall I whet it? Dear Liza,	With a stone! Dear Henry, Dear Henry,
With what?	A stone!
But the stone is too dry! Dear Liza, Dear Liza.	Then wet it! Dear Henry, Dear Henry, Dear Henry.
But the stone is too dry! Dear Liza,	Then wet it! Dear Henry, Dear Henry,
Too dry!	Wet it!
With what shall I wet it? Dear Liza, Dear Liza.	With water! Dear Henry, Dear Henry, Dear Henry.
With what shall I wet it? Dear Liza,	With water! Dear Henry, Dear Henry,
Wet it?	Water!
With what shall I fetch it? Dear Liza, Dear Liza.	With a bucket! Dear Henry, Dear Henry, Dear Henry.
With what shall I fetch it? Dear Liza,	With a bucket! Dear Henry, Dear Henry,
With what?	A bucket!

There's a hole in my bucket!

Virginia Maier, of Pittsburgh, PA, gave me a tape of this song from her collection.

Ever been through one of these discussions? I have, many of them – with grad students who didn't have a clear picture of what needed to be done!

picture of what should be done next. I decided that this was because I was trying to tell them what to do, whereas I should have been telling them where they were trying to go. I began to address the central issue: how could the student and I come to a meeting of the minds such that they would be able to solve all of their problems without coming to me?

This led to the practice of formulating the lab programs in terms of the questions we were trying to answer.

4.2 An Anchor and a Sieve

I view the motivating question as a sort of intellectual nail driven into the wall of the world and to which the student is tied by some sort of intellectual "rope." Once the motivating question has been really accepted (that is, thoroughly understood and accepted), the student is "tied" to that nail by her/his own rope of intent to finish. When students encounter an obstacle, they can pull themselves past it, always choosing a solution that moves them closer to the objective. This is a far different dynamic than the student trying to respond to my suggestions of what to do next. Once this approach was in place, I no longer felt I was trying to push ropes.[1]

I strongly recommend figuring out the "motivating question" for each experiment as a way of focusing attention on what must really be done. There are other advantages that accrue from working to answer a question as opposed to working to take data or study something. I hope they will become clear in this chapter.

4.3 Identifying the Motivating Question Clarifies Thinking

How can one identify the motivating question that drives an experiment, and who has the right to make this identification? Not the experimenter! Rather, it is the client who has this right because this is the person who is paying the bill for the experiment!

Identification of the motivating question is rarely easy. The urge to learn something may be strongly felt, but expressing exactly what you want to learn is very difficult. It is bad enough trying to talk about the subject when you have an hour or more to try to make clear your intentions. It is incredibly difficult to write down a concise description in such language that it cannot be misunderstood – and that is what we are trying to do. How often has the exercise of writing exposed our own sloppy thinking? How often have we felt despair at "trying to get it right" in writing? Take courage! The end result, your motivating question, is worth your effort to identify it early!

This chapter addresses how to formulate the motivating question and the advantage of working to a question you are trying to answer. If you experiment without a motivating question, you will take a series of steps – but when will you arrive?

4.3.1 Getting Started

When the urge strikes to run a new experiment, don't fight it. If the urge has struck you, then talk to yourself – takes notes on what you think you should do, what you think the rig might look like, or what you think might happen. Let this process run until

1 Upon my (Moffat) retirement from active teaching, my recent graduates presented me with a memorial plaque bearing a gold-plated piece of rope and the legend "You Can't Push a Rope." This saying is one of the first major lessons every engineer learns in the course Statics, when she/he draws forces on an object.

you feel you have really expressed yourself. Then stop and ask, "What question am I trying to answer?"

If the urge has struck someone else, and you are that person's sounding board, listen actively and take notes. Let the person keep talking until he or she runs dry – then the individual knows you have really heard the idea, entirely. Then ask, "If you do this, and you are successful, what question will you have answered?"

I think you must honor the enthusiastic urge by letting it have center stage until it plays itself out. I simply ask that the speaker (myself or my student) get reasonably specific right away. This is not intended in a critical sense but simply to detail the plan. I take notes about the proposed apparatus, the test conditions, and the proposed data, trying to absorb the real intent of the idea. I keep at this until the speaker (myself or the student) runs out of things to say about what to do.

At this point everything that could be said has been said. The pressure is off. We can get to work.

4.3.2 Probe and Focus

Now I raise the following five questions.

If we do this experiment and get all the data we have asked for:

Q1. What question will we have answered?
Q2. Is this question worth the cost of answering it?
Q3. Has it already been answered?
Q4. Is the proposed experiment the best way to answer it?
Q5. If we get what we asked for, will that solve the problem that led to this work?

Almost always, trying to answer Q1 honestly and carefully puts the issue in a new light.

Once attention is focused on *what we want to accomplish* instead of on *what we want to do*, we can admit the possible existence of other ways to accomplish the same objective. This often leads us to formulate a more important question than we first had in mind and to propose a quite different experiment.

It is a good idea to start by formulating several versions of what seems to be the motivating question, and then start critiquing them. The first versions will leave a lot unsaid, and a good "devil's advocate" approach will make some of them look downright naive.[2] Don't be reluctant to play devil's advocate – if the question allows a silly answer, the question is poorly formulated. Precision will soon start to emerge.

When you have a few good candidates, the test for identifying a really good motivating question is, "Which of these questions, if it alone is answered, will justify the cost of this experiment?"

There can be only one "top priority question." During the planning of the experiment, when a trade-off is required, the motivating question provides the decision criterion. The motivating question is the "mission statement" of the experiment.

2 On an exam in a graduate heat-transfer course, I once asked: "When can equation 8-16 be used?" One student responded, "On Thursday!" What could I say!

Sometimes, the question keeps on being refined, even while the program is in motion.[3] No harm in that.

4.4 Three Levels of Questions

In terms of experiment planning, it helps to distinguish three levels of questions that might be raised during the development of an experiment plan:

Level 1: What happens?
Level 2: By what means does it happen?
Level 3: What are the underlying physics?

It is important to identify the level of the motivating question and ensure that all following decisions are aimed at progress at that same level. The different levels lead to different experiments for a given situation.

For example, consider a program aimed at studying the effects of high free-stream turbulence u' on the value of the heat-transfer coefficient h on a flat plate.

The Level 1 question would be: "How does the heat-transfer coefficient h vary with free-stream turbulence u', all other factors remaining fixed?" This question would be answered by a series of experiments in which h was measured for different values of u' for some specified set of conditions.

A Level 2 question would be: "What changes in the structure of the boundary layer were responsible for the changes in h?" This question would be answered by measurements of the velocity and temperature distribution within the boundary layer, seeking a correlation between high h and some recognizable change in the boundary layer structure.

A Level 3 question would be: "How have the momentum and energy transports within the boundary layer been altered by the turbulence?" This question would be answered by measurements of the mixing length and turbulent Prandtl number or the turbulence intensity and dissipation length scale within the boundary layer.

Note that these different levels require entirely different types of measurements, so it is important to know which level is most important to the client, the ultimate customer.

There is an implicit presumption in modern fluid mechanics and heat transfer that low-order events can be predicted from high order, i.e. that knowledge of the turbulence transport mechanisms will allow calculation of the boundary layer structure, and that knowledge of the boundary layer structure will allow prediction of h at the surface.

This hierarchical assumption leads some experimenters to immediately start work at Level 2 or Level 3, seeking an elegant and powerful answer to the Level 1 question. In general, however, it seems to work best if one directly addresses the Level 1 question first over the entire range of the desired conditions before attempting to shift up to Level 2 and Level 3.

---

3 In one program, it took nearly a year to get the final form of the question worked out. The rig was nearly ready for shakedown runs before we were content. That question, written on a 3 × 5 card and thumbtacked to the wall in the office, guided our research through three PhD theses.

It is particularly important not to inadvertently mix levels in the same experiment plan. One may pursue two, or even three, levels of questions within one series of tests, and that is frequently done to save testing time, but one should remain aware that each level requires different data and that the different datasets are addressing separate objectives.

4.5 Strong Inference

A particularly powerful approach to selection of the "right" question is described by John R. Platt, professor of physics and biophysics at the University of Chicago (1953), as "strong inference." The essential features of strong inference are (1) the formulation of more than one alternative hypothesis concerning the major question, and (2) the execution of a set of experiments that have the possibility of disproving each hypothesis. The power of the method lies in the fact that, once a hypothesis has been disproved, then no further work need be done along that line. Note that if an experiment simply supports a hypothesis, not much progress has really been made, because the very next experiment may disprove it. To use strong inference, one must assemble systems of hypotheses that fully enclose the main question so that at least one hypothesis must finally survive its experimental test.

Platt lists the following steps:

i) Devise alternative hypotheses.
ii) Devise a crucial experiment (or several of them) with alternative possible outcomes, each of which will as nearly as possible exclude one or more of the hypotheses.
iii) Carry out the experiments so as to get clean results.
iv) Recycle the procedure, making subhypotheses or sequential hypotheses to refine the possibilities.

Identification of the motivating question corresponds to the step devising alternative hypotheses. The specific experimental objective is, then, to test these hypotheses by experiment.

Platt's contribution is his reminder to seek disproof for hypotheses, rather than support, as a more economic strategy.

4.6 Agree on the Form of an Acceptable Answer

Once the question has been determined, the next step is to agree on the form of an acceptable answer. A clear agreement on the form of an acceptable answer helps ensure that both parties have the same question in mind.

Thinking about the form of an acceptable answer may, in fact, lead to refining the question, especially if you play "devil's advocate" and look for "silly" answers. If "silly" answers are not precluded by the form of the question, then the question needs to be refined.

The question "What is the effectiveness of this heat exchanger at its design conditions?" needs only a single number for an answer: "The effectiveness is 0.86."

Another example, using instrument calibration, has more latitude. If the calibration question was: "Does this instrument meet its accuracy specification?" then a satisfactory answer would be a simple "Yes" or "No." If, however, the question was "How do the readings of this instrument compare with the true values, over its range?" then an acceptable answer would require a table of values, or a chart, listing the indicated value corresponding to each true value.

Note that a simple request to "Calibrate this instrument" does not establish which of these two answers would be acceptable and does not qualify as a desirable motivating question.

4.7 Specify the Allowable Uncertainty

The third point that should be dealt with is uncertainty. How much uncertainty can be tolerated in the answer to the motivating question?

For most heat-transfer situations, an uncertainty of ±5% is "Olympic-quality" data. Much of our present heat-transfer understanding was developed from historic and legacy databases with ±20% uncertainty. Often, the client already has some data or has a design approach that works reasonably well but wants to reduce the uncertainty. In such a situation, to produce more data with the same uncertainty as the existing dataset will not advance the art, nor will it satisfy the client.

The allowable uncertainty should be described in a manner consistent with current conventions. For most engineering work, the uncertainty is specified at the 95% confidence level and presumes a symmetric probability distribution, centered around zero error. For example, if the uncertainty is quoted as ±10% at the 95% confidence level, this means that 19 of 20 repeated trials of the same measurement act will yield a value within ±10% of the quoted value. The possible error is, by agreement, equally likely to be positive or negative.

In some cases, the client may wish to specify a "one-sided" tolerance, for example to demonstrate that a product is better than its competition. A one-sided test must be selected before the experiment is run. The one-sided test must be clearly indicated in the report so that the results can be interpreted accurately in terms of the one-sided error allowance. Otherwise, the experiment will be reckoned as a two-sided or "zero-centered" test.

4.8 Final Closure

From my experience, a written memo of understanding should be prepared that clearly states the motivating question, the form of an allowable answer, and the acceptable uncertainty. This should become a part of the experimental record.

It is difficult for some project managers to sign off on such a statement as, "The answer to this question, with no other benefit, will justify the cost of this experiment." Lurking in the back of most minds are hopes for "spin-off data," bits of knowledge they hope to get, or a general feeling that the experiment ought to be "kept loose, so we can steer it as we go." These urges usually arise out of reluctance to come to grips with the specific problem: what is it that we really must know?

I recommend organizing a "signing party" at which the people with a stake in the issue sign off on the question, the form of its answer, and the uncertainty. Too often, higher management pays little or no attention to these "philosophical issues" until too late. There is something about deliberately signing off that makes people pay attention.

Unless there is explicit agreement, early, on the objective of the experiment, a host of serious issues may ensue:

- Disagreements on "what to do next."
- Dissatisfaction with the outcome.
- "Second guessing" after the final report.
- A general whitewashing to cover up the lack of definite progress.

Too many experiments simply don't answer any question – they only contribute some facts that may or may not be useful.

Reference

Platt, J.R. (1953). *Address before the Division of Physical Chemistry*. American Chemical Society.

Homework

4.1 Propose at least three potential motivating questions that your experiment can answer.

4.2 Discuss with your client each of the potential motivating questions. Which one is preferred?

4.3 What questions does your client raise in light of the preferred motivating question?

4.4 How can this motivating question be refined? Can it be made more concrete?

4.5 Can your motivating question be clearly achieved?

4.6 Rinse and repeat from Exercise 4.1.

5

Choosing the Approach

5.1 Laying Groundwork

The "experimental approach" includes three parts: (i) the technique to be used, (ii) the data acquisition system (the instrumentation), and (iii) the data interpretation equations. Together these constitute an "approach." It is not possible to make a critical evaluation of one of the three without considering the other two at the same time.

A reasonable first step in choosing the approach would be to review the experimental literature: How has this measurement been handled in the past? What difficulties were encountered? How did the choice of method restrict the allowable range of variables?

The analytical literature will provide a source for another input: What assumptions are used in the accepted analyses? Can the necessary assumptions be satisfied in our experiment? Can our experiment cover a broader range of applicability by making fewer assumptions?

New experimental techniques and measuring systems are constantly being introduced. Have new techniques opened the way for a more direct approach to measure the desired results?

Computer data reduction allows greater sophistication in the handling of peripheral effects, such as variations in barometric pressure, humidity, etc., on flow measurement, changes in thermal conductivity with temperature, and so forth. Experiments that were too tedious to consider in the past due to the necessary data reduction can now be executed almost as easily as the simplest direct measurement.

Uncertainty analysis combined with computer data reduction is a powerful tool in evaluating proposed experimental methods. The object of an experimental program is to obtain data of as high a degree of reliability as possible, for a given amount of effort. Some approaches are inherently less sensitive to error propagation than others. Uncertainty propagation at constant probability is an accepted technique for determining the effect of uncertainties in the input values on the output. This technique was laid out by Kline and McClintock (1953, p. 3); we demonstrate uncertainty analysis in Chapter 10; Coleman and Steele (C&S; see preface) extensively present variations on the concepts in their text. Comparison of the uncertainties in the proposed methods, before the program is settled, will allow best use of the available resources by pointing out which methods are least susceptible to input perturbations.

Planning and Executing Credible Experiments: A Guidebook for Engineering, Science, Industrial Processes, Agriculture, and Business, First Edition. Robert J. Moffat and Roy W. Henk.

Companion website: www.wiley.com/go/moffat/planning

The uncertainty analysis should be built into the data-reduction program, and the uncertainty in each result displayed along with its value. Many tests are acceptably accurate in certain ranges of the variables and wildly uncertain in other ranges. During a years-long experimental program, it is not always easy to remember that the test must not be operated at certain combinations of the input parameters!

Critical consideration of the possible approaches will frequently point out extensions to the range or new interpretations addressed by the motivating question, which may have a bearing on the objectives of the program. After the approach has tentatively been chosen, the objectives and approach should be considered together and evaluated for compatibility. Can the desired accuracy be obtained over the desired range of variables, when the data are obtained in the proposed manner, and the results deduced with the proposed data-reduction program?

"Philosophical issues," such as identifying the motivating question, should be addressed early on to ensure that everyone has the same objective for the experiment.

5.2 Experiment Classifications

Clarify: what kind of experiment are we trying to run? Several general classifications of experiments are described in the following sections.

5.2.1 Exploratory

An exploratory experiment is often put together using whatever is at hand, to find out whether or not some process or phenomenon is worth studying carefully at a later date. Consequently, exploratory experiments often have large uncertainties in their data (20% or larger) and display a lot of scatter in their results. The results need only be accurate enough to answer the question "Is this area worth further study?" for such an experiment to be called a success.

For example, old rice cookers just boiled the rice until finished and then kept the rice warm. Then somebody explored "Can rice be cooked more deliciously?" It can be. High-tech rice cookers now follow a programmed temperature trajectory. Sometimes an exploratory experiment is purely qualitative in nature, like a visualization experiment conducted to learn the general features of a flow field.

5.2.2 Identifying the Important Variables

The relative importance of the variables in a new situation is usually determined using a statistical experiment plan, such as a two-level factorial or fractional factorial design or a screening design. The intent of such an experiment is simply to rank the candidate variables in order of importance. The incentive to do this, in turn, arises from the desire to simplify future experiments by restricting the number of variables that must be considered. Since excluded variables become contributors to uncertainty in the results, we wish to exclude only variables that cause least variation of the results or that are too expensive to measure. For our rice cooker example, food scientists included such candidate variables as time; temperature; white or brown rice; washed or unwashed; grain length short, medium, or long; origin of rice; and shape of temperature vs. time trajectory. A glance at the available settings of a modern rice cooker reveals which variables survived this experimental test.

5.2.3 Demonstration of System Performance

Most engineering experiments in industry are single-factor-at-a-time parametric studies aimed at documenting the performance of a system, or component, over its operating range. This is what most engineers think of when the word "experiment" is mentioned. A parametric experiment can be thought of as a partial differentiation of the process: vary "X" over some specified range while holding everything else constant and plot the results as a function of "X." The accuracy of measured numerical results is critical to the value of this type of experiment, documenting the performance of the equipment.

5.2.4 Testing a Hypothesis

Hypothesis-testing experiments come in two styles: supportive and refutative. A supportive experiment seeks evidence that the hypothesis is true, within the domain of the experiment, while a refutative experiment seeks evidence that the hypothesis is false. The distinction between supportive and refutative experiments is discussed by John R. Platt, highlighted in Section 5.6. Both supportive and refutative experiments are useful in engineering. Rarely do we seek invariant truths or universal constants, statements true without reservation or limitation. Usually we are content to examine truth over some specified range of conditions. An experiment that refutes a hypothesis makes a stronger statement than an experiment that supports it: a hypothesis need only be defeated once to be defeated entirely. Strictly speaking, most experiments don't prove or disprove anything. The most powerful statement that can be made based on most experiments is that over the range of conditions tested, the hypothesis appears to be true.

5.2.5 Developing Constants for Predetermined Models

Engineering analysis often deals with phenomena that are not fully understood at the fundamental physical level. This has made it customary to rely on plausible models that incorporate specially defined constants, often evaluated experimentally. The value of these constants depends both on the structure of the model and on the situation studied.

In our rice cooker example, manufacturers pursued experiments to determine the length of time (one constant) to maintain the rice at each temperature (another constant) through its cooking trajectory (the predetermined model).

Convection heat transfer is a prime example of a field that continues to improve models for engineers to use in their designs. Convection fluid flow remains the untamed field of classical physics, in contrast to conduction heat transfer and radiation heat transfer, where the physics is well understood. The addition of heat transfer or chemical reactions to flow convection compounds the complexity. Due to this complexity, engineers designing practical systems with fluid flow require useful models with predetermined coefficients. An example constant within convection heat transfer is labeled the "turbulent Prandtl number," which is defined as the ratio of the apparent turbulent diffusivities of momentum and heat. These two apparent turbulent diffusivities likewise are two other modeled constants deduced from temperature, velocity, shear stress, and heat flux distribution data. The turbulent Prandtl number is used to predict the temperature distribution once the velocity distribution has been found. The problem is that, given the same experimental data, two different expert modelers could easily deduce different distributions of the turbulent Prandtl number if their descriptions of the momentum diffusivity distribution were different. Both turbulent Prandtl number models, used appropriately with their respective momentum diffusivity models, might well predict the same temperature distribution in the boundary layer. However, the reported distributions of turbulent Prandtl number might be quite different. In short, measured values of such modeling constants cannot be tested for truth or falsity independent of the model for which they are intended: they are true if they work in the designated model and false if they don't.

This problem surfaces also when heat-transfer coefficients are "derived" from overall heat exchanger behavior and then compared with textbook values or computational fluid dynamics results. The "derived" heat-transfer coefficient incorporates all of the errors in the measured performance, filtered through the theoretical model presumed to describe the overall performance in terms of h.

5.2.6 Custody Transfer and System Performance Certification Tests

What is referred to here are tests whose procedures are rigidly prescribed, either by law, by contract, or by custom. Not only are the procedures fixed but also the allowable instruments and the method of data interpretation. For this class of tests, the accuracy of the results, in absolute terms, is of less importance than their repeatability. Units will be judged acceptable or not acceptable based on whether or not their indicated performance lies inside a specified band of values. Hopefully those bands will have been set based on broad, external criteria of acceptability. The acceptable behavior is then described in terms of the results from the standardized test procedures. An example is the test to check each fuel pump at a gasoline station. One does not lightly go about trying to improve the accuracy of such tests – once the criteria for acceptable performance have been established, no changes can be made in the test procedure unless a complete, overall recertification is contemplated. What is needed, in this type of experiment, is periodic evidence that the results are repeatable.

5.2.7 Quality-Assurance Tests

Quality-assurance testing is a combination of performance testing and sampling theory. From the experimental standpoint, absolute accuracy is not important, repeatability is. Once the criteria for acceptable performance have been described in terms of the output of standardized tests, the accuracy of the test procedures is no longer relevant. What is mainly needed here is, once again, a way of ensuring that the results are repeatable over long periods of time.

Quality assurance is a key focus in industries which emphasize quality manufactured products. The practices of such companies provide us extra guidance. In addition to product quality, these companies employ a variety of charts and tools to keep ongoing records of each piece of factory equipment and instrument. A partial list includes control chart, check sheet, cause-and-effect diagram, Pareto chart, flowchart, run chart, etc. The data collected and presented on these charts[1] are invaluable to your tests and

1 Quality assurance gained a solid foundation in industry notably through the work of W. Edwards Deming. Various industries in Japan, employing the methods of Deming, became known as top-quality manufacturers of cameras, automobiles, etc. at reasonable prices.

analyses. If your client lacks such charts for some of their equipment, by immediately adopting these charts they will boost your efforts to answer their Motivating Question.

5.2.8 Summary

Before getting too far into the plan for a specific experiment, it should be clearly agreed which classification of experiment is being considered. The criteria for a successful exploratory experiment are quite different from those for an archival experiment.

Regardless of the type, the credibility of your experiment will benefit by employing quality-assurance charts for each instrument and equipment. In remote locations, we have even used quality-assurance charts for batteries and light bulbs.

5.3 Real or Simplified Conditions?

This is a tough one. On the one hand, we are tied to reality in that whatever we try to measure must somehow be usable in field applications. On the other hand, it is difficult to get accurate measurements under field conditions. On the one hand, we are told "Let the engine vote."[2] On the other hand, we are trained to try to extract the significant elements from any situation and focus on the relevant physics.

This presents the experimentalist with a real challenge. Where, between reality and idealization, should the experiment be positioned? How much reality should be preserved in the proposed experiment? How do we "Separate the wheat from the chaff"?

Two anecdotes illustrate the extremes: Charles Lamb's "Dissertation upon Roast Pig" (Panel 5.1) and Joel Ferziger's "Consider a Spherical Cow" (Panel 5.2).

Both extremes of this dilemma are well buttressed by hosts of wise old sayings and by experience. Modern gas turbine testing is still plagued by the question of how much reality is needed in bench testing. For example, there is evidence that the heat-transfer coefficients measured on turbine blades and vanes in engines are higher than those measured in either large-scale rotating warm rigs or in stationary hot annular cascades. There is evidence that data from hot annular cascades show higher values than hot linear cascades, and that hot linear cascade data are higher than produced by cold linear cascades. But the data are not conclusive. The question remains, "How much reality do we have to retain?"

My view is that the less you know about the system, the closer you should stick to reality. But staying real is expensive. Cold linear cascade data are a lot cheaper than hot linear cascade data, and hot linear data are cheaper than hot annular data. And so on up the line to engine tests.

This, then, is one of the major challenges the experiment planner must face: how close to reality should the experiment be positioned?

5.4 Single-Sample or Multiple-Sample?

One decision that must be made early is whether to run a single-sample or a multiple-sample experiment.

2 W.A. Turunen, head, Gas Turbine Department, GM Research Labs, ca. 1955. Personal communication.

Panel 5.1 A Brief Summary of "Dissertation upon Roast Pig" (with Thanks and Apologies to Charles Lamb [1775–1834])

Charles Lamb was a fierce critic of British politics. One of his satirical essays published in England during the late eighteenth and early nineteenth centuries highlighted Parliament's inability to go after the root cause of a social problem.

In his essay, he described how roast pig was first introduced into London.

The London Mr. Lamb described was just like the real, turn-of-the-century London except that, in that particular era (his story went), no one had ever tasted cooked meat. All meat was eaten raw, although every kitchen was well equipped with stoves and ovens for cooking other dishes.

By accident, one family's house burned to the ground. Although all of the family escaped, the family's pet pig did not. It died in the fire, burned to death. The carcass of the poor beast, London's first roast pig, was discovered by the family while cleaning up the rubble.

At first, of course, they were saddened by their pet pig's demise, but they couldn't help noticing that the roast pig smelled pretty good. Being hungry, and opportunistic, they tasted it and were amazed at its fine flavor – far tastier than the raw pig they were accustomed to eating.

They shared this discovery with their neighbors, who were much impressed with the fine flavor of roast pig.

So impressed were the neighbors, in fact, that they burned down their own house and roasted their own pig.

Panel 5.2 Consider a Spherical Cow

Joel Ferziger, a colleague of mine at Stanford with a physics background, told of the theoretical physicist employed as a consultant by the California Milk Producers Association (CMPA) to increase the productivity of milk cows in California.

When first approached about this issue, he protested that he was a theoretical person, not a farmer. He doubted that he could contribute much. The CMPA was not to be put off. They were convinced that a sound, theoretically based approach would yield something of value. Unable to dissuade them that they were wrong, he accepted a provisional assignment.

He first demanded, and got, a private office with a good view, a new computer, a private secretary, and three months of solitude. At the end of three months he announced that he was, in fact, on the brink of discovering something important, but he was not yet ready to talk about it.

At the end of six months, the CMPA demanded a progress report.

The members assembled in the auditorium rented for the occasion, along with members of the press, alerted to the possible earth-shaking consequences of the announcement. The physicist stepped to the chalkboard, drew a perfect circle with a flourish and then, peering over his spectacles, began his lecture: "Consider a spherical cow."

The data from a single-sample experiment are processed one set at a time. Ten runs yield 10 results. In contrast, the data from a multiple-sample experiment are averaged first and then processed. Ten runs produce one result.

Performance assessment experiments and custody-transfer measurements are often multiple-sample experiments. The question addressed is, "Does the unit under test meet the performance guarantee?" Or "How much gas (or oil, or water, or steam) was delivered over this period?"

Averaging a large set of multiple-sample data taken during a period of stable operation and then calculating the result based on the average value of each measurand allows the random component of each measurement uncertainty to be quoted as a very small number. The alternative of processing each set of single-sample data individually and then averaging the result would also have yielded an average result and a similarly reduced estimate of uncertainty.

The issue to be aware of is that the average result found by processing the multiple-sample averages from the data is not the same as the average of the results calculated from the individual data sets.

I favor single-sample experiments over multiple-sample because single-sample experiments allow me to know what is going on. Averaging in multiple-sample experiments conceals some aspects of behavior that I wish to know about. For example, plotting the results from each set of observations as a function of time will often reveal something significant about the system: a periodicity that can be identified, a correlation between the results and time itself, or room temperature, or with opening and closing the door or ... whatever. If you can't look at the individual data, you can't see!

The two approaches, multiple-sample and single-sample, are equivalent only when the processes involved in the system are all linear and the experimental system was stable. In addition, the equations used in the data-reduction process must also be linear. In contrast, most of my world in thermo-fluids and convection heat transfer is nonlinear.

5.5 Statistical or Parametric Experiment Design?

The phrase "experiment design" has at least two distinctly different meanings. It may mean a statistical design (e.g. two-level factorial or fractional factorial) or it may mean a parametric design: a sequential, one-factor-at-a-time design. These two categories of experiment design have different objectives and play different roles. They are not interchangeable but are complementary.

Statistical experiment designs are generally used when a system sensitive to several variables operates at, or near, a single operating point, and the issue is "Which of these factors is most important"? The statistical experiment plan is used to determine which factors have the largest effect on the outcome of the process in the vicinity of that operating point.

Single-factor parametric experiments are used when a system operates over a wide range of only a few variables, as when calibrating a piece of equipment. The single-factor experiment is usually designed to hold constant all variables except one factor at a time, thus revealing the partial derivative of the result with respect to that factor over its entire range of possible values.

When approaching a new situation, it is often desirable to run a fractional factorial experiment design to identify the important factors and then a full factorial experiment design to identify interactions between those important factors. Then, finally, single-factor experiments fill in the response surface on each of the most important factors. These types of experimental plans work well together. They are complementary, not competitive.

In unfamiliar situations, it is common to apply both: first, a statistical plan to identify the important factors and then a parametric plan to determine how those factors affect the result over a wide range of conditions.

The difference between a statistical plan and a parametric plan is illustrated in Figure 5.1, for an example when there are only three variables, two factors adjustable by the operator and one resultant variable.

Statistical experiment plans are often used to rank order the importance of the factors in a new situation. In such experiments, a carefully chosen set of data points is used to identify the effects of each factor, centered on some single important operating point. The partial derivative of results with respect to each variable can be extracted from the measured effects. In effect, the results from a statistical plan describe the best tangent plane approximation to the operating surface.

Commercial statistical design of experiments packages offer several styles of full factorial or fractional factorial experiment plans. An alternative we recommend (and demonstrate) in chapter 9 is a powerful, free program for experiment design, developed at AT&T Bell Labs.

Parametric experiments, on the other hand, generally hold all but one variable (or parameter, such as Reynolds number) constant while varying the target variable over a wide range. These "one-factor-at-a-time" experiments are responsible for the largest fraction of output data generated by engineering experiments for the sake of calibrating industrial equipment or a system.

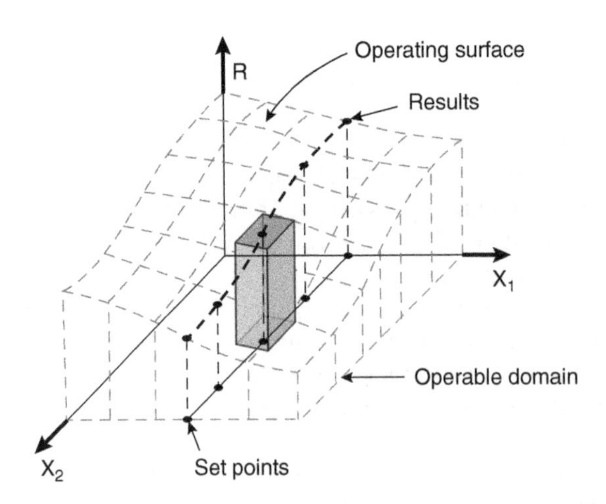

Figure 5.1 A statistical plan focuses on a region around the central point. A parametric plan traces a path across the operable domain.

5.6 Supportive or Refutative?

A strategy recommended by Platt (1953) is based on structuring experiments with the objective of refuting a hypothesis, rather than trying to support it. His thesis is that if you can disprove a hypothesis, you have accomplished something definite – that hypothesis can be dropped from the list of possible truths. Simply collecting supporting data may be only a momentary delay until someone else disproves it.

This is contrary to many engineering experiments that are aimed at amassing data in support of some notion.

Platt suggested the following strategy, which he named "strong inference":

i) Devise alternate hypotheses.
ii) Devise a set of experiments having alternative outcomes, each of which would conclusively defeat one of the hypotheses.
iii) Carry out the experiments with control of uncertainty to assure credible results.
iv) Recycle the process, making subhypotheses or sequential hypotheses to refine the possibilities remaining until one survives.

This strategy seems more likely to be applicable when strongly idealized tests are planned. It is a powerful concept. It may not be applicable to "reality-based" tests, with many uncontrolled variables. When it can be applied, it tends to lead to definite results in a small number of tests.

5.7 The Bottom Line

Once these general issues have been dealt with, it is time to get on with what is generally thought of as "picking the approach": selecting the test method and data interpretation equations.

The first step is a literature search: What methods have been used in the recent past to investigate situations like the one you have in mind? What ranges did they cover, what uncertainty did they yield?

Instrumentation capabilities change rapidly, and methods that were tedious and "of academic interest only" a few years ago may be accurate, reliable, and inexpensive now. For example, consider particle image velocimetry (PIV) to measure fluid motion. PIV was originally done by tracking particles in movie film, frame by frame, with the tracking done by graduate students. Now PIV is done by correlation analysis from high-speed video records. Mapping surface temperature distribution that has long been done using arrays of thermocouples is now done by full-field infrared or liquid crystal imaging.

Look both to journals that report advances in instrumentation and those that report on the physics of the events.

Sometimes local issues rule. If your organization has established policy to fill their database by a particular method, you will be judged prudent if you use that method. If you plan to introduce a new approach, you will have to baseline the new results against the old.

References

Kline, S.J. and McClintock, F.A. (1953). Describing the uncertainties in single-sample experiments. *Mechanical Engineering* 75: 3–8.

Platt, J.R. (1953). *Address before the Division of Physical Chemistry*. American Chemical Society.

Homework

5.1 Classify: which kind of test or experiment is preferred by the client?

5.2 List important factors and variables that can (or might) affect the results of the experiment.

5.3 List important factors and variables that can (or might) affect the operation of the experiment.

5.4 Which conditions can be realistic?

5.5 Which conditions must be simplified or idealized?

5.6 Plan all the factors that must be recorded. How will the values be observed? How will the values be recorded?

5.7 What forms and charts already exist for your targeted process? Take inventory of process charts and quality-control charts that are in use or must be implemented. Examples include Pareto charts, check sheets, control charts, cause-and-effect, and flowcharts. What charts are used to keep track of the performance of each piece of equipment? Each piece of equipment might become its own targeted experiment (if fault or failure).

5.8 Who records each value?

5.9 How can your Motivating Question be refined?

5.10 For beginning researchers, daily practice data collection. For your daily commute, use the commute chart (in online resources) to keep track of time, path, mode, and other factors. How would you improve it? For drivers, use the car operating record shown below to keep track of fuel and service. As an added benefit, such a chart typically increases the resale value of the vehicle.

Car operating record.

Date	Mileage	Gal.	Fuel $	Oil, qt.	Where	Remarks, repairs, other service

This table guides data collection practice from the daily life of student researchers. Your Motivating Question: How does my vehicle typically perform and can I detect when it changes?

Task 5.10a: Build a spreadsheet file where one page appears like this. You may add additional columns if you need.

Task 5.10b: Print the page and use it in your vehicle. (Note, this table often increases the resale value of the vehicle.)

Explanation of car operating record data.

Column name	Values	Units	Description
Date	Quantitative	mm/dd/yy	Standard date entry
Mileage	Quantitative	Mile	Reading on car odometer
Gal.	Quantitative	Gallon	Reading on fuel pump at service station
Fuel $	Quantitative	Dollar	Cost of fuel
Oil, qt.	Quantitative	Quart	Amount of oil added to engine on date
Where	Categorical	Location	Location of incident; service performed or fuel received
Remarks	Categorical	Text	Remarks, full/not, repairs, alerts, tire condition, octane, % ethanol

This table guides data collection practice from the daily life of student researchers.

Task 5.10c: Add a separate tab to the same spreadsheet file to create a sheet with this codebook. The codebook elaborates on the columns and values that appear in the data sheet.

6

Mapping for Safety, Operation, and Results

6.1 Construct Multiple Maps to Illustrate and Guide Experiment Plan

Before conducting your experiment, you may construct maps as a guide to obtain your treasured results safely and effectively. One map ensures that your data cover the necessary domain to get the new data that motivate your experiment and to check their quality. Another map keeps you, and other operators of the facility, within safe operating bounds. A third type of map anticipates the treasured results via a response surface to efficiently collect data while minimizing uncertainty.

The term "parametric mapping" describes the process of constructing a visual display of the domain of an experiment, using coordinates that fit your needs at the moment. Three types of maps are illustrated in this section, but there are a number of variants that you may find helpful. Such maps are helpful in planning the sequence of data points to be taken.

Verbal descriptions of an operable domain are not very reliable indicators of the combinations of conditions that can really be reached – some combinations of values may not be attainable, even though each parameter individually might be able to reach the desired value. A map is unambiguous.

The following sections illustrate the more common types of maps. These illustrations were made freehand, not using a graphics package, to show that anyone can work in three dimensions, artistic or not! When it comes to understanding what is going on in the lab, a three-dimensional isometric, sketched approximately on the back of an envelope RIGHT NOW, is more useful than an accurate rendition that comes out of the computer an hour from now.

6.2 Mapping Prior Work and Proposed Work

Parametric maps may be used to visualize the extent of prior work, as shown in Figure 6.1.

This is the operable domain plot that I (RM) generated to help define the scope of my doctoral thesis experiment. It shows the regions covered by published studies of transpired boundary-layer heat transfer for low-velocity, constant properties flow. The large dashed-line rectangle was the domain I chose to study.

Planning and Executing Credible Experiments: A Guidebook for Engineering, Science, Industrial Processes, Agriculture, and Business, First Edition. Robert J. Moffat and Roy W. Henk.

Companion website: www.wiley.com/go/moffat/planning

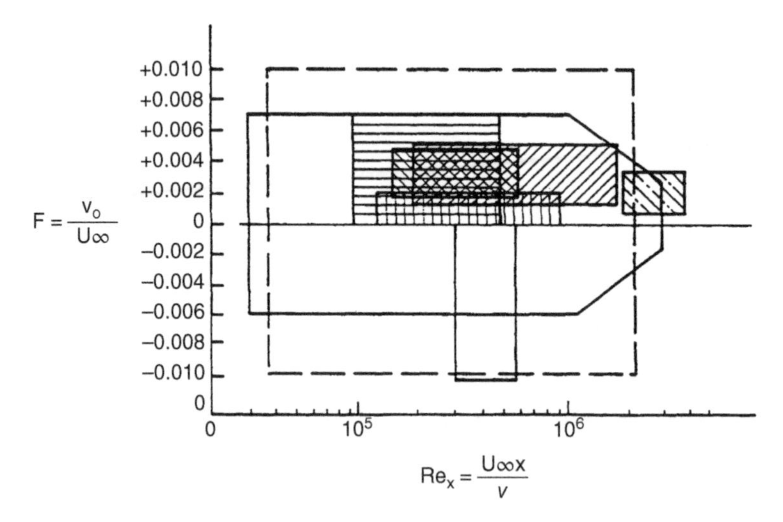

Figure 6.1 Mapping the prior art shows where to work and where to look for a baseline value.

My situation had two independent variables: x-Reynolds number and blowing fraction, F. The various enclosed regions show the reported work by each different investigator.

Note that some domains were rectangular, others were not. For each researcher, the characteristics of the apparatus used in the researcher's study determined the shape of its operable domain.

One of the benefits of this map was that it helped me select the combination of parameters that I should use to baseline my experiment. Five different experiments had reported data in the vicinity of $Re_x = 3 \times 10^5$ and F = 0.002. In order to compare my data with previous research, I was careful to include this data point within the operable domain of my apparatus.

Operable domain maps may also be useful in selecting the best design for a new apparatus. For an example we will look at a separate experiment. Figure 6.2 shows the operable domain map used to select the design parameters for a boundary-layer roughness study. Although this research also used wind tunnels, it differed from my experiment described in Figure 6.1. One rig for this experiment already existed, with roughness elements of 0.2 mm diameter. How should a second research tunnel be designed? The task addressed by this map was, "What should be the parameters of the second tunnel?" The challenge was to get the largest possible range of variables from the combined contribution of both tunnels, with validation and hand-off credibility from the first rig to the second.

The mapping parameters were x-Reynolds number and roughness Reynolds number. The objective of the research program was to cover as wide a range of each parameter as possible. The independent variables to be manipulated for these maps were roughness element size, test section length, and free-stream velocity. Free-stream velocities less than 3 m/s were not allowed, in order to assure a fully developed turbulent boundary layer within 0.3 m (1 ft) of the entrance of the test section. The upper velocity of 76 m/s (250 fps) was chosen to assure that high-velocity effects, such as heating due to viscous dissipation in the boundary layer, were negligibly small. Test section maximum

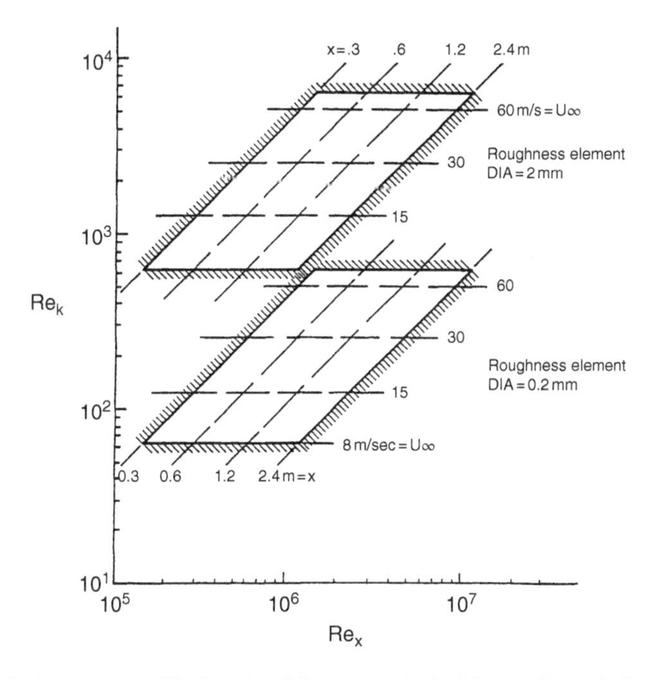

Figure 6.2 Positioning the operable domain of the second rig in this way allowed a broad range of conditions to be covered.

working length of 2.4 m (8 ft) was dictated by the space available. The minimum working length was taken to be 0.3 m (1 ft) to assure a fully developed boundary layer.

With these constraints, several domain maps were constructed with roughness size (k) as a free parameter until a second domain was identified that provided an optimum apparatus. Note that the final design for the combined pair of facilities allows a vertical line at a constant x-Reynolds number of 1×10^6 to cover a range of roughness Reynolds numbers from 60 to 6000. Likewise, the combined facilities allow a horizontal line at a roughness Reynolds number of 600 to cover an x-Reynolds number range from 1.5×10^5 to 1.5×10^7.

Such operable domain maps are useful very early in the design stages of an experiment, when the hardware is being selected. It is at that point that the flexibility and range of the apparatus are determined.

6.3 Mapping the Operable Domain of an Apparatus

One of the more common events in an engineer's life is to "inherit" a system designed and built by someone who has "moved on." Often, there is little or no hand-off period. The first issue is, "What can this system really do?"

A parametric map can be used to describe the operable domain of an apparatus, by mapping its constraints into the desired coordinates. This mapping can be done in either engineering coordinates or in operator's coordinates. Again we consider a wind tunnel, but this example shows the sequence of steps followed in mapping the operable

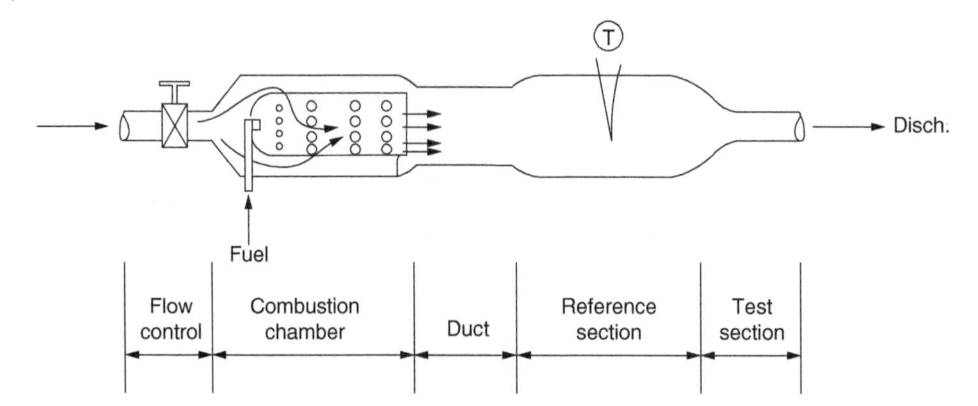

Figure 6.3 The hardware components of the applied thermometry rig.

domain into the practical and familiar coordinates of Mach number and unit Reynolds number. Once available, the map can be to evaluate the range of the apparatus as well as to select the most productive trajectories through the domain for test scheduling.

The following transformations illustrate a step-by-step mapping from hardware geometry into engineering coordinates, based on a few simple calculations.

The problem was to find the operable domain of the hot-air tunnel shown in Figure 6.3, which was proposed for a study of the effects of Mach number on convective heat transfer. The intent was to run a series of points at constant Reynolds number but varying Mach number and vice versa. The coordinates of interest are thus Mach number and Reynolds number. The question is, "What combinations of Mach number and unit Reynolds number can this apparatus reach?"

The apparatus shown in Figure 6.3 is a direct-combustion wind tunnel, burning natural gas in a gas turbine combustion chamber. The purpose of this rig was to provide a stable, high-temperature test chamber for high-temperature heat-transfer and temperature measurement studies.

Air enters from a compressor capable of delivering 1.36 kg/s (3 lbm/sec) and leaves through a convergent nozzle discharging to a plenum chamber at atmospheric pressure. The throat area is 0.0046 m^2 (0.05 ft^2).

The first step in the mapping was to determine the minimum and maximum temperatures at which the system could run. Test cell records indicated that the system could not operate reliably above 1000 °C (about 1800 °F). That set the upper burn-out limit on the operable domain. The lower limit was set by the temperature of the air delivered by the compressor, about 100 °C (212 °F).

The next step was to determine, by a few careful tests, that the combustion chamber had a lean blowout line that considerably restricted low-temperature running. Beginning in the middle of the operable domain, the rig was fired up and gradually throttled down, at constant air temperature until something forced a shutdown. It was the combustion chamber. As shown in Figure 6.4, the combustion chamber had a blowout limit line that prevented operation below certain velocities for a given temperature. A few tests established the line with sufficient accuracy to plot the excluded domain as the left cross-hatched area of Figure 6.4.

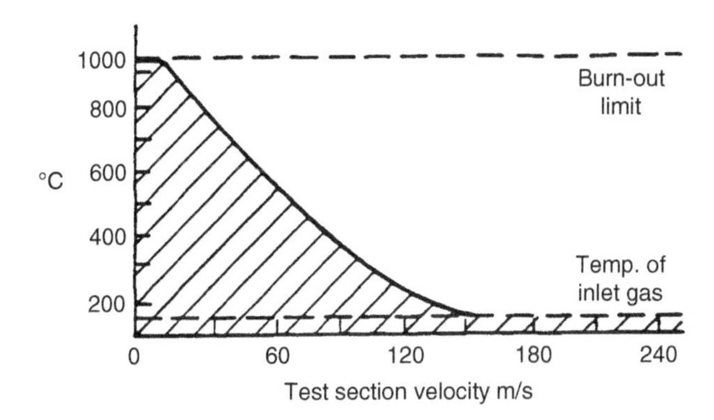

Figure 6.4 The first description of the operable domain, in the simplest coordinates.

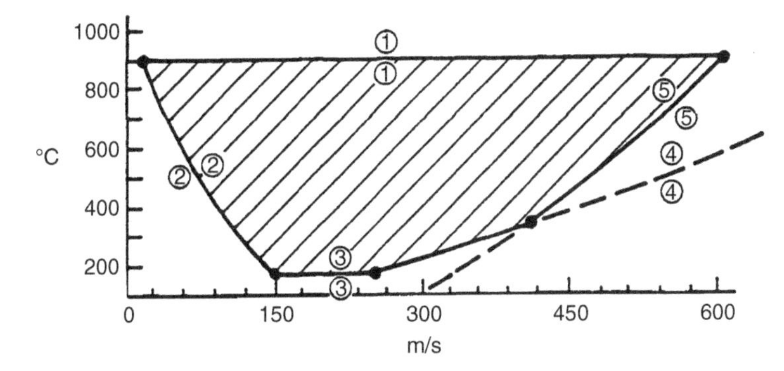

Figure 6.5 The domain is closed by two constraints: the mass flow limit line and the acoustic velocity line.

In Figure 6.5, the view changes so that the accessible domain is now shown by the cross-hatched area. The maximum flow limitation (1.36 kg/sec) could be expressed as a line in temperature versus velocity coordinates because the pressure is known to be atmospheric, using a perfect gas equation of state for air. The pressure, flow rate, and area are fixed, so the attainable values of V are linearly related to T. This constraint appears as line 4 in Figure 6.5.

$$\dot{M} = \rho A V \tag{6.1}$$

Using the ideal gas law, $P/\rho = RT$, we obtain

$$V = \left(\frac{\dot{M}R}{AT}\right)T \tag{6.2}$$

The convergent nozzle limits the maximum velocity in the throat to less than or equal to the speed of sound. Since the acoustic velocity is a function of temperature, the velocity is

$$V \le \sqrt{g\gamma RT} \tag{6.3}$$

Line 5, the acoustic velocity line, closes the domain, as shown in Figure 6.5. The boundaries are:

i) Upper temperature burn-out limit.
ii) Lean blowout line.
iii) Compressor delivery temperature.
iv) Mass flow limit.
v) Acoustic velocity limit.

We can now convert the ordinate from "temperature" to "unit Reynolds number," because:

$$\frac{Re}{D} = \frac{\rho V}{\mu} = \frac{P}{R}\frac{V}{\mu T} \approx 39.7\frac{V}{\mu T} \tag{6.4}$$

Each corner of the domain was mapped into the new engineering coordinates, along with the center points of each line segment. The curves were then completed to the desired resolution.

The resulting domain, in unit Reynolds number versus velocity, is shown in Figure 6.6. Note that the transformations to this point have inverted the view of the domain. What was the upper temperature limit in the first coordinate set of physical coordinates now forms the lower limit of the domain in engineering variables. Compressor discharge temperature, which defined the lower bound in the first map, now forms the upper bound. Lines 4 and 5 (the mass flow limit and the acoustic velocity limit) are nearly coincident.

The final coordinates can be found by calculating the Mach numbers corresponding to several points around the boundaries of Figure 6.6 and identifying line of constant

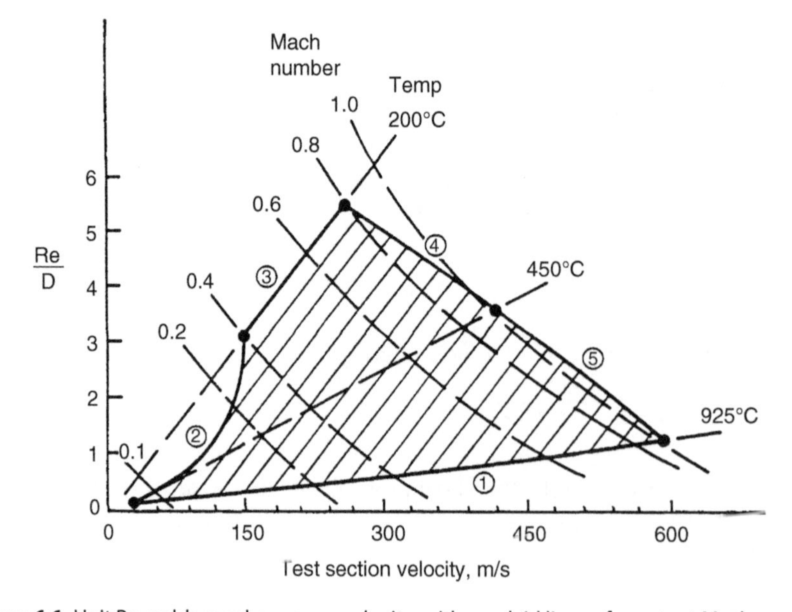

Figure 6.6 Unit Reynolds number versus velocity, with overlaid lines of constant Mach number.

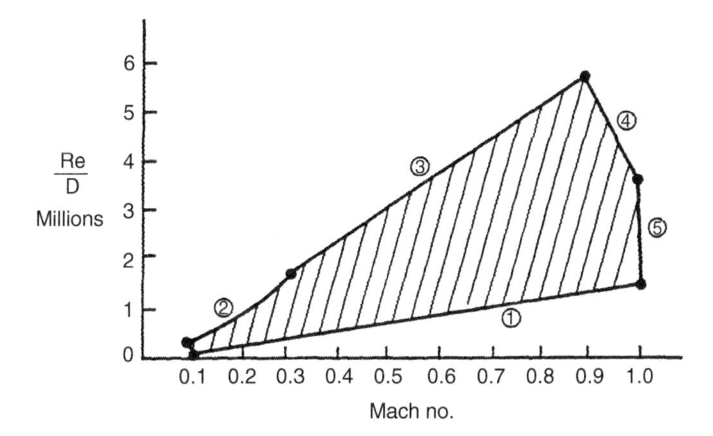

Figure 6.7 The final form of the operable domain map, in unit Reynolds number versus Mach number.

Mach number on the domain, by trial and error and interpolation, because:

$$M \equiv \frac{V}{a} = \frac{V}{\sqrt{g\gamma RT}} = \frac{V}{49.1\sqrt{T_0/1.2}} \tag{6.5}$$

The factor 1.2 allowed use of the stagnation temperature T_0 in the denominator rather than the static temperature. The resultant overlay is shown in Figure 6.6.

The final form, in Re/D versus M, was extracted by cross plotting from Figure 6.6 into the desired dimensionless coordinates.

Figure 6.7 shows the operable domain in the coordinates of interest: unit Reynolds number versus Mach number. As shown in Figure 6.7, although the apparatus is capable of operating between Mach numbers of 0.1 and 1.0 and unit Reynolds numbers from almost zero up to 5.5×10^6, there is only a very restricted range within those limits that can actually be attained.

Having a map showing the attainable test conditions, it is easy to plan effective experiments. The endpoints of each data trajectory are known in advance, data points can be nested near the endpoints, and tests can be planned (evoking a "block design") to minimize wait time (by operating at a sequence of points that are nearly at the same temperature).

6.4 Mapping in Operator's Coordinates

In addition to their usefulness in program planning and equipment evaluation, parametric maps can be used during the execution of tests as "road maps" to guide the operator in running the apparatus.

It is particularly important in heat-transfer work to know where you are in the domain map at all times. Thermal systems tend to be slow in response to changes in power or flow. Without a guide as to what combinations of conditions to set, the rig can move from one set point to another only by "feeling its way along." Attempts to "hurry up" the process by abrupt changes in power or flow often result in catastrophic failure of the rig.

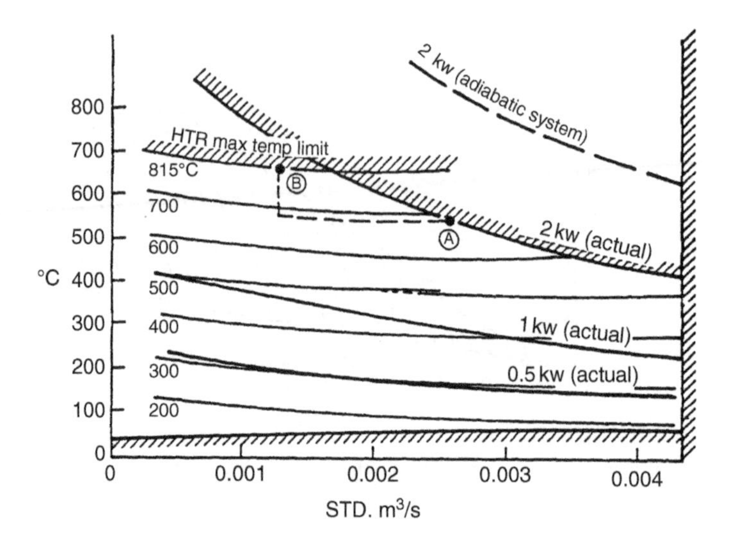

Figure 6.8 A map in operator's coordinates allows for speedy and safe changes in condition.

An operator's domain map is particularly useful when certain combinations of control settings might lead to failure (i.e. burn-out, vibration, excessive loads, etc.). Such a map is illustrated in Figure 6.8, which represents the operable domain of an electrically heated air supply system for an experiment concerning catalytic action at high temperatures. Physically, the apparatus consists of an air tank fitted with air inlet and exhaust lines and equipped with an electric heater. Air flow is limited to 0.0044 m^3/s (or 9.2 SCFM), and electrical power is limited to 2.0 kW by a circuit breaker. The maximum allowable heater temperature is 815 °C (1500 °F).

The operable domain is displayed in terms of the surveillance data available to the test operator: exit air temperature, air flow rate, and electric power.

The most subtle and most dangerous constraint is that represented by the line showing conditions that bring the heater to its maximum safe operating temperature.

For flows less than 0.0017 m^3/s (3.5 SCFM), the heater will burn out at less than 2.0 kW, even though the exit air temperature is far below the heater burn-out temperature.

The first estimate of the general shape of this domain map can be found using simple heat-transfer and energy conservation principles. A second cut, using a rough estimate of heat loss from the enclosure to its surroundings, is sufficient to yield an estimate of the actual 2 kW energy balance portion of the boundary. Once the 2 kW line and the heater burn-out line are established, the interior lines of half power and quarter power can be sketched in. Finally, a rough estimate of the radiant and convective heat transfer from the heater shows that there may be trouble at low flows and high temperature.

This "first draft" map was then used to guide exploratory experiments to accurately locate the most critical boundaries and to locate interior operating lines. The final map represents experimental findings, interpolated based on analytical considerations.

The boundaries of the domain have different characteristics. The "zero flow" boundary, the "maximum flow" boundary, and the "compressor delivery temperature" boundary are all "hard" boundaries: the operator cannot cross them, even if trying to intently.

The 2.0 kW boundary is a "nuisance boundary" because attempting to cross it will drop out the circuit breaker and shut down the system. This, in turn, will require re-establishing the steady-state test condition, and some time will be lost.

The "heater burn-out" boundary is a "penalty boundary" because attempting to cross it may cause the heater to fail and require a complete rebuild of the system.

A domain map in operator's coordinate is particularly helpful in training a new operator. Consider the problem of changing the test rig set point from point A, on the energy-balance limited boundary, to point B, on the heater burn-out line. The only paths between A and B that are safe are those that lie entirely within the operable domain. It would not be safe, for example, to decrease air flow at constant power, which would cause the heater to cross the burn-out boundary. The fastest way from A to B would, perhaps, be a simultaneous reduction in power and flow aimed at following a path of essentially constant heater temperature down to the desired flow, followed by a step increase in power to boundary point B. If an operator's domain map like Figure 6.8 is available, it helps train new technicians to operate the equipment safely.

Operating maps are also useful in organizing test programs for minimum time consumption. In thermal systems, much time is lost waiting for thermal equilibrium. Approximately five times the characteristic time of the apparatus must elapse after each change in set point before a new, stable thermal equilibrium is reached. A test series at constant flow with varying power or constant power with varying flow will require much more time to run than a series at constant temperature. Combinations of flow and power required to attain constant temperature can easily be read from an accurate operating map.

Test operators can benefit from operating maps in which the domain is described in terms of the quantities available for direct control (pressures, flows, and temperatures) rather than engineering coordinates of Mach number, Reynolds number, etc.

With the advent of computer graphic displays, it seems reasonable to display the operating point continuously, on an operating map, so the operator can keep track of the condition of the equipment.

6.5 Mapping the Response Surface

There are a variety of ways to present the results of an experiment.

6.5.1 Options for Organizing a Table

The tabular form shown in Table 6.1 provides a compact format showing exact details of the operating domain and the response (the desired results) from my experiment that began this chapter. The format style of Table 6.1 is familiar because most reference manuals in physics, chemistry, and material properties (including equations of state) are presented in this compact form. This form saves paper and bookshelf space and so historically has been the choice for hundreds of years. During an oral presentation to the target audience, you may wish to keep handy the results in Table 6.1 form; your knowledge of exact details would enhance your personal credibility.

Table 6.1 Stanton number as a function of blowing fraction, F.

Ordered runs	Reynolds number	F = 0.000	F = 0.001	F = 0.002	F = 0.004	F = 0.008
1–5	34,700	0.00215	0.00156	0.00105	0.00057	0.0042
6–10	104,100	0.00111	0.00081	0.00166	0.00290	0.00216
11–15	173,500	0.00095	0.00325	0.00335	0.00245	0.00163
16–20	243,000	0.00089	0.00340	0.00271	0.00192	0.00114
21–25	312,000	0.00090	0.00286	0.00231	0.00171	0.00098
26–30	382,000	0.00126	0.00236	0.00202	0.00150	0.00069
31–35	451,000	0.00321	0.00229	0.00191	0.00151	0.00067
36–40	520,000	0.00332	0.00223	0.00178	0.00129	0.00056
41–45	590,000	0.00305	0.00213	0.00170	0.00125	0.00056
46–50	659,000	0.00285	0.00205	0.00170	0.00116	0.00055
51–55	719,000	0.00270	0.00202	0.00155	0.00116.	0.00049
56–60	798,000	0.00264	0.00195	0.00161	0.00110	0.00053
61–65	867,000	0.00246	0.00185	0.00151	0.00099	0.00043
66–70	937,000	0.00242	0.00180	0.00142	0.00108	0.00037
71–75	1,006,000	0.00236	0.00176	0.00143	0.00106	0.00039
76–80	1,075,000	0.00229	0.00175	0.00132	0.00095	0.00035
81–85	1,145,000	0.00229	0.00179	0.00136	0.00103	0.00036
86–90	1,214,000	0.00225	0.00179	0.00139	0.00089	0.00034
91–95	1,284,000	0.00220	0.00168	0.00129	0.00101	0.00038
96–100	1,353,000	0.00218	0.00179	0.00127	0.00082	0.00027
101–105	1,422,000	0.00215	0.00159	0.00129	0.00077	0.00031
106–110	1,492,000	0.00212	0.00156	0.00121	0.00071	0.00031
111–105	1,561,000	0.00211	0.00153	0.00130	0.00087	0.00032
106–110	1,631,000	0.00211	0.00163	0.00120	0.00084	0.00037

The format of Table 6.1, however, hinders statistical analysis of the data by computer. Why? There are multiple values for the response in each row and in each column. But there is a more excellent way to save experimental results for the purpose of credible analysis.

Table 6.2 is the preferred format to organize and store data for statistically analyzing the results by computer, but it would not be suitable for publication. Notice that each combination of operating conditions occupies a single row, with the measured response values. Plan to save your data so that all operating conditions, variables, and results are stored in a format like Table 6.2. Factors that are saved can be analyzed for how they impact results; variables not saved at the time the experiment is conducted cannot be credibly included in the analysis. In order to fit the table on one page of this book, dummy values are placed in many columns. The values for F (transpiration) and Reynolds number were derived from the operating conditions set by the operator (who). The value for Stanton number (St) is the derived response of the experimental results.

Please notice in Table 6.2 that good experimental practice advises randomizing the order of operating conditions. Reasons for randomization will be discussed in later chapters. Plan to randomize even within a "block design," which is typically chosen when an operating factor (for example, system temperature) requires inordinate (compared to other factors) time to alter. On a page separate from the table, create a codebook (an example codebook is shown in Chapter 5, Exercise 5.10 and Chapter 10, Figure 10.10) to explain units, dimensions, ranges, and levels of each variable.

6.5.2 Options for Presenting the Response on a Scatter-Plot-Type Graph

When presenting results, your audience will likely prefer a graph more than tables. A conventional two-dimensional graph of the results is shown in Figure 6.9. The x-Reynolds number is shown on the abscissa (horizontal axis). The response of the experiment is the Stanton number on the ordinate (vertical axis). An individual curve is drawn for each value of F, the transpiration parameter. Although it is relatively easy to visualize the operable domain of a two-variable problem (the domain is two-dimensional), a three-variable problem requires a bit more artistic ability, but not so much that you should avoid sketching it.

Let's demonstrate by referring back to our original graph of the operating domain. When the operating domain is two-dimensional, we use it as a base region in the ($x1$, x2) plane. Then the response can be shown graphically in the third dimension as a surface lying above that plane with an elevation in the $x3$ direction. For my research heat-transfer experiment at the beginning of this chapter, the two parameters of the operable domain were x-Reynolds number and transpiration value F. The results of the Stanton number are presented in a three-dimensional response surface representation in Figure 6.10. For most persons, Figure 6.10 provides an immediately understandable terrain.

Figure 6.10 shows that the values of the Stanton number form a response surface consisting of a curved ridge with steep slope on the low Reynolds number face, and a gradual descent on the high Reynolds number side. The crest of the Stanton number ridge rises as F increases and simultaneously shifts to lower Reynolds numbers. The low Reynolds number valley is the minimum of laminar behavior, while the steep slope is the transition to turbulence region. The gradual descent on the high Reynolds number

Table 6.2 Data format for computer and statistical analysis: text example.

Run true	Date	Time	Main flow	Blow flow	Power	Where	Who	Temp	F	ReyNum	St
1	d1	t1	f1	b1	w1	z1	R1	k1	0.002	797 942	0.00161
2	d2	t2	f2	b2	w2	z2	R1	k2	0.000	242 852	0.00089
3	d3	t3	f3	b3	w3	z3	R2	k3	0.004	1 491 804	0.00071
4	d4	t4	f4	b4	w4	z4	R1	k4	0.008	1 630 576	0.00037
…											
17	d17	t17	f17	b17	w17	z17	R2	k17	0.000	936 714	0.00242
…											

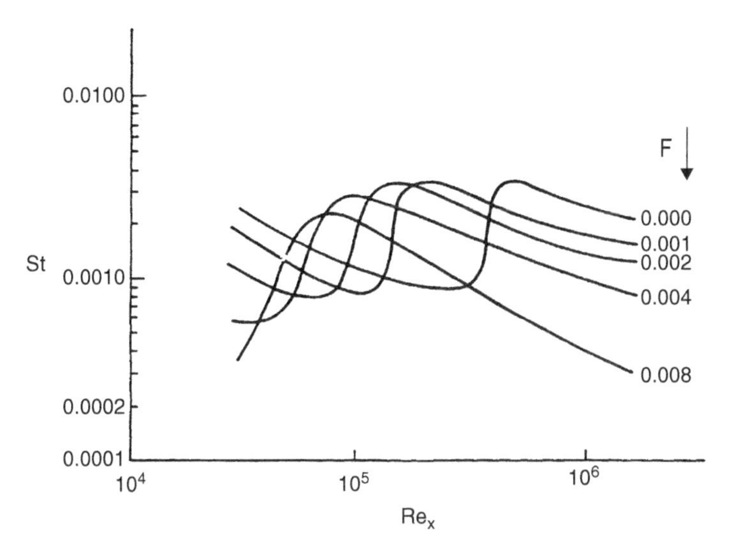

Figure 6.9 The data from Table 6.1, plotted in a conventional manner, are not much more informative than looking at the table itself.

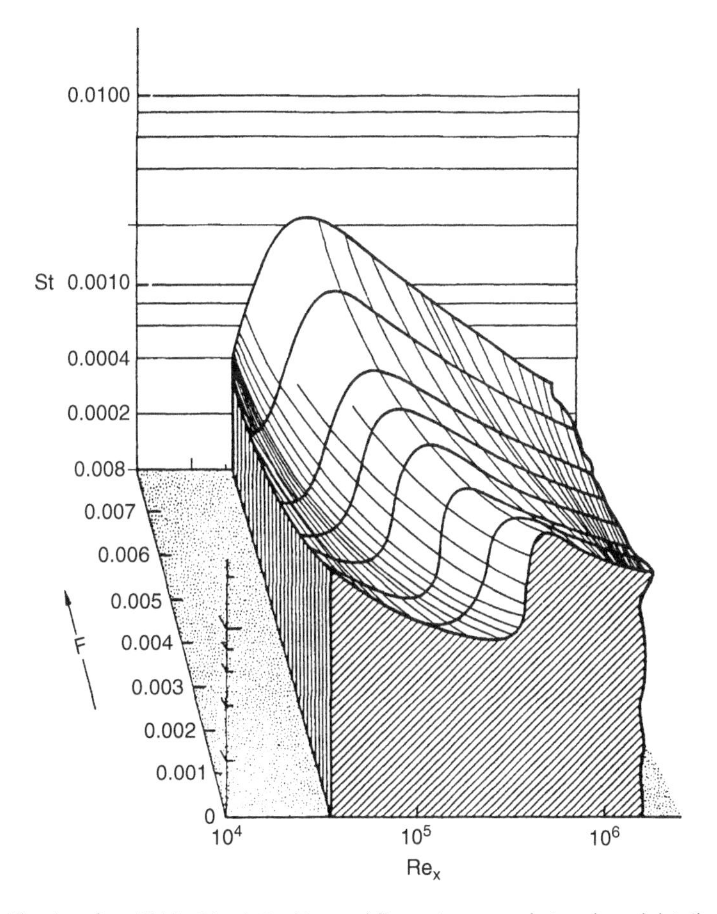

Figure 6.10 The data from Table 6.1, plotted in an oblique view, reveals trends and details of the flow physics via the topography of the response surface.

side of the crest is the turbulent regime. Clearly the figure is easier to understand by sight than to explain with words. The tables, although numerically precise, did not convey physical understanding.

During the planning phase of an experiment, there are advantages to viewing expected results of an experiment in terms of its response surface representation. One immediate advantage is the ease with which spatial relationships can be grasped when presented in response surface form.

Let's consider how Figure 6.10 shows many important features of the physics of this system and guides in planning the next series of measurements. One option is to take data along paths where each path consists of a series of runs at constant Reynolds number, varying the blowing fraction, F. If such a path were followed in the transition region, the resulting data would be likely to show considerable scatter, due to the steep slope of the response surface perpendicular to the path. Small differences in the Reynolds number on successive data points would result in considerably different Stanton numbers. If those data were subsequently treated as though they had all been taken at exactly the same Reynolds number, the average value, the Stanton number variations associated with those small differences in Reynolds number would appear as scatter in the Stanton number. A second, better path option would fix the blowing fraction and vary the Reynolds number.

A better option improves upon the second option by explaining more Stanton number variation and by minimizing Stanton number scatter via collecting data along trajectories that follow the "path of steepest descent," or the "fall line," to use a skier's terminology.

Response surfaces may be as simple as a single surface when there are only two operational variables. If a third operational variable acts, the response may be a nested set of surfaces.

As a personal challenge, imagine the following: if the operating domain contains more than three factors, how can the response be shown? Clever options do exist!

MM&AC (see preface) is a useful reference for this chapter.

Homework

6.1 Construct a map of the domains that (where results) have already been reported in the literature.

6.2 Construct a map of the proposed contribution by your experimental test. Where does your experiment overlap extant results? Where does your experiment extend knowledge (science)? Where does your experiment reduce uncertainty?

6.3 Construct a map of the operable domain for your equipment.

6.4 Construct a map of the safe domain for your equipment.

6.5 Draw a response surface for expected behavior of each type of result across your proposed domain.

6.6 How can your motivating question be improved?

7

Refreshing Statistics

We limit our scope to practical statistics for finite amounts of data:

1) To refresh basic statistics concepts.
2) To illustrate the most fundamental data distributions.
3) To adjust conclusions when experimental data are few.
4) To construct a simple model to explain results using free open-source software.

This chapter aims to provide a sufficient background of key concepts. We harness statistics in order to explain "variation in the context of what remains unexplained" (Kaplan 2012). To provide a context for these concepts, we adopt an economic experiment. Consider a small shoe store in an ideal location, but a nearby competitor is larger. The owner wants to satisfy 19 out of 20 young customers who desire the Moffat fishing shoe. The small store cannot stock dozens of every size, so the owner asks us to do an experiment to fit the store's inventory to its potential customers.

7.1 Reviving Key Terms to Quantify Uncertainty

7.1.1 Population

The *population* specifies all possible objects which could be included in the data collected. Our population is not infinitely large, so here we emphasize finite statistics. In contrast, classical statistics assumed an infinite population with a Gaussian (or "normal") distribution, which was useful in theory but never realistic – even the number of atoms in the universe, about 10^{80}, is much less than infinity. For each experiment, it is wise to specify a realistic population.

Our economic example offers several candidate populations. One candidate is the total number of Moffat shoes manufactured by the factory; its inventory has an exactly known mean of 24.5 cm and Standard Deviation of 2.2 cm. A second candidate population for our shoe store experiment describes all possible customers who will seek for the Moffat shoe locally during the next year. Since the second population is relevant to our client, we adopt it.

The virtue of an experiment is to understand the population and answer the motivating question without measuring every member of the population. Instead we only measure a subset of the population called the *sample*.

Planning and Executing Credible Experiments: A Guidebook for Engineering, Science, Industrial Processes, Agriculture, and Business, First Edition. Robert J. Moffat and Roy W. Henk.

Companion website: www.wiley.com/go/moffat/planning

7.1.2 Sample

A *sample* is the set of cases selected from the population and measured in the experiment. Each member of the sample is one case. For many science and engineering experiments, if the sample size is close enough to infinity, then a Gaussian distribution may be assumed. We take the view that most experiments have finite sample sizes far from infinity. We will soon show how to adjust small-sample results to achieve an acceptable answer to our motivating question.

Results improve when the objects in the sample are drawn randomly from the population. Randomization is a key strategy for counteracting bias in an experiment. Randomizing improves data quality, for example, by converting bias drift into (unexplained) noise. Be diligent to employ randomization wherever possible in each stage of experiment design.

The sample for our example experiment will be 12 customers, chosen at random. For this text, the shoe size (centimeter) was randomly "measured" by rolling a dice and adding 20. A six-sided dice has a theoretical mean of 3.5. The data of our random sample are shown in Table 7.1. In the bar chart (Figure 7.1), the observed sample is compared with a model of what we expected from the dice.

A small sample of truly random events rarely looks like the infinite distribution.[1] In our example with dice, the distribution does not look uniform. Likewise a small sample of normal data rarely looks Gaussian. It is fair to ask: how confidently can we understand a population if the experimental sample size is small? The answer requires concepts later in Section 7.3.

Table 7.1 Economic experiment example for shoe store.

Case	Age	Size = Y	Dice
1	11	22	2
2	13	26	6
3	12	24	4
4	12	22	2
5	11	22	2
6	11	22	2
7	12	23	3
8	13	24	4
9	12	23	3
10	13	23	3
11	11	21	1
12	11	22	2

1 Make these ideas tangible. Do your own experiment with a single six-sided dice. Before rolling the die 12 times, predict the number of events for each side. A fair die has a uniform distribution for infinite rolls, so we expected each side to appear twice in 12 rolls. Now roll and record. How did your prediction compare?

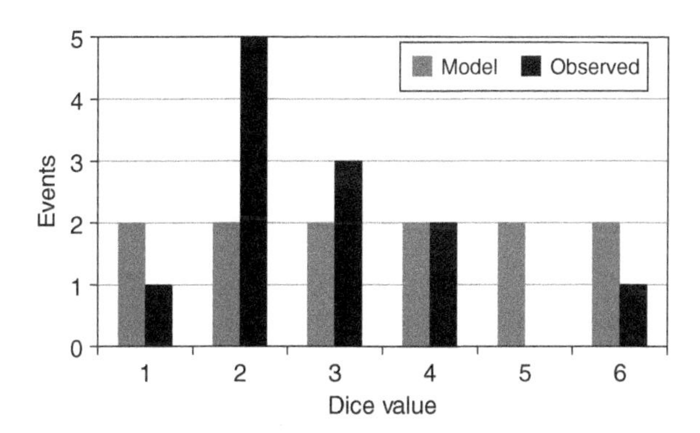

Figure 7.1 Model prediction and experimental results.

We next use the above example experimental results to illustrate the definitions of central value, mean, variance, Standard Deviation, Uncertainty, and Degrees of Freedom.

7.1.3 Central Value

The *central value* is a single number that best represents the sample. Three popular candidates for the central value are mean, median, and mode. Data must be sorted to find the median and mode. The mode is the most frequent value within the sample – for our example, the mode is 22 cm. The median is the value for which an equal number of cases are above and below the median – for our data the median is between 22 and 23 cm, thus the median is chosen to be 22.5 cm. The third and most popular candidate for the central value is the mean or average.

7.1.4 Mean, μ or $\overline{Y}$

The most familiar goal in statistics is to estimate the mean of a population, which is the central value about which the distribution is equally weighted. The population mean will be designated μ and is called the *true mean*. Will the mean of the sample $\overline{Y}$ provide a credible, accurate estimate of the population mean? Although the true mean is rarely known, we happen to know that the true mean for a single dice is 3.5; thus the true mean for the artificial "dice+20" shoe size in our example was 23.5 cm. Now compare the sample mean to the true mean. The sample mean is calculated from the following equation. For our data the sample mean was 22.83 cm from the general equation

$$\overline{Y} = \frac{1}{n}\sum_{i=1}^{n} Y_i = \frac{1}{Df}\sum_{i=1}^{n} Y_i \tag{7.1}$$

7.1.5 Residual

Now that we expect data to be near the central value, how much does an actual datum deviate from the expected value? The approved term to quantify this deviation is the *Residual*. Commonly the expected value is the mean; then the *Residual* remains after subtracting the mean from an actual data value. More broadly, the Residual is the deviation of each datum from its predicted model value. The Residual for the ith datum Y_i is calculated as

$$\text{Residual} = Y_i - \bar{Y} \tag{7.2}$$

7.1.6 Variance, σ^2 or S^2

The average dispersion (squared Residual) of the whole sample about model values is the variance. The variance of the sample Y, when the mean $\bar{Y}$ is the model value, is calculated by

$$S_Y^2 = \frac{1}{(n-1)}\sum_{i=1}^{n}\left(Y_i - \bar{Y}\right)^2 = \frac{1}{Df}\sum_{i=1}^{n}\left(Y_i - \bar{Y}\right)^2 \tag{7.3}$$

7.1.7 Degrees of Freedom, *Df*

Before any calculations are made from sample data, the Degrees of Freedom equals the number of data points. Consequently, the equation for the mean used $Df = n$ as its denominator. In contrast, the equation for the variance used the sample mean value $\bar{Y}$, which we just derived from the data. By reusing $\bar{Y}$, we lost one Degree of Freedom. Thus the definition of variance required $Df = (n - 1)$ in the denominator. This is a statistical principle: each time a value is derived from the data and reused, the Df decreases by one.[2] For example, if data are modeled by three coefficients (rather than just the mean) then when variance is calculated, the denominator becomes $Df = (n - 3)$.

7.1.8 Standard Deviation, σ_Y or S_Y

The Standard Deviation of a sample of Y values is simply the square root of the variance of the sample,

$$\sigma_y \approx S_Y = \sqrt{S_Y^2}. \tag{7.4}$$

In theory the population Standard Deviation is designated σ_Y, while the sample Standard Deviation is S_Y. For our data the sample variance was 1.788 cm^2. This means that the sample Standard Deviation was $S_Y = 1.337$ cm around the sample mean $\bar{Y} = 22.83$ cm.

Our sample Standard Deviation differs from what theoretical distributions predict. The Standard Deviation for dice with a purely uniform distribution is $\sigma_Y = 1.97$ around the true mean.

Our small sample resembles a Gaussian, but is it? Let's check. For a purely Gaussian distribution, 68% of the cases lie within 1σ of the true mean of the population and 95% of the cases lie within 2σ. For our data, 10/12 = 83% of the sample cases lie within 1S_Y of the sample mean. Likewise 11/12 = 92% of the data lie within 2S_Y of the mean. Since our sample was not Gaussian, the shopkeeper would be wise to reject our advice. Our faulty

2 Ponder: must there be more than one data point to find variance? Answer: there is one exception. In the special case of n = 1, if the true mean μ is used instead of the sample mean in the variance equation, then *Df* equals 1.

advice? Assuming Gaussian variance, we advised that the inventory of the small shop should focus on shoe sizes between 20 and 25.5 cm. To give more accurate advice from a small sample size, we must adjust our estimated values for the mean and the Standard Deviation according to the "t-distribution" (Section 7.3).

7.1.9 Uncertainty of the Mean, $\delta\mu$

For a credible experiment, we must report an Uncertainty level along with results. How certain can we be that the sample mean $\bar{Y}$ estimated the population mean μ? The Standard Deviation helps describe the Uncertainty of the sample mean. The Uncertainty $\delta\mu$ is smaller than the Standard Deviation, because σ is divided by the square root of the sample size, as in

$$\delta\mu = Multiplier \frac{\sigma}{\sqrt{n}}, \tag{7.5}$$

where the t-distribution provides the multiplier at the confidence level selected by the client. We then report the range of the mean as

$$\bar{Y} - \delta\mu < \mu < \bar{Y} + \delta\mu. \tag{7.6}$$

In our example, we find $\frac{\sigma}{\sqrt{n}} = \frac{1.337}{\sqrt{12}} = 0.3860$. We choose the multiplier = 2.20 since $Df = 11$ on the t-distribution. By simplifying $22.83 - 2.2 \cdot 0.386 < \mu < 22.83 + 2.2 \cdot 0.386$, we claim with 95% confidence that the true mean is in the range $21.98 < \mu < 23.68$; recall that the true mean $\mu = 23.5$.

By taking multiple-samples of a population, means are distributed normally. Just as the means tend to a Gaussian distribution, the Standard Deviation has its own typical distribution, called χ^2.

7.1.10 Chi-Squared, χ^2

How well does a model predict reality (the population)? We shall heed George Box who remarked "All models are wrong. Some models are useful" (BH&H; see preface). The χ^2 value is useful for testing the goodness of a model. A χ^2 value closer to zero indicates a better fit. Let's check our example.

The equation for χ^2 is

$$\chi^2 = \sum_{i=1}^{N} \left(\frac{O_i - E_i}{E_i} \right)^2 \tag{7.7}$$

where

O_i = the number of observed events in a bin,
E_i = the number of expected (predicted by model) events in a bin,
N = the number of data bins or intervals.

Our model yielded $\chi^2 = \left(\frac{1-2}{2}\right)^2 + \left(\frac{5-2}{2}\right)^2 + \left(\frac{3-2}{2}\right)^2 + \left(\frac{2-2}{2}\right)^2 + \left(\frac{0-2}{2}\right)^2 + \left(\frac{1-2}{2}\right)^2 = 4.0$, but a Gaussian had $\chi^2 = \left(\frac{1-E_1}{0.23}\right)^2 + \left(\frac{5-E_2}{1.68}\right)^2 + \left(\frac{3-E_3}{4.09}\right)^2 + \left(\frac{2-E_4}{4.09}\right)^2 + \left(\frac{0-E_5}{1.68}\right)^2 + \left(\frac{1-E_6}{0.23}\right)^2 = 27.7.$

The χ^2 test affirms that our uniform model (Figure 7.1) fitted this small sample better than a Gaussian-shaped model did. Note: expected events need not be an integer.

Although the χ^2 equation appears similar to variance, notice the differences. Variance is the *average* dispersion. In contrast, the χ^2 equation divides each term by its expected vicinity before squaring, capturing how well the model matches the amount of data expected. Notably, cousins of the χ^2 equation reappear in the Uncertainty estimate, t-distribution, etc.

7.1.11 p-Value

The p-value estimates the probability of an event occurring. The range of the p-value is between zero and one, inclusive. If the p-value = 1, then the event has a 100% probability of occurring. It is certain. If the p-value = 0, then the event has zero probability of occurring. It will never happen – for example, two fermions occupying the same quantum state, violating the Pauli Exclusion Principle. In experimental physics, an event was reckoned impossible if the p-value is less than 1×10^{-50}; some rare nuclear decay events of slightly higher probability are known to occur. In Section 7.1.10, we encountered the p-value (multiplied by total events) in the values used for E_i. In everyday engineering and science, we often employ a target p-value = 0.05, which is equivalent to 1 in 20 odds. The p-value takes a major role in the Null Hypothesis.

7.1.12 Null Hypothesis

Every experiment begins with a question and a hypothesis, as does the scientific method. Why then must we consider a "Null Hypothesis," when "null" encompasses meanings of "nothing, empty, void, vain, worthless"? We are looking for significant answers, aren't we? The Null Hypothesis offers many benefits that make the analysis of an experiment more credible. For one, a Null Hypothesis focuses the experimenter on the Motivating Question.

Credible experiment analysis applies the Null Hypothesis to every factor that might affect the results – significant enough to be recorded and possibly included in an explanatory model. The Null Hypothesis takes the view "this factor is unlikely to be important" and then challenges us to try to prove the Null Hypothesis wrong. Logically, this tactic helps because it is easier to prove an assertion wrong than it is to prove an assertion correct.

Beware. The Null Hypothesis arises everywhere, often unintended. Whenever we do not record a specific factor during an experiment, by default we have decided that factor is null. Why? We will not be able to use any unrecorded factor to explain the results. We have already assigned unrecorded factors to the null heap.

In our shoe store example, the only factors recorded in the datatable were age, shoe size, and dice. In order to demonstrate the issues of random, small samples, we rolled a die and recorded its value, then added 20 for the size. Could boy/girl be a factor? It was not recorded.

Statistics uses the χ^2 equation to find the probability (p-value) that a Null Hypothesis is true. To find significant factors with 95% confidence, we seek a p-value less than 0.05 (5%) so that we can "reject the Null Hypothesis." One Null Hypothesis made big news in 2012: after the p-value was less than 0.000006 that the "Higgs boson was not found," then the Higgs was announced. How does the wording make sense? In contrast, a high p-value suggests the null might be true, which in the cautious language of statistics becomes "cannot reject the Null Hypothesis." Any factor with a high p-value begs scrutiny because no benefit comes from including a null factor in an explanatory model. Due diligence advises an analyst to exclude null factors from a model and then recalculate. To merely omit a factor without recalculating constitutes statistical malpractice.

7.1.13 F-value of Fisher Statistic

The Fisher statistic is quantified in the F-value. Though rarely covered in elementary statistics, it deserves a mention here. The definition, description, and usefulness of the F-value will be discussed later in this chapter. When selecting significant contributing factors for a model, we consider the F-value to be superior to the p-value of the Null Hypothesis.

7.2 The Data Distribution Most Commonly Encountered

The Normal Distribution for Samples of Infinite Size

The bell-shaped curve shown in Figure 7.2 is called the *Normal Distribution* or *Gaussian Distribution*. This distribution is useful because it typifies results for an infinite number of repeated trials. If a normal distribution does not fit your process, other options are available.

The normal distribution has a central value, μ, about which the distribution is symmetric, so positive and negative excursions from the mean are equally likely. The height at one Standard Deviation from the mean is about half the height at the mean. Note also

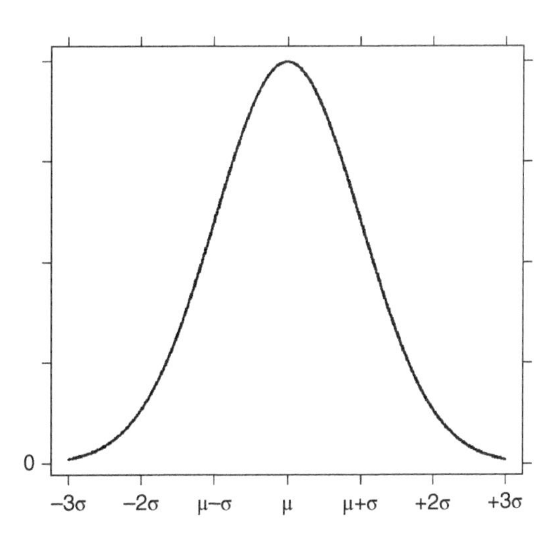

Figure 7.2 The Normal Distribution.

that the "tails" approach zero asymptotically, thus deviations far from the mean are rare but not prohibited.

The dispersion about the mean is given by a shape parameter σ, the Standard Deviation. For a pure Gaussian distribution, 68% of the members of the population lie within 1σ of the mean of the population, i.e. between $(\mu - \sigma)$ and $(\mu + \sigma)$. Likewise, 95% of the population lies within 1.96σ of the true mean. Appendix A discusses more details about this distribution.

The normal distribution does not apply to all measurements nor to all experiments. One example is a logNormal distribution of income. It also would not apply to the set of ball bearings which have fallen through a hole of a certain size, thereby excluding large sizes.

7.3 Account for Small Samples: The t-Distribution

As we have already noted, the distribution of a small sample size rarely matches any theoretical distribution, whether normal, uniform, or other. However, when a set of several small samples is taken from the same population, the distribution of the means fits the t-distribution. Consider a group of samples; each sample consisting of n items has an exact mean. We collect all of the sample means in the group to estimate the population mean. The t-distribution tells us how to account for a small number of items in the sample.

The values of "t" are almost normal. The "almost" refers to the fact that the t-distribution has longer tails, depending upon the sample size. Table 7.2 supplies the multiplier to stretch the tails, which are stretched for 95% certainty when the sample size is small. Glance at the bottom of the table – the t-distribution for an infinite population echoes the result for a pure Gaussian distribution, that 95% of the distribution lies within 1.96σ of the true mean. At the top of the table, the Degrees of Freedom Df equals 1 when only two data points existed to calculate the Standard Deviation. When $Df = 1$, the multiplier is 12.70σ rather than 1.96σ, exceeding by six times a pure Gaussian.

Our example had a small sample size, $n = 12$. The sample mean was $\overline{Y} = 22.83\text{cm}$. Using the sample mean, the Degrees of Freedom decreased by one, $Df = n - 1 = 11$. The estimate of the sample Standard Deviation was $\sigma_Y = 1.34\,\text{cm}$. The t-distribution 95% factor in the table is 2.20, so the range is $\overline{Y} \pm 2.20\sigma_Y = 22.83 \pm 2.20 \times 1.34$. To satisfy 95% of visiting customers, we now advise that the inventory of the small shop should focus on shoe sizes between 19.5 and 26 cm. For our sample, 12/12 = 100% of the sample lies within 2.20σ of the mean. Through this example, we can demonstrate to the shopkeeper that our technique is credible before collecting any data. We may recommend that an actual experiment collect other factors which explain shoe size (such as sex, foot width, ...) and that a larger sample size will reduce Uncertainty. The shopkeeper can weigh the benefits against the cost of each data point collected, in order to decide the extent of your experiment.

The main point of the t-distribution is that, when the sample size is small, we must adjust our understanding of what the experiment tells us about the population. The t-distribution makes the adjustment predictable. Appendix A discusses more details.

Table 7.2 t-Distribution.

Df	0.95 Factor
1	12.70
2	4.30
3	3.18
4	2.78
5	2.57
6	2.45
7	2.36
8	2.31
9	2.26
10	2.23
11	2.20
12	2.18
15	2.13
20	2.09
30	2.04
40	2.02
60	2.00
120	1.98
∞	1.96

7.4 Construct Simple Models by Computer to Explain the Data

A quick, easy, and accurate way to analyze statistics is available via the free open-source R language. We stored our example shoe results in the file "diceShoe.csv," of which the top five lines are shown below:

```
age,size,dice
11,22,2
13,26,6
12,24,4
12,22,2
...
```

7.4.1 Basic Statistical Analysis of Quantitative Data

Let's analyze our example using the R language. We will find all of the values discussed above.

```
> a=read.csv('diceShoe.csv')  # Read the csv format data file into an R datatable.
> head(a)                     # Review the top six cases in the R datatable.
  age size dice
1  11  22  2
2  13  26  6
3  12  24  4 ...
> mean(a$size)                # Calculate the sample mean of our example dice shoes.
  22.833                      # This value is identical to the discussion in the text.
> sd(a$size)                  # Calculate the sample Standard Deviation.
  1.3371                      # This value is identical to that in Section 7.1.
> mod=lm(size~1,data=a)       # Perform a linear model based on zero explanatory factors.
> summary(mod)                # Display R analysis of the linear model.
lm(formula = size ~ 1, data = a)
Coefficients:
              Estimate Std.Error t-value Pr(>|t|)
(Intercept)   22.833     0.386    59.16 3.98e-15
Residual Standard Error: 1.337 on 11 Degrees of Freedom
```

Notice that summary gave the results we previously found: the Intercept from the linear model is the mean; the Standard Error of the Intercept is $\sigma/\sqrt{n}$, which we used for the Uncertainty of the mean; the Residual Standard Error is the Standard Deviation; the Degrees of Freedom reduced by one because one coefficient was reused to calculate the Residual, that is $12 - 1 = 11$.

A well-chosen graph will present information cogently to help the client/reader understand the answer to the motivating question. R offers many informative options to inspect results (Figure 7.3) of which R-commands are listed in Sec. 7.5.1.

The histogram shows the data distribution but lacks direct information of the mean or Standard Deviation. The density plot mimics the histogram, yet as a smooth curve. The boxplot offers a view of the data more familiar to economics but reveals direct information useful in science.

The boxplot emphasizes the median and the central 50% of data between the first and third quartiles. The InterQuartile Range (IQR) quantifies the difference between the first and third quartiles. Any data point located beyond 1.5 IQR above the third quartile (or 1.5 IQR below the first quartile) is identified as an "Outlier." The whisker locates the

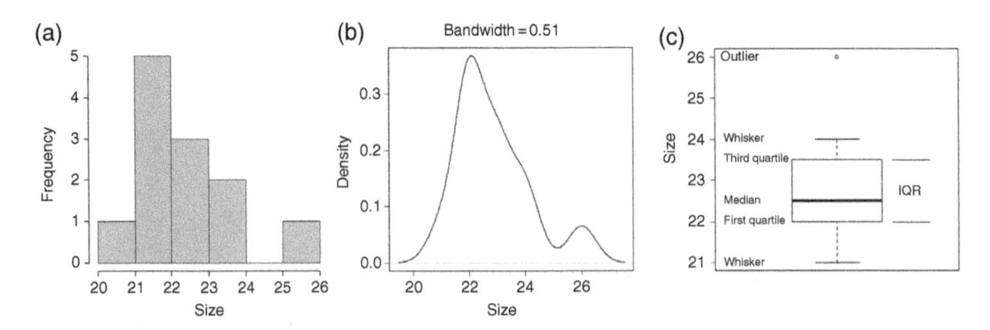

Figure 7.3 Various plots to inspect results of the shoe store example experiment. (a) Histogram, (b) density plot, and (c) boxplot.

furthest data point which is not an outlier. If the shopkeeper is only interested in the central 50% of potential customers, the boxplot gives best guidance.

```
> anova(mod)                        # Display R Analysis of Variance (anova) of the model.
Analysis of Variance Table
Response: size
          Df Sum Sq Mean Sq
Residuals 11 19.667  1.7879
```

Analysis of Variance (anova) yielded the value "Mean Square" which equaled our variance.

The analysis above of a single variable (shoe size) was sufficient to answer the motivating question demanded by the shopkeeper: what shoe sizes would satisfy the majority of customers? We did not try to explain how shoe size depended on other factors. In contrast, if our client was the shoe manufacturer, the motivating question may be more complex: how is shoe size related to age, foot length, width, sex, time of day, or other factors? The power of the R language shines when analyzing data with many variables.

7.4.2 Model Data Containing Categorical and Quantitative Factors

Next we use the R language to analyze a different dataset where electrical resistance varies according to two factors: temperature and material alloy. These data are stored in the file "hiloy.csv," which is listed in Appendix A and also available for download at the book website.

```
> b=read.csv("hiloy.csv")  # Read the csv format data file into an R datatable.
> head(b)                  # Review the top six cases in the R datatable.
  temp   ohm alloy         # The response is resistance, measured in ohm.
1  106 22.69     C         # One explanatory factor is temperature, in degrees.
2   40 18.73     N         # One explanatory factor is type of alloy, by category.
3   85 24.13     N ...     # Notice that temp and alloy were randomized.
> length(b$ohm)            # Display the number of cases in the file.
  44
> mean(b$temp)             # Calculate the sample mean of one explanatory factor
  78.8636                    # to avoid extrapolation and to reduce Uncertainty.
> b$tm=b$temp-78.86        # Transform temperature to tm, centered about the mean.
> head(b)                  # Notice that a new factor appears in the R datatable
  temp   ohm alloy     tm   # while the original data are unaltered by R analysis.
1  106 22.69     C  27.14
2   40 18.73     N -38.86  ...
> lm(ohm~ -1+alloy,data=b) # Display averages for each alloy, one explanatory factor.
> mod=lm(ohm~1+alloy+tm,data=b) # Fit a linear model based on two explanatory factors.
> summary(mod)             # Display R analysis of the linear model.
lm(formula = ohm ~ 1 + alloy + tm, data = b)
Coefficients:
            Estimate Std.Error t-value Pr(>|t|)
(Intercept) 19.4225  0.4114    47.214  < 2e-16
alloyN       5.5121  0.5831     9.454 7.43e-12 # Reject the Null Hypothesis for these
tm           0.1492  0.008541  17.470  < 2e-16 # factors: for tm, p-value < 2 × 10^-16.
Residual Standard Error: 1.925 on 41 Degrees of Freedom
```

7.4.3 Display Data Fit to One Categorical Factor and One Quantitative Factor

```
> plot(ohm~temp,data=b)
> points(fitted(mod1)~temp,pch="+",data=b)
```

Figure 7.4 shows the data in open symbols and the fitted model in solid symbols. The Intercept was 19.42 Ω at `tm = 0` for the type C alloy; temp = 78.86 °C. The Intercept at `tm = 0` for the type N alloy differed from type C alloy by 5.51, which adds up to $19.42 + 5.51 = 24.93\,\Omega$. A single slope was fitted for both alloys as 0.1492 [Ω/K]. Other values in the summary statement will be explained later in this chapter. Although this model is far from perfect, it explains alloy behavior better than averages did. When the Residual for this model was calculated, the expected value was not the mean – rather the expected value was the model value, plotted as solid symbols. Likewise the variance was the sum of the squared Residuals from the model.

The model function shown in Figure 7.4 is written explicitly as

$$ohm = 19.42\Omega + \begin{bmatrix} 5.51, \text{if N} \\ 0.00, \text{if C} \end{bmatrix} \Omega + 0.1492 \left[\frac{\Omega}{K}\right] (temp - 78.86) \left[{}^{\circ}\text{C}\right]. \tag{7.8}$$

7.4.4 Quantify How Each Factor Accounts for Variation in the Data

```
> anova(mod1)
Analysis of Variance Table
Response: ohm
           Df  Sum Sq Mean Sq F-value    Pr(>F)
alloy       1  227.64  227.64   61.42 1.113e-09
tm          1 1131.13 1131.13  305.20 < 2.2e-16
Residuals 41  151.95    3.71
```

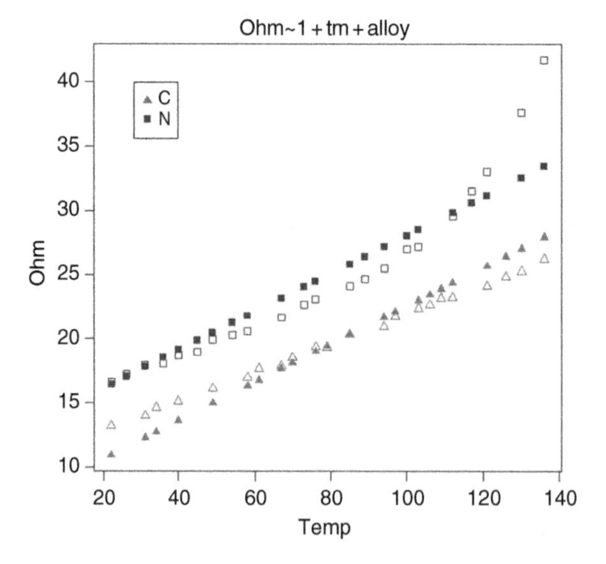

Figure 7.4 Initial model of hiloy data.

Analysis of Variance (`anova`) reveals how much variance in the response ohm is captured by each explanatory factor. The two alloys explained 227.64 Ω^2 of the total variance. The slope with respect to temperature explained 1131.13 Ω^2 of the total variance. We conclude that temperature explained more variance of the response than the type of alloy; our conclusion is confirmed by `tm`'s larger value of the Fisher statistic (F-value). The Fisher statistic measures the benefit of a factor against the cost of including that factor. In the anova table, we see that the temperature and alloy are both significant for explaining the response. A leftover 151.95 Ω^2 of the total variance remains unexplained by this choice of model. The leftover variance is then divided by the remaining Degrees of Freedom (*Df*) to give average variance. As the square root of the average variance, the Residual Standard Error (by `summary`) is akin to the Standard Deviation. Notice that the *Df* reduced from 44 to 41 because three data-derived coefficients were reused to calculate the expected values used in the equation for Residuals (as in Section 7.1.5).

The Fisher statistic for the whole model is the average of the F-values listed in the anova table. A large value for the Fisher statistic is preferred. The whole-model Fisher statistic is F = (ratio of sample size to model size) × (ratio of explained to unexplained) or

$$F=\left(\frac{n-m}{m-1}\right)\left(\frac{R^2}{1-R^2}\right). \tag{7.9}$$

Figure 7.4 shows us that the model is not perfect, reaffirming "All models are wrong, but some models are useful" (George Box). Viewing the graph may guide you how to devise an improved model, for example, seeking individual slopes for each alloy. We will inspect and evaluate more model options in Section 7.5. Try them out. By using R to execute and examine each model option suggested, you can tap the power of R for statistical analysis.

7.5 Gain Confidence and Skill at Statistical Modeling Via the R Language

7.5.1 Model and Plot Results of a Single Variable Using the Example Data diceshoe.csv

```
a=read.csv('diceShoe.csv') # Read data file into R.
head(a)                    # View the initial cases in the R datatable.
mean(a$size)               # Calculate the sample mean. Up arrow ↑. Edit command.
sd(a$size)                 # Calculate the Standard Deviation of the selected variable.
mod0=lm(size~1,data=a)     # Perform a linear model based on zero explanatory factors.
summary(mod0)              # Compare results: mean, Uncertainty, Standard Deviation.
anova(mod0)                # Perform Analysis of Variance.
boxplot(a$size,ylab="Size",cex.lab=1.4)          # Draw and label a boxplot.
hist(a$size,xlab="Size",main="",breaks=
  c(20,21,22,23,24,25,26))                       # Histogram.
plot(density(a$size))      # Draw and label a density plot.
rug(a$size)                # Add marks where data points occur.
xCum=sort(a$size)          # Sort in order to draw a cumulative plot.
yCum=seq(0,1,length=length(xCum))
plot(yCum~xCum,type='b',xlab='Size',ylab='Portion',main='Cumulative Plot')
```

7.5.2 Evaluate Alternative Models of the Example Data hiloy.csv (Sections 7.5.3 Through 7.5.8)

7.5.2.1 Inspect the Data

```
b=read.csv('hiloy.csv',na.strings='')   # Read data file into R; detect blank entries.
head(b)                                  # View the initial cases in the R datatable.
length(b$ohm)                            # Display the number of cases in the datatable.
```

7.5.3 Grand Mean

```
mean(b$ohm)                 # Calculate the sample mean of the selected variable.
mod1=lm(ohm~1,data=b)       # Perform a linear model based on zero explanatory factors.
mod1                        # A model with no explanatory factors is the Grand mean of all data.
summary(mod1)               # Complete results: compare mean, Uncertainty, Standard Deviation.
```

7.5.4 Model by Groups: Group-Wise Mean

```
mod2a=lm(ohm~ -1+alloy,data=b)  # Prevent Intercept; model on a categorical factor.
mod2a                   # The result displays each group mean. Up arrow ↑↑. Edit command.
mod2=lm(ohm~alloy,data=b)     # Include default Intercept, model on the same factor.
mod2                    # Compare mod2 to mod2a. The results are identical but presented
                        # differently; to obtain the other group mean, add the coefficients.
summary(mod2)           # Compare results: coefficients, uncertainties, mean of Residuals.
plot(ohm~temp,data=b)                          # View original data.
points(fitted(mod1)~temp,pch='&',data=b)       # Show Grand mean for all data.
points(fitted(mod2)~temp,pch='+',data=b)       # Show means for each group.
boxplot(ohm~alloy,data=b,ylab='Ohm',cex.lab=1.4)  # Compare boxplots for each
                                                  # group.
```

7.5.5 Model by a Quantitative Factor

```
mod3=lm(ohm~temp,data=b)  # Model on a single quantitative factor.
mod3                      # Notice the slope and the Intercept extrapolated to temp=0.
plot(ohm~temp,data=b)                          # View original data.
points(fitted(mod3)~temp,pch='+',data=b)       # Show least-squares fit of all data.
mean(b$temp)              # Calculate the mean of a quantitative factor and use it to
b$tm=b$temp-78.86           # avoid extrapolation to reduce Uncertainty.
head(b)                   # Notice how cases now appear in the R datatable.
mod3a=lm(ohm~tm,data=b)   # Model after centering on a quantitative factor.
mod3a                # Notice the slope and the Intercept interpolated at the mean temp.
summary(mod3)        # Complete results: the Std.Error of the extrapolated Intercept is large.
summary(mod3a)       # Compare results: the Std.Error of the interpolated Intercept is minimized.
```

7.5.6 Model by Multiple Quantitative Factors

```
mod4=lm(ohm~tm+alloy,data=b)  # Model on two factors; earlier example with order reversed.
plot(ohm~temp,ylim=c(10,45),data=b)          # View original data, scale y axis.
points(fitted(mod4)~temp,pch='+',data=b)     # The model for both groups has same slope.
summary(mod4)   # Compare results: the slope is identically the same for each group.
anova(mod4)     # Compare Section 7.4.4: this order of factors explains variance most to least.
```

7.5.7 Allow Factors to Interact (So Each Group Gets Its Own Slope)

```
mod5=lm(ohm~tm+alloy+tm:alloy,data=b)    # Model with interaction. The temperature
                                          # slope can differ for each alloy group.
summary(mod5)         # Interaction improved the model. The Std.Errors and Residual
                      # decrease.
Coeff:          Estimate Std.Error  t-value  Pr(>|t|)
(Intercept)     19.5371  0.3256     60.011   < 2e-16   # Reject the Null Hypothesis.
tm               0.11372 0.00971    11.708   1.69e-14
alloyN           5.5039  0.4603     11.956   8.77e-15
tm:alloyN        0.06852 0.01349     5.077   9.25e-06
Residual Standard Error: 1.52 on 40 Degrees of Freedom
plot(ohm~temp,ylim=c(10,45),data=b)
points(fitted(mod5)~temp,pch='+',data=b)    # Each group has own least-squares
                                            # fit slope.
mod5a=lm(ohm~tm*alloy,data=b)               # This shortcut notation exactly
                                            # mimics mod5.
summary(mod5a)        # Compare to mod5: identical.
anova(mod5)           # Explained more than mod4.
```

The model mod5 derived the best possible straight-line fits, with individual slopes for each alloy, shown in Figure 7.5. A straight line fits alloy C data well with small Residuals over the whole temperature domain. In contrast, a line does not not fit alloy N data over the whole domain, resulting in large Residuals for alloy N. At low temperatures, the resistance of alloy N behaved linearly, but a solid phase change caused the behavior to change at higher temperatures.

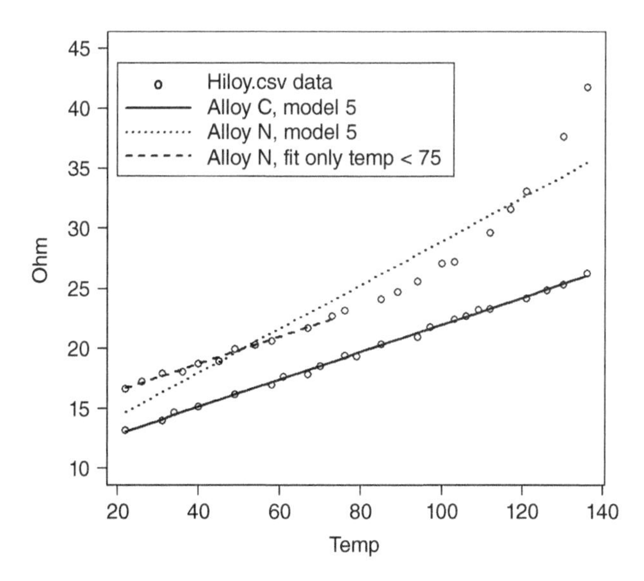

Figure 7.5 Model 5 provided best "least-squares" straight line fits for each alloy.

7.5.8 Include Polynomial Factors (a Statistical Linear Model Can Be Curved)

```
mod6=lm(ohm~alloy*(tm+I(tm^2)),data=b)   # Model with polynomial and interaction.
summary(mod6)                             # Inspect results of quadratic fit.
plot(ohm~temp,ylim=c(10,45),data=b)
points(fitted(mod6)~temp,pch='+',data=b)  # Each group has its own second-
                                          # order polynomial fit.
```

This model gives no improvement for alloy C. This model equation allows us to predict the electrical resistance of alloy N with greater accuracy than previous models. Notice, however, that the modeled N alloy resistance at low temperatures reaches a minima and increases again for even lower temperatures. This more accurate model with lower uncertainty defies physics. Do we use it? The Motivating Question guides. The client decides.

7.6 Report Uncertainty

A credible experiment must report values along with their uncertainties. Recall Section 7.1 described the Uncertainty of the mean. In Section 7.4 the `summary` command yielded the Uncertainty of the mean under the heading "Residual Standard Error." Now please review the results of each `summary` command in Section 7.5. The second column gives the `estimate` of each coefficient in the linear model and the third column gives the `Standard Error` of the same coefficient. From the Standard Error, we obtain the uncertainties that we must report. Figure 7.6 shows the Uncertainty for each modeling term at 95% confidence. The total Uncertainty is a hyperbolic curve. Let's see how the Uncertainty was calculated from summary(mod5) values.

Notice that the Degrees of Freedom reduced from 44 to 40, because four coefficients were reused to calculate the Residuals. For 95% confidence and $Df = 40$, the t-factor is 2.02. We view only the alloy C group. The Intercept Standard Error causes the modeled

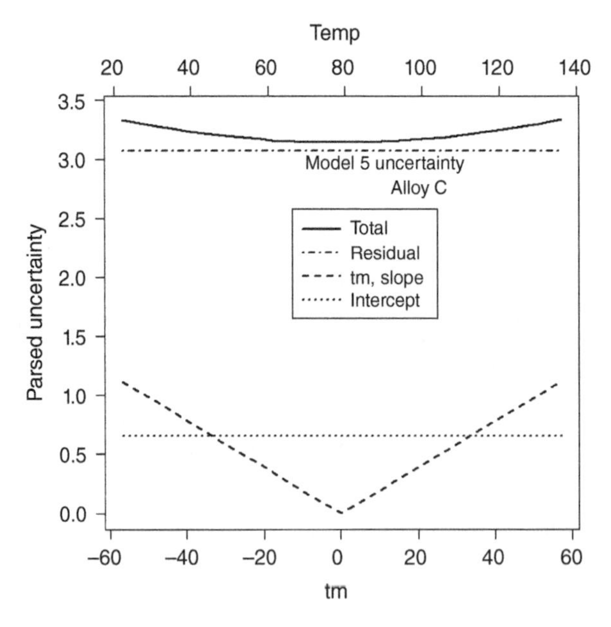

Figure 7.6 Contributions to Uncertainty at 95% confidence for model 5 terms.

constant Uncertainty to be $2.02 \cdot 0.3256 = 0.658$. The tm Standard Error gives a slope Uncertainty of $2.02 \cdot 0.0097 = 0.0196$. The Residual Standard Error yields an unexplained (nonmodeled) Uncertainty of $2.02 \cdot 1.52 = 3.07$. Since the nonmodeled Residual is orthogonal to the modeled Intercept Uncertainty, the constant uncertainties combine $\sqrt{\left(0.658^2 + 3.07^2\right)} = 3.14$. Thus the complete mod5 equation including Uncertainty for alloy C is

$$ohm = \left(19.54 \pm 3.14\right)\Omega + \left(0.1137 \pm 0.0196\right)\left[\frac{\Omega}{K}\right]\left(temp - 78.86\right)[^\circ C]. \tag{7.10}$$

It is clear that the dominant contributor to the total Uncertainty in Figure 7.6 and the model equation is the Standard Error of the Residuals. How can the Standard Error be reduced?

7.7 Decrease Uncertainty (Improve Credibility) by Isolating Distinct Groups

We recommend two techniques to credibly and honestly decrease Uncertainty (NO fudging or ignoring data, which is another form of statistical malpractice):

1) Avoid extrapolation by translating each quantitative factor to a reasonable zero. For our example hiloy data, we zeroed the temperature explanatory factor about its central value and modeled using tm. Recall the models 3 (using temp) and 3a (using tm). When we compared the Std.Error of the Intercept, model 3a (tm) had much smaller Standard Errors.
2) Reduce the number of groups in a model. Isolate and model each distinct group based on combinations of the categorical factors.

Our example data had only one categorical factor "alloy" with two levels C and N. We use the `subset` command to isolate each group. Then we analyze each group separately.

```
bC=subset(b,b$alloy=='C') # Create a subset R datatable with only the C alloy group.
head(bC)                  # View the initial cases in the subset R datatable.
length(bC$ohm)            # Display the number of cases in the datatable.
mod7=lm(ohm~tm,data=bC)   # Fit a linear model using the centered explanatory factor.
summary(mod7)             # Compare to mod3a, our previous best. Standard Errors
                          # are reduced.
anova(mod7)               # Compare how this model explains variance.
```

Let's write the mod7 equation with 95% confidence. For $Df = 20$ the t-factor is 2.09. The Intercept Uncertainty is $2.09 \cdot 0.0373 = 0.0780$, slope is $2.09 \cdot 0.00111 = 0.00233$, and due to Residual Std.Error is $2.09 \cdot 0.1742 = 0.364$. Since the Residual is orthogonal to the Intercept Uncertainty, they combine $\sqrt{\left(0.078^2 + 0.364^2\right)} = 0.372$. Thus the complete mod7 equation for alloy C is

$$ohm = \left(19.54 \pm 0.372\right)\Omega + \left(0.1137 \pm 0.0023\right)\left[\frac{\Omega}{K}\right]\left(temp - 78.86\right)[^\circ C]. \tag{7.11}$$

Let's plot these results along with confidence intervals:

```
bC$un7=sqrt(0.372^2 + (0.00233*bC$tm)^2)  # Calculate mod7 Uncertainty for
                                           # C alloy.
bC$un5=sqrt(3.14^2 + (0.0196*bC$tm)^2)    # Calculate mod5 Uncertainty for
                                           # C alloy.
plot(ohm~temp,ylim=c(10,30),data=bC)       # View original alloy C data.
points(fitted(mod7)~temp,pch='+',data=bC)          # View linear fit of alloy C.
points(fitted(mod7)+un7~temp,pch='*', data=bC)     # Add +mod7 Uncertainty.
points(fitted(mod7)-un7~temp,pch='-', data=bC)     # Add –mod7 Uncertainty.
points(fitted(mod5)+un5~temp,pch='v', data=bC)     # Add +mod5 Uncertainty.
points(fitted(mod5)-un5~temp,pch='^', data=bC)     # Add –mod5 Uncertainty.
help(plot)                  # Access online help for the plot function (or any R function).
```

Figure 7.7 displays Uncertainty bands for our two best models of alloy C. Isolating the alloy C group (as we did for mod7) resulted in a narrower band which appears appropriate for the data. In contrast model 5 has wide band – a consequence of the large Residuals of alloy N. This plot shows the advantage of isolating groups when selecting a preferred model. Since the Residual Standard Error is a dominant contributor to Uncertainty, improved models aim at reducing Residuals.

Referring back to Figure 7.6, the total Uncertainty band has a slight hyperbolic curve, narrowest at the zero reference of the quantitative factor. We created the variable tm, centered about the mean temp 78.86 °C, in order to minimize the total Uncertainty band and to avoid extrapolation. Extrapolation causes wide Uncertainty bands.

7.8 Original Data, Summary, and R

Rather than use an original datafile for statistical analysis, credible practice reckons the original data as sacrosanct. After the data are collected and recorded, the original datafile is never changed. Defend yourself against data corruption or charges of scientific malpractice.

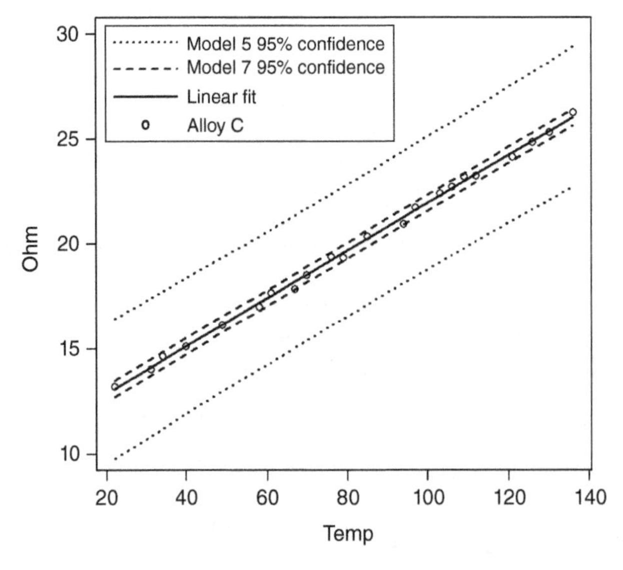

Figure 7.7 Compare Uncertainty bands for two models of alloy C.

Files containing data are typically saved in comma-separated variable (csv) format, but other formats are available. You may record your data in a spreadsheet, such as open document (ods) or excel (xls) format. The first row of the spreadsheet contains the labels of each factor or variable, including response. It helps to refrain from punctuation in factor labels. Each subsequent row contains a single case with the factor value in its own column. Extra sheets/tabs in the spreadsheet may be used as a codebook to explain ranges and categories for each factor, along with physical dimensions. Before R analysis, export original data from the spreadsheet into csv format. Within R, load the data with the command `read.csv()` as we did in examples.

The R language enables a history of graphs to be saved in individual files (png, pdf, jpg, etc.) for import into other documents. We recommend the free open-source R statistical language, but many other tools are available for statistical analysis. The commands in this text work in the commercial S language. Other commercial packages have similar capabilities, though commands may differ. To obtain the R language, search for "r cran" or "r language." Download and install the base package that pertains to your operating system (linux, apple, or ms). See Appendix D.1 for details.

Since the R function for linear modeling performs a least-squares fit,[3] the Standard Error for each term and for the Residual are orthogonal (that is, independent) to each other and add together similar to the Pythagorean Theorem. Uncertainty bars are calculated by the root of the sum of variances (squared Std.Error) as discussed by Kline and McClintock (1953).

Having refreshed basic statistical concepts, we hope that you can confidently perform basic statistical analysis of up to three variables. The next chapter introduces wisdom from statistics to collect data more effectively as our skill extends to dozens of variables.

References

Kaplan, D. (2012). *Statistical Modeling: A Fresh Approach*, 2e. Project Mosaic Books.

Kline, S.J. and McClintock, F.A. (1953). Describing the uncertainties in single-sample experiments. *Mechanical Engineering* 75: 3–8.

Homework

We recommend that you try Exercises 7.1 through 7.6 by hand or with minimal use of a calculator (spreadsheet). Of course it is possible to use spreadsheet or computer language functions to determine the "sample" values requested in Exercises 7.1 through 7.6. However, for the "true" values requested in Exercises 7.1 through 7.6, no such built-in spreadsheet function (or calculator function) exists. The true values in each exercise may be calculated by hand, by calculator, or by creating your own spreadsheet macro.

3 The "least square-root sum-of-residuals-squared method" for fitting lines and models is commonly known as the "least-squares." Who would not be thankful for this brevity?

7.1 Add these two new data to the set in Table 7.1. Taking the roll of a single standard fair six-sided dice as the source, determine the true mean of the dice and the sample mean of the augmented data (all 14 samples). Repeat for the shoe size.

Case	Age	Size = Y	Dice
13	12	26	6
14	13	23	3

7.2 For the augmented set of Exercise 7.1, determine the true Residual (based on true mean) and the sample Residual. How do these values differ between the dice and the shoe size?

7.3 For the augmented set of Exercise 7.1, determine the Degrees of Freedom for subsequent data analysis based on the true mean. Repeat for the Degrees of Freedom based on the sample mean.

7.4 For the augmented set of Exercise 7.1, determine the Standard Deviation and the sample Standard Deviation. Furthermore determine the variance and the sample variance. How do these values differ between the dice and the shoe size?

7.5 For the augmented set of Exercise 7.1, determine the Uncertainty of the true mean. Furthermore determine the Uncertainty of the sample mean. How do these values differ between the dice and the shoe size?

7.6 For the augmented set of Exercise 7.1, consider the Chi-Squared, χ^2. Determine the model expected values for the dice. Furthermore determine the model expected values for the shoe size. Then determine the Chi-Squared, χ^2, for the dice. Repeat to determine the Chi-Squared, χ^2, for the shoe size.

Chapter 7 Modeling Exercises

Remember: statistics explains variation in the context of what remains unexplained.

7.7 For the augmented data of Exercise 7.1, follow the steps in Section 7.4.1 to create a simple model of purely quantitative data. Compare the values in the R-language summary to your answers in Exercises 7.1–7.6.

To create an augmented data csv file for analysis by the R-language, do the following:

a) Use a spreadsheet (LibreOffice calc or ms Excel) to open the file `diceShoe.csv`.
b) Immediately save into a new filename `diceShoeAug.csv`. Use File→Save As.
c) Add two rows of data in the appropriate columns.
d) Save the augmented data into the new filename.
e) Close the spreadsheet.
f) Follow the text commands and steps in Section 7.4.1 for file `diceShoeAug.csv`.

7.8 Follow the steps in Section 7.5.1 to do more advanced statistical analysis and plots on the augmented file `diceShoe.csv`.

7.9 For the data `hiloy.csv`, redo the analysis in Section 7.4.2 on the original `temp` factor rather than the centered tm variable we created. Compare new values provided in the `summary()` command with the values in 7.4.2. Note which values are the same. Note which values differ. Why? Hint: where is the value for the `Intercept` evaluated? Remark on the relative percentage that the Standard Error (`Std.Error`) has changed.

7.10 Following the steps in Section 7.4.2, create two linear models of shoe size as explained by the factor age. Create one model from the original values in the `diceShoeAug.csv`. Create a second model for a dynamic factor `agem`, which is centered on the mean age. Compare values provided in the `summary()` command. What other factors in real life besides age would help explain shoe size?

7.11 In Section 7.4.2, the analysis resulted in the following model function:

$$ohm = 19.42\Omega + \begin{bmatrix} 5.51, \text{ if N} \\ 0.00, \text{ if C} \end{bmatrix} \Omega + 0.1492 \left[\frac{\Omega}{K}\right] (temp - 78.86) \; [°C].$$

Write this equation as two separate equations, one for each alloy.

7.12 For each of the two equations in Exercise 7.11, write an equation that includes and displays the Standard Error, found in the `summary()` command. Refer to Section 7.6 for an example. Each coefficient shows ± one Standard Error. Recall that for a normal distribution, one Standard Error accounts for 68% of the data. For a normal distribution, two Standard Errors accounts for 95%. Write both modeling equations including 95% confidence of the coefficients.

7.13 Do the same as Exercise 7.12 for the modeling equation (95% confidence) that results from your answer to Exercise 7.9.

7.14 Do the same as Exercise 7.12 for the modeling equation (95% confidence) that results from `mod2` in Section 7.5.4.

7.15 Do the same as Exercise 7.12 for the modeling equation (95% confidence) that results from `mod3a` in Section 7.5.5.

7.16 Do the same as Exercise 7.12 for the modeling equation (95% confidence) that results from `mod4` in Section 7.5.6.

7.17 Do the same as Exercise 7.12 for the modeling equation (95% confidence) that results from `mod5` in Section 7.5.7.

7.18 Do similarly to Exercise 7.12 for the modeling equation (95% confidence) that results from `mod7` in Section 7.7.

7.19 Create a model `mod8`, similar to the steps in Section 7.7, isolating alloy N. Do similarly to Exercises 7.12 and 7.18 for the modeling equation (95% confidence) that results from `mod8`.

7.20 Project: work through a series of models on your own experimental data.

8

Exploring Statistical Design of Experiments

8.1 Always Seeking Wiser Strategies

This chapter shows how to reduce the number of trials while extracting as much information as possible out of an experiment. This requires planning experiments statistically, to maximize the information derived and minimize the work required. There are many excellent references available concerning the statistical design of experiments. This chapter has a more limited scope:

i) To overview full-factorial designs and fractional-factorial designs.
ii) To set up a well-designed 12-run Plackett–Burman (PB) screening design in order to identify the three or four most important factors from a list of up to 11 candidate factors.
iii) To quantify the most important factors measured in either task (i) or (ii) above.
iv) To reinforce wise strategies for experiment design.

With a goal of collecting reliable data in an experiment, it helps to be aware of common pitfalls that reduce the quality of collected data. Thankfully the field of statistics has provided solutions to avoid these pitfalls. By employing them, you may be spared the grief of time wasted taking futile data. Every experimentalist we know has her/his stories of such grief. We all wish that we had learned from these wise strategies for taking data earlier in our careers.

8.2 Evolving from Novice Experiment Design

Novice experimental plans often run "one factor at a time," stepping across the range of each variable in turn while holding all others constant. This kind of experiment is second nature – for example with a radio we first select AM or FM, then adjust the frequency dial to catch a signal, and finally adjust the sound volume dial. For the hiloy test in Chapter 7, we might select only the N alloy, pick a temperature, measure resistance and record, increase the temperature, and so on. In contrast, randomizing factors, as we did for the hiloy test, are counterintuitive – it would be like toggling the radio AM/FM switch and adjusting dials simultaneously. A single-factor test favors our ability to

Planning and Executing Credible Experiments: A Guidebook for Engineering, Science, Industrial Processes, Agriculture, and Business, First Edition. Robert J. Moffat and Roy W. Henk.

Companion website: www.wiley.com/go/moffat/planning

detect trends; randomization thwarts it. From a math viewpoint, a single-factor test yields the partial derivative with respect to the changed variable. Problems with this approach when there are more than two factors lead us to statistical design.

Consider, for example, a three-factor experiment with five points to span the range of each factor. Mapping the complete operating domain requires $5^3 = 125$ cases. The time and cost may be acceptable if your client has already established that the proposed experiment is worthwhile – it might be tedious, but not impractical. Then there are other concerns: how repeatable is the set value after a factor is changed? Does any value drift after the factor is set? Hassles changing the setup tempt us to measure more points for one factor rather than change other factors. Furthermore, in the early stages of most investigations, a long list of factors might be of interest. There is neither time nor money available to run the whole set of possibly significant factors if the plan is to investigate each factor one at a time.

The key mission of this chapter is to urge that plans for data collection be launched through a Plackett–Burman screening design. A 12-run PB design can economically identify the most important factors from a list of up to 11 candidates. One need not be skilled in statistical analysis to pursue a 12-run PB design – many readers start right off with Section 8.5 and only read other sections as their value becomes clear. In order to lay a foundation for PB, we introduce concepts of statistical experiment design via a two-level factorial design.

8.3 Two-Level and Three-Level Factorial Experiment Plans

The term "statistical design of experiments" typically refers to a method of wisely selecting the experimental points which will be run. One may plan a reduced *sequence* of measurements to obtain minimum run time for an acceptable confidence, trading away higher confidence for the savings in experimental time.

The principal impetus for statistical treatment of experiments is that the classic "one factor at a time" experiment plan becomes prohibitively expensive when there are more than three factors. If the response surface of an experiment is reasonably well behaved (i.e. continuous derivatives inside the domain), then one need not measure every possible combination of independent factors in order to accurately assess the shape of the response surface. The statistical treatment aims at choosing the data cases so as to get the most information for the effort exerted.

The fundamental basis upon which the statistical design of experiments rests is that the response surface is relatively "smooth" but not necessarily flat. Let's consider a small region of the response surface which could be well described with a Taylor's series expansion about a midpoint containing terms of second order. This assumption is valid for any continuous surface in a small enough domain and sometimes over large domains. Statistical treatment of the data will reveal the best estimate of the slope, second derivative, and first mixed derivatives of the result with respect to each factor. This will be the estimate consistent with the assumption that the higher-order derivative terms don't significantly affect the behavior of the result.

In order to achieve the best possible estimate of the shape of the response surface, it helps to move quickly to the boundaries of the operable domain when choosing data points. This "bold" strategy covers the corners of the operable domain completely,

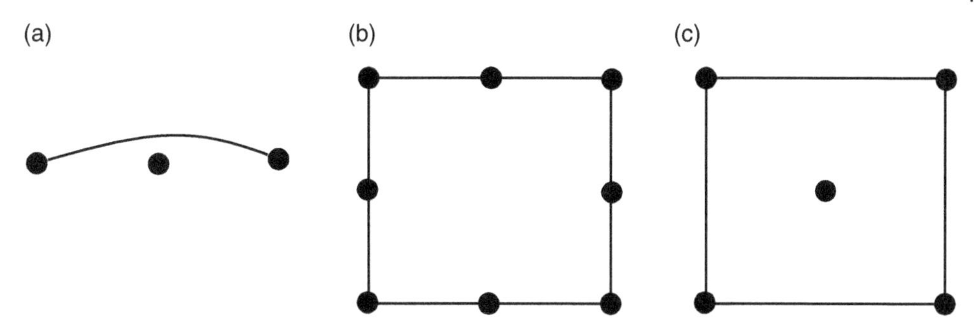

Figure 8.1 Sampling locations.

setting the maximum range for each independent factor. Considering a single factor as in Figure 8.1a, values at two extreme points and one center point suffice to obtain the mean level of the response and to estimate the first derivative and second derivative. Considering a two-factor problem, the three points on four sides become eight points as in Figure 8.1b, treating each side as two endpoints and a central point. Using locations as in Figure 8.1c, four corner points plus a point in the center of the operating domain, likewise allows estimation of the level, slope, and second derivative using only five data points instead of eight. Yet the trade-off is that the five points cannot distinguish which factor impacts the second derivative. All these lines of reasoning have given rise to the subject known as "statistical design of experiments."

The term "statistical" also invokes the notion of randomness. Although statistical design of experiments is a very systematic procedure, the experimental cases must be executed in a random order. The intent of randomizing the sequence of cases is to break up any correlation which might exist between the result of the process and time. Two factors which are perfectly correlated cannot be separated from one another with respect to their effect on an experiment, since whenever one factor changes, so does the other. Moreover, time is a factor in every test, always increasing. Thus any monotonic factor will correlate with and be confounded by the time factor; a notable yet oft unrecorded factor is that the skill of the experimentalist improves over time.

8.4 A Three-Level, Three-Factor Design

Consider an experiment with three independent factors, x_1, x_2, and x_3. The operable domain of this experiment can be visualized as a cube if one imagines that the range of each independent factor is appropriately scaled. The unit cube is shown in Figure 8.2 with the corner points identified by their values of x_1, x_2, and x_3.

To organize a two-level factorial experiment, it is convenient to establish first the range of each factor and to code it, i.e. identify its largest value by (+1) and its smallest value by (−1). For a third level, a central value is identified as zero. An illustrative example sets a context for the following discussions: a three-level, three-factor experiment tests the degree of polymerization (response) as a function of the pressure, temperature, and process time. The codebook for factors and response is listed in Table 8.1.

The three explanatory factors (x_1, x_2, and x_3) become the axes of the scaled cubic operable domain shown in Figure 8.2. The corners of the cube represent the combination of

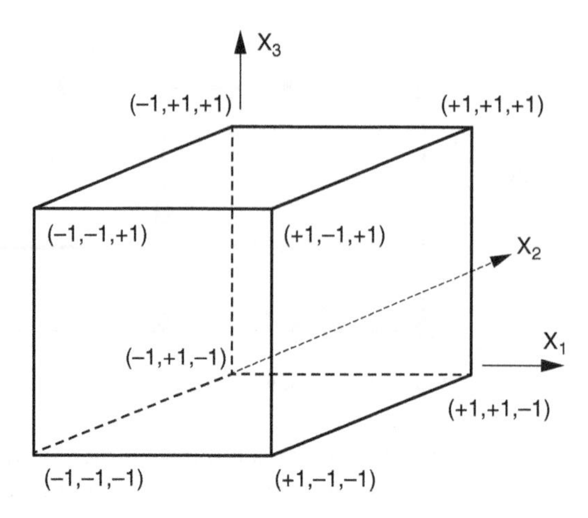

Figure 8.2 The coded unit cube with eight corner points.

Table 8.1 Example 1, Part 1: codebook of test conditions for a three-level, three-factor experiment.

	Factor Coding	–	0	+	
Label	Factor	Low	Mid	High	Units
X1	Pressure	20	50	80	kPa
X2	Temperature	240	260	280	°C
X3	Time	30	40	50	min
	Response				
Y	Polymerization	0		300	

experimental parameters at which data will be taken. Each trial will be identified by its appropriate triad of (+) and (−) signs. The calculation scheme we use requires that each corner of the operable domain be designated by a specific "trial number," according to Table 8.2, and this nomenclature must be preserved. These identifications will be preserved for optional worksheet instructions.

Full-factorial experiment designs anticipate taking data at every corner of the domain. Table B.2 shows full two-level factorial designs up to five factors. For our three-level example, Table 8.2 added just a central point. BH&H (see preface) offer other options for three-level designs.

Rather than just run the nine trials listed in Table 8.2, the client opted for 20 case runs as shown in Table 8.3. The purpose for extra cases was to reduce uncertainties at the eight corner points, sampling twice. The center was sampled four times to derive a pooled variance (Appendix A). In Table 8.3, the testing sequence is given with the experimental response shown in column Y.

Please notice the randomized order in which the trials were actually made. The randomizing step is important in order to thwart possible correlations between time

Table 8.2 Example I, Part 2: coding and assignment of trial identification numbers.

	Coded			Physical		
Trial ID	X1	X2	X3	X1	X2	X3
1	–	–	–	20	240	30
2	+	–	–	80	240	30
3	–	+	–	20	280	30
4	+	+	–	80	280	30
5	–	–	+	20	240	50
6	+	–	+	80	240	50
7	–	+	+	20	280	50
8	+	+	+	80	280	50
9	0	0	0	50	260	40

Table 8.3 Example 1, Part 3: the randomized trial schedule and test results.

Test Order	Trial ID	X1	X2	X3	Y
1	9	50	260	40	63
2	1	20	240	30	9
3	5	20	240	50	91
4	1	20	240	30	11
5	3	20	280	30	14
6	2	80	240	30	30
7	9	50	260	40	67
8	7	20	280	50	155
9	6	80	240	50	125
10	3	20	280	30	20
11	6	80	240	50	129
12	5	20	240	50	99
13	2	80	240	30	36
14	9	50	260	40	60
15	4	80	280	30	38
16	8	80	280	50	177
17	7	20	280	50	155
18	4	80	280	30	48
19	8	80	280	50	179
20	9	50	260	40	72

and the process under study. The reader may use a ready-made list of randomized order, such as Table B.1. For a 20-trial experiment, select one column from the 32 trials and then delete numbers larger than 20. A handy, randomly arranged list of 1–20 trials remains.

As you design an experiment, there are trade-offs to consider. A full-factorial design tests every combination of factors. The main benefit is that every case contributes to every model coefficient of the response. This "concealed replication" enhances the confidence level. One trade-off, Fractional Factorial Design,[1] exploits the replication to reduce the number of trials and save time. Another trade-off, done here with the blessing of our client, is to do more tests than a full factorial specifies in order to refine uncertainties. Our aim is to equip you to face the trade-offs, employing advances in statistical design, so that you may plan for credible results at lowest cost of experiment time.

Just as we saw in the Chapter 7, the free open-source R language provides quick, easy, and accurate analysis of a statistically designed experiment. The data in Table 8.3 have been stored in file mh7cube.csv with one row for each observation of the experiment. The key commands are:

```
> a=read.csv("mh7Cube.csv")          # Read the csv file into R.
> mod1=lm(Y~X1*X2*X3, data=a)        # Perform a linear model including interaction
                                     # terms.
> summary(mod1)                      # Display coefficients for all model terms.
> anova(mod1)                        # Display Analysis of Variance to identify dominant
                                     # terms.
```

Table 8.4 shows an Analysis of Variance to determine how each factor and all cross terms contributed to the response variable Y. Only four factors survived the Null

Table 8.4 Example 1, Part 4: Analysis of Variance of the modeling terms.

Term	Df	Sum Sq	Mean Sq	F-value	p-Value
X1	1	2704	2704	29.264	0.00016
X2	1	4096	4096	44.329	2.33E-005
X3	1	51076	51076	552.771	2.10E-011
X1:X2	1	9	9	0.097	0.76033
X1:X3	1	9	9	0.097	0.76033
X2:X3	1	2209	2209	23.907	0.00037
X1:X2:X3	1	36	36	0.390	0.54419
Residuals	12	1109	92		

1 In Section 8.5, we introduce the PB method, which is easy to explain without a computer. N.J.A. Sloane at AT&T developed software that expands PB and full-factorial designs. It optimizes flexible low-cost designs for strong confidence levels. The power of the open source Gosset program will be highlighted later in this chapter.

Hypothesis (low p-value) to comprise a final model. Ranked according to dominance, the factors are X3 (time), X2 (temperature), X1 (pressure), and X2:X3 (temperature: time).

Section 8.6 explains complete details of how to perform this analysis by evaluating alternative model options. Reasons for the final preferred model then become clear.

8.5 The Plackett–Burman 12-Run Screening Design

The main event is the Plackett–Burman 12-run Orthogonal Experiment Plan. As a fractional factorial plan, PB dramatically reduces the number of cases below a pure full-factorial design.

The PB screening designs are a family (eight run for up to seven factors, 12 run for up to 11 factors, 16 run, etc.). The PB class of experiment designs is used to identify which of several candidate factors impact results without executing a full two-level factorial experiment. As an example of the economy which can be achieved, consider an experiment with 11 factors. A full two-level factorial experiment requires $2^{11} = 2048$ runs, but the PB can screen all 11 factors with only 12 runs. The economy of PB is apparent, but PB yields only main factor coefficients along with an estimate of their significance.

You won't get as much information from 12 runs as from 2048, but you may get enough! What is sacrificed by the PB design is the possibility of learning about interactions between the factors. A PB design will identify the main effects of each of the factors but will not speak to the interaction question. Hence a good strategy, when confronted with a multifactor problem, is to screen it first with a PB design to order to identify which factors impact results most. Subsequently, run a full-factorial experiment (described in Section 8.3) on the reduced parameter list – this approach saves time and cost.

As a demonstration of PB 12-run design, we consider material production with six factors. The codebook of natural factors for Example 2 along with their types and their codings is shown in Table 8.5. Note that the PB design allows for discrete, categorical,

Table 8.5 Example 2, Part 1: Codebook of factors and codings for a six-factor, 12-run PB design. The codebook also explains the Response variable(s) data.

Label	Factor	–	+	Units	Factor Type
X1	Tension control	Manual	Automatic		Discrete
X2	Temperature	200	250	°C	Continuous
X3	Throughput	10	20	kg/min	Continuous
X4	Mixing	Single	Double		Discrete
X5	Machine used	#1	#2		Discrete
X6	Moisture	20	80	%	Continuous
Y	**Response**	Strength (etc.)		MPa	

Table 8.6 Example 2, Part 2: 12-run PB design for a six- parameter problem.

	Coded						Natural						
Trial	X1	X2	X3	X4	X5	X6	X1	X2	X3	X4	X5	X6	Y
1	+	+	−	+	+	+	Auto	250	10	Dbl	2	80	85
2	+	−	+	+	+	−	Auto	200	20	Dbl	2	20	114
3	−	+	+	+	−	−	Man	250	20	Dbl	1	20	67
4	+	+	+	−	−	−	Auto	250	20	Sgl	1	20	64
5	+	+	−	−	−	+	Auto	250	10	Sgl	1	80	56
6	+	−	−	−	+	−	Auto	200	10	Sgl	2	20	68
7	−	−	−	+	−	+	Man	200	10	Dbl	1	80	13
8	−	−	+	−	+	+	Man	200	20	Sgl	2	80	108
9	−	+	−	+	+	−	Man	250	10	Dbl	2	20	90
10	+	−	+	+	−	+	Auto	200	20	Dbl	1	80	22
11	−	+	+	−	+	+	Man	250	20	Sgl	2	80	130
12	−	−	−	−	−	−	Man	200	10	Sgl	1	20	23

nonquantitative variables![2] The six factors shown in Table 8.5 are assigned "high" and "low" values and placed in the first six columns of Appendix B, Table B.3. This layout defines the conditions to be used for each trial number. The Response is explained.

The procedure for executing a 12-run PB design is essentially the same as the procedure for the two-level factorial design.

- Assign the factors to high or low states for the trial numbers using Table B.3.
- Randomize the running order of the test cases, perhaps by using Table B.1.
- Run the 12 test cases and tabulate the data.
- Calculate the 11 factor effects using R, or the worksheet patterned in Table C.5.
- Assess the significance of the factor effects and identify the main factors.

The results of this experiment are shown along with the coded factors in Table 8.6.

Since we selected the 12-run PB, we must make all 12 runs even with less than 11 factors. Had you noticed that we could have screened six factors with an eight-run PB design?

The next step would be to choose the order in which the tests would be run. The randomization could be done using Table B.1. After arranging the trials in random order, the 12-run test plan was executed. The responses for each trial number are listed in nonrandomized order in column Y of Table 8.6. The data are ready for analysis (whether saved in random order or not).

2 As an example of a nonscientific screening, a PB design could be used to investigate the reasons for a difference in opinion between a school board and the admission committee as to which students should be admitted: different weight functions on traits yield different factor effects.

Table 8.7 Example 2, Part 3: Analysis of Variance as parsed for each model term.

Label	Df	SumSq	MeanSq	F-value	p-Value
X1	1	40	40	0.216	0.66181
X2	1	1728	1728	9.244	0.02874
X3	1	2408	2408	12.883	0.01572
X4	1	280	280	1.500	0.27527
X5	1	10 208	10 208	54.610	0.00071
X6	1	12	12	0.064	0.81007
Residuals	5	935	187		

```
> apb=read.csv("mh7PB.csv")            # Read the csv file into R.
> head(apb)                            # Inspect the top six lines of the file.
> mod2=lm(Result~X1+X2+X3+X4+X5+X6, data=apb) # Linear model the main factors.
> summary(mod2)                        # Display analysis for coefficients of main factors.
> anova(mod2)                          # Display Analysis of Variance for main factors.
```

Table 8.7 to the right shows an analysis of the above data to find how each X factor contributes to the variance in response Y. The F-value is a ratio of the benefit of a factor divided by the cost of including the factor. Analysis of Variance reveals three dominant terms by the F-value. In order of dominance, the factors are X5, X3, and X2.

By looking at the coefficients for these dominant factors (cf. Sections 8.6.6 and 8.6.7), the results reveal that Machine 2 produced better strength than Machine 1 (X5), fast throughput was better than slow (X3), and higher temperature improved strength (X2).

Thus the 12-run PB experiment design screened all the proposed factors contributing to product strength. The most dominant factor was Machine 2, a discrete categorical factor, which considerably outperformed Machine 1. Experiment designs are not limited to continuous variables. The subsequent two-factor (X3, X2) full factorial experiment design only used Machine 2.

Section 8.6 explains how to analyze full-factorial and PB results by free open-source software. Appendix C explains the equivalent worksheet analysis by hand, for practice. The analysis is done using a worksheet similar to that used for the two-level factorial design, but simpler.

8.6 Details About Analysis of Statistically Designed Experiments

8.6.1 Model Main Factors to Original Raw Data

Let's methodically build the analysis of a full-factorial experiment by reconsidering Example 1 in Section 8.4. The quickest, easiest, and most accurate analysis of a statistically designed experiment is available through the free open-source R language. First analysis only looks at the main factors.

```
> a=read.csv("mh7Cube.csv")          # Read the csv file into R.
> head(a)                            # Inspect the top six lines of the file.
> mod3=lm(Y~X1+X2+X3,data=a)         # Perform a linear model just on main factors.
> summary(mod3)                      # Display analysis with coefficients for main factors.
Call: lm(formula = Y ~ X1 + X2 + X3, data = a)
Coeffs:        Estimate Std.Error t-value  Pr(>|t|)
(Intercept)-376.7667    49.8375   -7.560  1.15e-06   ***
X1             0.4333     0.1210    3.582  0.002492   **
X2             0.8000     0.1815    4.409  0.000439   ***
X3             5.6500     0.3629   15.568  4.37e-11   ***
Residual Standard Error: 14.52 on 16 Degrees of Freedom
```

The coefficients for each main factor appear along with the Standard Error, which is akin to the "Standard Deviation" of the coefficient. All of the main factor coefficients appear to be significant. Since we now have four dimensions, the three factors plus the response, a Response Surface Map is not easy to plot graphically (yet clever graphical solutions do exist).

Recall that in Chapter 7, we found it was advantageous to center all quantitative variables about their mean in order to avoid extrapolation and to reduce Standard Errors.

8.6.2 Model Main Factors to Original Data Around Center of Each Factor

```
> x1m=mean(a$X1); x1m
> x2m=mean(a$X2); x2m
> x3m=mean(a$X3); x3m
> a$X1c=a$X1-x1m
> a$X2c=a$X2-x2m
> a$X3c=a$X3-x3m
> head(a)                            # Inspect the top six lines of the dynamic datatable.
> mod4=lm(Y~X1c+X2c+X3c,data=a)
>summary(mod4)
Call:  lm(formula = Y ~ X1c + X2c + X3c, data = a)
Coefs:        Estimate  Std.Error  t-value  Pr(>|t|)
(Intercept)   78.9000     3.2461   24.306  4.64e-14   ***
X1c            0.4333     0.1210    3.582  0.002492   **
X2c            0.8000     0.1815    4.409  0.000439   ***
X3c            5.6500     0.3629   15.568  4.37e-11   ***
Residual Standard Error: 14.52 on 16 Degrees of Freedom
> anova(mod4)
```

Compared to model mod3, the mod4 Standard Error of the Intercept decreased dramatically. In essence, the relative Standard Error decreased from 13% of the estimate in mod3 to just (3.24/78.9) = 4% of the estimate in mod4. Since the Standard Error is equivalent to Standard Deviation, we have substantially reduced Uncertainty by centering original data. All the other results were identical, unaffected by data centering.

8.6.3 Model Including All Interaction Terms

Next, we include all the interaction terms as well as the main factors. This is similar to the analysis in Section 8.4 except that now the factors are centered about their mean (nonextrapolated).

```
> mod5=lm(Y~X1c*X2c*X3c, data=a)  # Perform a linear model including interaction
                                  # terms.
> summary(mod5)                   # Display analysis with coefficients for all model
                                  # terms.
> anova(mod5)                     # Display Analysis of Variance to identify dominant
                                  # terms.
```

ANOVA results for mod5 are identical to Table 8.4, Example 1, Part 4. All of the main factor coefficients plus just one of the interaction terms survive the Null Hypothesis. We sort the surviving factors on dominance based on the F-value from highest to lowest.

8.6.4 Model Including Only Dominant Interaction Terms

A more advanced model for the mh7cube data looks only at dominant factors, including just one interaction factor, in the order of dominance.

```
> mod6=lm(Y~X3c+X2c+X1c+X2c: X3c, data=a) # Model only dominant terms in
                                          # rank order.
> summary(mod6)              # Display coefficients and Std.Errors for dominant terms.
Call: lm(formula = Y ~ X3c + X2c + X1c + X2c:X3c, data = a)
Coeffs:       Estimate Std.Error t-value  Pr(>|t|)
(Intercept) 78.90000   1.96876  40.076   < 2e-16 ***
X3c          5.65000   0.22011  25.669  8.27e-14 ***
X2c          0.80000   0.11006   7.269  2.75e-06 ***
X1c          0.43333   0.07337   5.906  2.88e-05 ***
X3c:X2c      0.05875   0.01101   5.338  8.28e-05 ***  # Interaction term.
Residual Standard Error: 8.805 on 15 Degrees of Freedom
```

For mod6, the Standard Errors of all coefficients have decreased, as has the Residual Standard Error. In contrast with R-language analysis, hand analysis is unable to estimate the Uncertainty of the coefficients.

```
> anova(mod6)                 # Display Analysis of Variance on dominant terms.
```

Analysis of Variance Table

```
Response: Y
           Df Sum Sq Mean Sq F-value     Pr(>F)
X3c         1  51076   51076 658.875 8.266e-14  ***
X2c         1   4096    4096  52.838 2.745e-06  ***
X1c         1   2704    2704  34.881 2.884e-05  ***
X3c:X2c     1   2209    2209  28.496 8.283e-05  ***     # Interaction term.
Residuals  15   1163      78
```

8.6.5 Model Including Dominant Interaction Term Plus Quadratic Term

Estimating second derivative effects, as described in Section 8.3, was the motive for including the center point in the 20 trials but has not yet been modeled. With only a single center point, as in Figure 8.1c instead of 8.1b, it is impossible to discern which factor caused curvature; the data are ambiguous. In order to analyze second-order effects, we must do extra work by creating a squared factor (similar to the curved fit in Section 7.5.8). Our final model adds the second-order term.

```
> a$X2c2=a$X2c^2         # Create second-order factor from main factor X2c.
> head(a)                # Inspect the top six lines of the growing dynamic datatable.
> mod7=lm(Y~X3c+X2c+X1c+X2c:X3c+X2c2, data=a)  # Model adds the second-order
                                               # factor.
> summary(mod7)          # Display analysis of linear quadratic model with coefficients and
                         # Std.Errors.
Call: lm(formula = Y ~ X3c + X2c + X1c + X2c2 + X2c:X3c, data = a)
Coeffs:       Estimate  Std.Error  t-value  Pr(>|t|)
(Intercept) 65.50000    2.17535    30.110  3.97e-14  ***
X3c          5.65000    0.10877    51.946  < 2e-16   ***
X2c          0.80000    0.05438    14.710  6.59e-10  ***
X1c          0.43333    0.03626    11.952  9.85e-09  ***
X2c2         0.04188    0.00608     6.887  7.48e-06  ***  # second-order term.
X3c:X2c      0.05875    0.00544    10.803  3.56e-08  ***  # interaction term.
Residual Standard Error: 4.351 on 14 Degrees of Freedom
```

For mod7, all of the factors survive the Null Hypothesis. The Standard Errors of all coefficients decreased as well as the Residual. We add Analysis of Variance:

```
> anova(mod7)                       # Display Analysis of Variance on modeled terms.
Analysis of Variance Table
Response: Y
           Df  Sum Sq  Mean Sq   F-value     Pr(>F)
X3c        1    51076    51076  2698.355   <2.2e-16  ***
X2c        1     4096     4096   216.393  6.586e-10  ***
X1c        1     2704     2704   142.853  9.846e-09  ***
X2c2       1      898      898    47.431  7.478e-06  *** # second-order
                                                           term.
X3c:X2c    1     2209     2209   116.702  3.560e-08  *** # Interaction term.
Residuals 14      265       19
```

The addition of the second-order term in mod7 caused the Fisher statistic (F-value) to increase dramatically for all other factors (which is desirable). We must caution that the analyst selected X2 for the second-order term based on experience; it provided a coefficient for curvature and improved the model. The only way to impartially discern which factors impacted curvature is to perform more tests, but not at the exact center, nor at corners. In Chapter 9 we discuss issues to guide the selection of test points beyond factorial designs. The final model equation is

$$
\begin{aligned}
Y[polymerized] = 65.50 + 5.65\left[\frac{1}{min}\right](t-20.0)[min] + 0.800\left[\frac{1}{°C}\right](Temp-260.)[°C] \\
+ 0.433\left[\frac{1}{kPa}\right](P-60.)[kPa] + 0.0419\left[\frac{1}{°C^2}\right](Temp-260.)^2\left[°C^2\right] \\
+ 0.0588\left[\frac{1}{kPa°C}\right](P-60.)[kPa]\cdot(Temp-260.)[°C].
\end{aligned}
\tag{8.1}
$$

8.6.6 Model All Factors of Example 2, Centering Each Quantitative Factor

Next let's reconsider Example 2 in Section 8.5 and methodically build the analysis of a PB screening test. The data have been stored in the mh7PB.csv file with one row for each case of the experiment. Having learned our lesson, we center all the quantitative factors. Please note below that a warning occurred when asking for the mean of a categorical factor; only quantitative factors have a mean. The first analysis only looks at the main factors.

```
> apb=read.csv("mh7PB.csv")          # Read the csv file into R.
> mean(apb$X1)
  [1] NA
  Warning message:
  In mean.default(apb$X1) # argument is not numeric or logical: returning NA
> x2m=mean(apb$X2); x2m
> x3m=mean(apb$X3); x3m
> x6m=mean(apb$X6); x6m
> apb$X2c=apb$X2-x2m
> apb$X3c=apb$X3-x3m
> apb$X6c=apb$X6-x6m
> head(apb)                          # Review the top six lines of the file.
> mod8=lm(Result~X1+X2c+X3c+X4+X5+X6c, data=apb) # Model the main
                                                  # factors.
> summary( mod8)                     # Display analysis for coefficients of main factors.
Call: lm(formula = Result ~ X1 + X2c + X3c + X4 + X5 + X6c, data = apb)
Coeffs:       Estimate  Std.Error  t-value  Pr(>|t|)
(Intercept) 34.16667    7.89374    4.328   0.007511  **
X1 Man       3.66667    7.89374    0.465   0.661812
X2c          0.48000    0.15787    3.040   0.028738  *
X3c          2.83333    0.78937    3.589   0.015719  *
X4 Sgl       9.66667    7.89374    1.225   0.275270
X5 M2       58.33333    7.89374    7.390   0.000714  ***
X6c         -0.03333    0.13156   -0.253   0.810074
Residual Standard Error: 13.67 on five Degrees of Freedom
> anova(mod8)                        # Display Analysis of Variance for main factors.
```

Table 8.7 showed the Analysis of Variance for Example 2, Part 3. Only three of the main factor terms appear to be significant. We sort the surviving factors on dominance based on the F-value from highest to lowest.

8.6.7 Refine Model of Example 2 Including Only Dominant Terms

Our final model for the mh7PB data looks only at dominant factors, in the order of dominance.

```
> mod9=lm(Result~X5+X3c+X2c, data=apb)    # Linear model only significant factors.
> summary(mod9)               # Display analysis with coefficients for main factors.
Call: lm(formula = Result ~ X5 + X3c + X2c, data = apb)
Coeffs:        Estimate  Std.Error  t-value  Pr(>|t|)
(Intercept)    40.8333     5.1384    7.947  4.58e-05   ***
X5 M2          58.3333     7.2667    8.027  4.26e-05   ***
X3c             2.8333     0.7267    3.899   0.00455   **
X2c             0.4800     0.1453    3.303   0.01081   *
Residual Standard Error: 12.59 on Eight Degrees of Freedom
```

The Intercept value pertains to Machine 1. It is clear that Machine 2 yields much stronger products than Machine 1. Analysis of Variance revealed how variance in the response was captured by the dominant factors.

```
> anova(mod9)                 # Display Analysis of Variance for dominant factors.
Analysis of Variance Table
Response: Result
           Df    Sum Sq   Mean Sq   F-value     Pr(>F)
X5          1   10208.3   10208.3    64.440   4.26e-05   ***
X3c         1    2408.3    2408.3    15.203   0.004551   **
X2c         1    1728.0    1728.0    10.908   0.010815   *
Residuals   8    1267.3     158.4
```

The final model equation for the PB screening test of Example 2 is

$$Y[strength] = 40.83[MPa] + \begin{bmatrix} 0.00, \text{if M1} \\ 58.33, \text{if M2} \end{bmatrix}[MPa] + 2.833\left[\frac{MPa}{\text{kg/min}}\right](flow - 15.)[\text{kg/min}] + 0.480\left[\frac{MPa}{°C}\right](Temp - 225.)[°C]. \tag{8.2}$$

Refer to Chapter 7 to write the model equation including uncertainties with 95% confidence.

The first notable decision from the PB test of Example 2 was to use only Machine 2 for production; Machine 1 could then be assigned to other use. Subsequent three-level, full-factorial tests of factors X2 and X3 were done to refine the model equation.

8.7 Retrospect of Statistical Design Examples

This chapter used two examples to introduce elementary concepts of statistical experiment design. Planning experiments statistically maximizes the information that can be derived while reducing the amount of experimental work required. The foremost and essential strategy is randomization. The second strategy we recommend is to set up a well-designed Plackett–Burman 12-run screening design in order to identify the most important factors from a list of up to 11 candidate factors. The third strategy we used was a factorial design on the reduced set of dominant factors to achieve promised results (answer the Motivating Question) at the required Uncertainty. The consequence of statistically designed experiments is to reduce the number of trials while extracting as much credible information as possible.

The Plackett–Burman algorithm encompasses a family of screening designs. If each experimental run is expensive and there are less than three factors, use PB for four runs. The pattern follows PB eight runs for up to seven factors; 12 runs for up to 11 factors; ($4n$) runs for up to ($4n$-1) factors. Please remember that PB accommodates categorical (nonquantitative) factors as well as quantitative factors. PB provides guidance to efficiently identify dominant factors.

The R statistical language provides a quick, powerful way to process experimental results.

If the experiment you are planning has $m \leq 11$ candidate factors, and if your factors scale into an m-dimensional cube, and if your anticipated Response Surface is flat or gently curved, then this chapter may suffice for you. Plan a PB screening as in Example 2, identify $j \leq m$ dominant factors, then randomize runs in the j cube as in Example 1.

8.8 Philosophy of Statistical Design

Deliberately randomize.

The Factorial Design Cube, tabulated in Appendix B, Table B.2, selects sampling points at the extremes of each combination of factors. In essence it covers the volume of the Sampling Map with one sample at each surface vertex. For an experiment with n factors, the Factorial Design Cube designates 2^n samples. As experimentalist, your client needs to know that as the number of factors increases, the required experimental runs increase exponentially. One solution is Fractional Factorial Design.

A Fractional Factorial Design Cube, as befits its name, chooses a smaller portion of surface vertices to cover the volume of the Sampling Map.

In our Example 1, the mh3cube was our modification to the Factorial Design Cube. Our two modifications were (i) to sample at the center of the volume and (ii) to sample each location multiple times. The purpose of sampling multiple times was to estimate the uncertainties (via a Standard Deviation) for each location and for the total experiment. Two samples at a location is the bare minimum for estimating a Standard Deviation and requires the t-test multiplier 12.7 to determine local Uncertainty. At the center of the volume, four samples required the t-test multiplier 3.18 times the center Standard Deviation. Since our Example 1 only had three factors, there were a total of 20 samples. The full analysis can be done by hand calculator or spreadsheet. For practice and confidence, the reader is welcome to work through Appendix C. Then compare with R analysis of the same data. You will find that the Standard Error and subsequent Uncertainty, as calculated by R, is less than the Uncertainty calculated by hand. The improved Uncertainty by R is a consequence of R using inherent information of the entire dataset.

The Plackett–Burman algorithm is an efficient special case of Fractional Factorial Design. PB requires only m samples, where m is some multiple of four, rather than 2^n samples for n factors. Since $m \geq n+1$, PB or mimic algorithms are ideal to efficiently screen for dominant factors. We recommend PB.

Beyond factorial design, we advise sampling the interior of the Sampling Volume Map. Chapter 9 provides guidance for selecting high-impact sample points.

Organize your data by recording a single observation for each selected sample of combined factors.

Linear models are quickly and accurately calculated by statistical software if data are recorded individually, as in the previous sentence. We recommend R, since it is powerful, widely used in science, and has thousands of specialized packages of algorithms contributed by fellow researchers. Furthermore, R is open source and free.

Refresh your aim to answer the Motivating Question within the specified Uncertainty.

Diligently randomize.

8.9 Statistical Design for Conditions That Challenge Factorial Designs

Not all experiments fit within the straight segments which bound the Sampling Volume map of an n-dimensional Factorial Design Cube, e.g. Figure 8.2. Just as the blind spot of your eye prevents imaging part of your field of vision, allowed ranges of your devices and equipment may limit your operating map. As we saw in Figures 6.4–6.7, the operating map for that experiment had curved boundaries and limited safety conditions. Curved boundaries and limits often do prevent scaling to the straight edges of the Factorial Cube. There are advanced statistical design tools available to help us.

Real life is not always smooth. Physics and life do have sudden shocks and discontinuities, e.g. waves crest. Then limiting oneself to sampling only at the boundary, specifically at the vertices of an n-dimensional Factorial Design Cube, may fail to capture the behavior of system physics within the Sampling Volume. For heat transfer from a surface with blowing, as we saw in Figure 6.10, the Response Surface Map can be far from a tame parabolic curve. A short list of discontinuities and challenges encountered in physics, chemistry, materials science, agriculture, and economics with interesting and contorted Response Surface shapes includes:

i) Phase transition.
ii) Shock waves.
iii) Superconductivity, semiconductors.
iv) Catastrophic events, explosion, extinction.
v) Stock market collapse, bee colony collapse.
vi) Discontinuities, cresting waves, cusps.
vii) Bifurcations.
viii) Hysteresis.
ix) And so on …

There are advanced statistical design tools to help us.

8.10 A Highly Recommended Tool for Statistical Design of Experiments

A single software tool, Gosset, is able to guide all of the steps we highlighted for statistically designed experiments. We highly recommend Gosset:

i) Gosset can design a minimal number of screening runs, rivaling PB.
ii) Gosset selects sampling points on the surface and interior to optimally fill a Sampling Volume.
iii) Gosset simultaneously ensures randomization.

The Gosset program was developed by N.J.A. Sloane and R.H. Hardin at AT&T Labs. Gosset is open source and free. Recently Sloane and Hardin put Gosset in the public domain. The Gosset code is flexible and powerful. To date, Gosset only runs on Linux and Unix systems. Gosset requires study and skill to use. Gosset comes with substantial documentation and numerous examples. Gosset is available for free from the website for this text or from URL NeilSloane.com/gosset.

Gosset capably handles a variety of Sampling Volumes. Of course, it handles the standard factorial cubic domain. It can also handle multidimensional factorial spherical domains. If constraints truncate the cube or sphere, Gosset can select sample points within the reduced Sampling Volume. Gosset cannot cook breakfast or mow the lawn.

Gosset can help solve math optimization problems. Gosset can optimize Type I, Type D, and Type E experiment designs. Gosset can select new sampling points after data collection has commenced. Please read and study Gosset documentation for more capabilities. Did we mention Gosset is free?

8.11 More Tools for Statistical Design of Experiments

The Taguchi Method provides an alternative strategy by targeting component quality rather than attempting to reduce experimental Uncertainty (our aim). The Taguchi method is commonly used on industrial production lines. This important method requires a whole textbook unto itself. We mention Taguchi in case the reader would like to investigate an alternative technique.

In the popular press, there are a variety of memes which vie for the label statistical experiments. These include Big Data, Internet of Things, Data Science.

Methods for data acquisition and for filtering will be discussed in later chapters.

In this text, we advocate open-source software, R and Gosset, which we have found to equal or surpass the power of commercial software. Commercial packages are available and do compete.

There are many excellent references available concerning the statistical design of experiments.

8.12 Conclusion

When planning initial data collection, we urge that you consider a screening design such as Plackett–Burman to select initial sampling points, followed by analysis to identify dominant factors. Subsequent sampling, on a reduced factor space, can then be guided by standard statistical design such as a Factorial Cube or Fraction-Factorial Design or by Gosset software. Be diligent to randomize sampling.

Further Reading

Box, G.E.P. and Draper, N.R. (1987). *Empirical Model Building and Response Surfaces.* Wiley.
Kaplan, D. (2012). *Statistical Modeling: A Fresh Approach,* 2e. Project Mosaic Books.
Montgomery, D.C. (2017). *Design and Analysis of Experiments,* 8e. Wiley.
Myers, R.H. and Montgomery, D.C. (2016). *Response Surface Methodology,* 4e. Wiley.
See list of titles under search term "Taguchi methods" in the Wiley OnLine Library (https://onlinelibrary.wiley.com/).

Homework

Statistics explains variation in the context of what remains unexplained (D Kaplan, 2012).

8.1 Refer to Section 8.4 to analyze data collected on a three-dimensional Factorial Cube. Follow the R-language commands in Section 8.4 to analyze the data in `mh7Cube.csv`. Based on your R-language skill developed in Chapters 7 and 8, how complicated would it be to write the modeling equation from the coefficients provided in the results in `summary(mod1)`?

8.2 Continue the analysis of Exercise 8.1 for the three-dimensional Factorial Cube. Perform "Analysis of Variance" on the first model using the command `anova(mod1)`. Note which values duplicate values reported in `summary(mod1)`. A new value that appears in anova is the Fisher statistic in the column F-value. For experiment decision-making, focus on high F-values.

In this text, we recommend choosing terms with high F-values. A term with a high F-value, positive far from zero, has a high benefit-to-cost ratio; a term with a high F-value explains a significant portion of variation in the results. Note that these high F-values correspond to low p-values.

Statistically speaking, the p-value is the probability that the Null Hypothesis may be valid. The p-value has high visibility in publications. Analysts often prefer to choose a modeling term where they can reject the Null Hypothesis, that is, terms with low p-values. Your report will include p-values.

8.3 Refer to Section 8.6.1 to analyze only the main linear terms for the three-dimensional Factorial Cube. Based on your R-language skill, write the modeling equation from the coefficients provided in the results in `summary(mod3)`. This equation is much simpler than Exercise 8.1.

8.4 Refer to Section 8.6.2 to analyze each of the main linear terms centered about their means. Write the modeling equation from the coefficients provided in the results in `summary(mod4)`. Compare with Exercise 8.3 the coefficients and Standard Errors for each term. Calculate and compare the relative Standard Error. Perform the `anova(mod4)` and compare values with the `summary(mod4)`. Furthermore, compare the F-values with the value in Exercise 8.2.

8.5 Refer to Section 8.6.3 to analyze the full set of factors and interactions centered about means.

Compare with Exercises 8.1 and 8.2 the coefficients and Standard Errors for each term. Perform the `anova(mod5)` and compare values with the `summary` and `anova` of `mod1`. Remark on the decisions you would make based on your analysis of mod5. Which terms dominate based on F-values and survive for further modeling?

8.6 Refer to Section 8.6.4 to analyze the full Factorial Cube including only the centered terms that survived the full analysis and screening of Exercise 8.5. Compare with Exercise 8.4 the coefficients and Standard Errors for each term. Perform the `anova(mod6)` and compare values with earlier models, especially `mod4` and `mod5`. Remark on the improvements based on your modeling decisions in Exercise 8.6.

8.7 Recall in Section 8.3 that a central point was included in the experiment design so that quadratic curvature could be included in the model. Refer to Section 8.6.5 to analyze the full Factorial Cube with all of the terms in Exercise 8.6 plus one quadratic term. Compare with Exercise 8.5 the coefficients and Standard Errors for each term. Perform the `anova(mod7)`. Based on the F-values, does the quadratic term survive along with all the terms in `mod6`? Remark on the improvements in this, our final model. Based on your R-language skill, write the modeling equation from the coefficients provided in the results in `summary(mod7)`.

8.8 For the modeling equation based on `mod7` in Exercise 8.7, write an equation that includes and displays the Standard Error in the `summary()` command. Refer to Section 8.6 for an example. Each coefficient shows ± one Standard Error. Write a second modeling equation including 95% confidence of the coefficients (± 2 * Standard Error).

8.9 Refer to Section 8.5 to analyze data collected for the PB 12-run screening design.

Follow the R-language commands in Section 8.5 to analyze the data in `mh7PB.csv`. The model includes the three quantitative factors purely linearly. The model also includes the three categorical two-level factors. Determine the coefficients and Standard Errors for all terms from `summary(mod2)`. Determine the F-values for all terms from `anova(mod2)`. Remark on the decisions you would make based on your analysis of `mod2`. Which terms dominate based on F- values and survive for further modeling?

8.10 Based on your results in Exercise 8.9, write the modeling equation from the coefficients provided in the results in `summary(mod2)`.

8.11 For the modeling equation based on mod2 in Exercise 8.10, write an equation that includes and displays the Standard Error in the summary() command. Refer to Section 8.6 for an example. Each coefficient shows ± one Standard Error.

8.12 For the modeling equation based on mod2 in Exercise 8.10, write an equation that includes and displays the Standard Error in the summary() command. Refer to Section 8.6 for an example. Write this modeling equation including 95% confidence of the coefficients (± 2 * Standard Error).

8.13 Based on Exercise 8.9, your analysis of mod2 led you to decide to use an improved model with three surviving dominant explanatory factors: X2, X3, and X5. Two factors are quantitative continuous (X2 and X3), and X5 is categorical (discrete two-level). Create a linear model, mod8, from these three factors using the PB 12-run screening design data mh7PB.csv. Determine the coefficients and Standard Errors for all terms from summary(mod8). Determine the F-values for all terms from anova(mod8).

8.14 Based on your results in Exercise 8.13, write the modeling equation from the coefficients provided in the results in summary(mod8).

8.15 For the modeling equation based on mod8 in Exercise 8.14, write an equation that includes and displays the Standard Error in the summary() command. Refer to Section 8.6 for an example. Write this modeling equation including 95% confidence of the coefficients (± 2 * Standard Error). Has mod8 improved the predictive power of your modeling equation?

8.16 Repeat Exercise 8.13, for the same three factors. Now for mod9, first center the quantitative modeling factors, each about their own mean. Create a linear model, mod9, using the PB 12-run screening design data mh7PB.csv. Determine the coefficients and Standard Errors for all terms from summary(mod9). Determine the F-values for all terms from anova(mod9). Compare these results with those of Exercise 8.13.

8.17 Based on your results in Exercise 8.16, write the modeling equation from the coefficients provided in the results in summary(mod9).

8.18 For the modeling equation based on mod9 in Exercise 8.16, write an equation that includes and displays the Standard Error in the summary() command. Refer to Section 8.6 for an example. Write this modeling equation including 95% confidence of the coefficients (±2 * Standard Error). Has mod9 improved the predictive power of your modeling equation? Compare to your answer in Exercise 8.16.

8.19 Summarize your key lessons learned concerning statistical design of experiments.

9

Selecting the Data Points

At this point, your Motivating Question has been identified – you know what you are trying to accomplish. You know what question you are trying to answer, you know the form of the answer you seek, and you have decided the allowable Uncertainty.

You are about to run a parametric experiment. You know which variables appear in the defining equation for the answer to the Motivating Question. The task is to measure those variables with acceptable Uncertainty and to take whatever additional data you need to establish the credibility of those measurements.

9.1 The Three Categories of Data

There are three categories of data that should be planned for:

Output data	These are the data needed to calculate the desired resultant and answer the Motivating Question.
Peripheral data	These data are necessary and sufficient to establish the credibility of the output data – to show that they are representative and correct.
Backup data	These data will be needed if the main hypothesis (or objective) is found to be "not true" or "not attainable" or if something really "weird" happens.

9.1.1 The Output Data

No item on the "output list" is optional. Work must continue until all of the data you have named on this list have been obtained.

Your report shall include, along with the recorded data value of each factor, the following: its Uncertainty estimate along with the predetermined confidence level, as in:

$$x_i = x_i \pm \delta x_i (20:1) \tag{9.1}$$

Omitting (i) the Uncertainty estimate derived from your data and (ii) the confidence level you chose in advance *is not an option*. The credibility of your report as well as the journal and scientific community demand it.

Planning and Executing Credible Experiments: A Guidebook for Engineering, Science, Industrial Processes, Agriculture, and Business, First Edition. Robert J. Moffat and Roy W. Henk.

Companion website: www.wiley.com/go/moffat/planning

9.1.2 Peripheral Data

The purpose of the peripheral data is to establish the credibility of the output data and to defend it against reasonable challenges. The planners should ask "What questions might I face?" under the presumption that the data will be challenged.

Following is a list of questions the experimenter should be prepared to answer and the issues they address. The key word in each question is "representative." Identifying the peripheral data needed to defend the output may well lead to extending the instrumentation requirements (to accomplish the required checks and balances) and the test schedule (to confirm results against other representative or baseline data from accepted sources).

Calibration	Is the output of each instrument representative of the value of its measurand?
Uniformity	Is the value of the measurand at the location of the instrument representative of the values in the claimed region? If the sensors had been located differently, would the same result have been found?
Repeatability	Are the results representative of the time-averaged behavior? Do the results repeat?
Scaling laws	Does the hardware of the experiment represent the hardware of the full-scale system?
Test procedures	Do the boundary conditions of the test represent the boundary conditions of the full-scale system?
Baseline data	Has the apparatus been qualified by tests at a baseline condition?
Checks/balances	Are the data internally consistent? Are the fundamental checks and balances satisfied (mass, momentum, energy conservation)?

9.1.3 Backup Data

In some tests, there may exist one set of results which will be judged "successful" and another (or all others) which will be judged a "failure." The peripheral dataset is, as we have seen, an insurance policy which defends the successful result against challenge. The backup dataset is the experimenter's insurance policy against a total failure.

Prudence dictates that the possible failure modes be identified as clearly as possible and a sparse (but usefully complete) set of data taken to provide guidance for the follow-on experiment, which would then be necessary. Since one cannot afford the luxury of providing for every possible outcome, the principal failure modes should be covered.

9.1.4 Other Data You May Wish to Acquire

There are frequently more candidates for the "necessary and sufficient" list of output data than can be kept in that highest-priority category. Those candidates should be reviewed after the output and peripheral list have been finished. Frequently, a minor increase in instrumentation can be sneaked on board to provide fringe benefits without jeopardizing the main line objective.

Beware. One must be careful, though, to avoid turning the equipment into the proverbial porcupine, spiked with extra electrical wires! No instrumentation should be added unless someone makes a convincing case that it would be a good investment.

9.2 Populating the Operating Volume

9.2.1 Locating the Data Points Within the Operating Volume

In parametric plans, the data points may be distributed along paths through the operable domain called "data trajectories." Naturally, our goal is collect data that will be most densely nested in the regions of most interest, or near locations in the operable domain where abrupt changes in the Response Surface are expected. It seems clear that the more you know about the shape of the Response Surface, the better you are able to site the data points.

9.2.2 Estimating the Topography of the Response Surface

Experimentalists appreciate the clarity of hindsight even more than do their theoretical counterparts – everything is easier the second time around. How nice it would be if one always knew in advance the approximate shape of the Response Surface before each new experiment!

In fact, your training in your expertise frequently enables you to estimate the shape well enough for planning purposes, by combining the results of previous experience (your own or from a literature search) with some elementary knowledge. After all, one rarely encounters a totally new problem. Most work progresses like dendrites – growing in a new direction but starting from a known base. That means that usually at least one boundary of the proposed domain has already been documented in the literature. With one known boundary and some basic engineering judgment, you can usually propose a plausible shape for the Response Surface.

9.3 Example from Velocimetry

The process is illustrated in Figures 9.1 and 9.2, which show an attempt at estimating the Response Surface for an experiment on hot-wire anemometers. Visualize in this instance that the whole of the operating volume is represented by two factors in the ($x1$, $x2$) plane. The Response Surface (your resulting data from the experiment) floats above the ($x1$, $x2$) plane.

Let's assume that the question we are trying to answer is, "How do the two factors, length and thermal conductivity of the sensing wire, affect the apparent turbulence measured in a specified flow field by a hot-wire anemometer?"

The sensor itself is a slender wire, perhaps 5 μm in diameter, attached to the tips of two relatively heavy support prongs. The active portion of the wire is typically 1–5 mm long, and that is the length whose effect we wish to investigate.

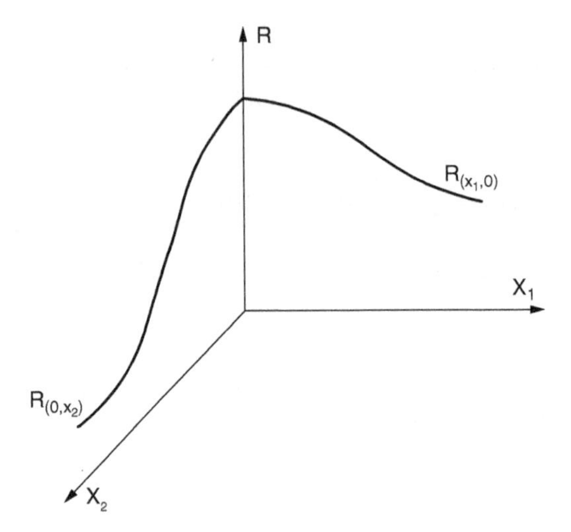

Figure 9.1 We generally can estimate the shape of the Response Surface at one or two of its boundaries.

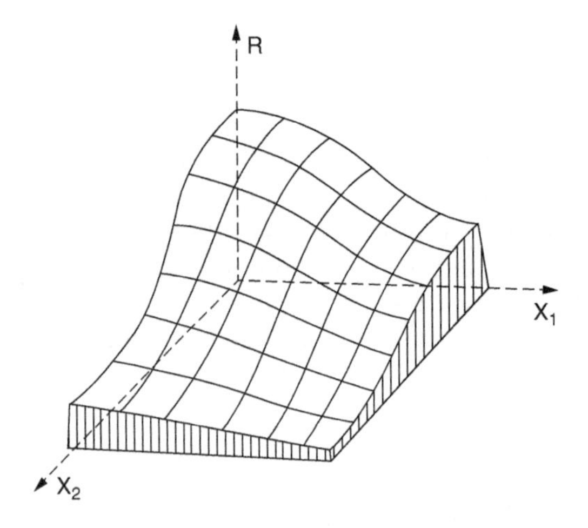

Figure 9.2 An estimated Response surface.

Figure 9.1 shows what is presumed to be known at the beginning of the study: the effect of wire length (x_1) on perceived turbulence, for zero conductivity, and the effect of conductivity (x_2) at a given wire length. These estimates need not be based on data. For each factor ask, "How does the curve look as the factor approaches zero and as it approaches infinity?"

9.3.1 Sharpen Our Approach

We can argue on first principles that a wire of zero length would sense even the smallest scales of turbulence and would, therefore, always yield the highest value of the apparent turbulence.

As the wire length increases from zero, the apparent turbulence will remain constant until the wire length exceeds the size of some small, representative energy-bearing structure. Hence the curve has zero slope as the wire length approaches zero length. The signal from a longer wire could be interpreted as the average of the signals from a series of point sensors along its length, each responding to its own local velocity. As the wire becomes very long, the apparent turbulence must approach zero, since the apparent turbulence is defined in terms of the mean and fluctuating signal from the instrument. Furthermore, its average output represents the average velocity along its length, the steady value. Again the curve must approach zero with zero slope.

Similar arguments can be made about the effect of conductivity. The sensing wire of the hot-wire anemometer acts as a fin, attached to its support prongs at each end. At the point of attachment, the wire must be at the temperature of the prongs. The extent of the wire affected by the temperature of the prongs depends on the conductivity of the wire. A wire of infinite conductivity would be at "prong temperature" for its entire length. The prongs which support the hot-wire sensing wire are massive, compared to the sensing wire, and do not respond to turbulent fluctuations at all. Hence, for infinite conductivity, the apparent turbulence would always be zero. At zero conductivity, the sensing wire would be unaffected by the temperature of the support prongs, hence would show the highest value of apparent turbulence for any wire length. By these two arguments, the traces shown in Figure 9.1 were constructed: apparent turbulence as a function of wire length, at zero conductivity, and apparent turbulence as a function of wire conductivity, at zero length.

The Response Surface expected for the combined experiment is shown in Figure 9.2. It was constructed using the assumption of a smooth, slowly varying surface joining the two wall traces. The profile at each interior section is assumed to be the product of two (appropriately scaled) functions. The process is related to the notion of separation of variables used in solving partial differential equations. This process is simple but works.

In Chapter 1, we quoted Abraham Lincoln: "Give me six hours to chop down a tree and I will spend the first four sharpening the axe." You have developed your expertise. Some call your ability to estimate a Response Surface as "making an educated guess." We rather say your ability is "using your sharpened axe."

9.3.2 Lessons Learned from Velocimetry Example

Note the benefit of the assumption that the Response Surface is a simple, continuous, low-order function with continuous derivatives.

Estimations like this should be done early in the planning process, when they can have maximum effect on the experiment design. For example, if Figure 9.2 is accepted as a reasonable description of the expected Response Surface, then it would seem logical to take the "front" corner of the domain as the first data point. That first data point would be at the joint maximum – highest conductivity, longest length. If that combination produces a very low value of apparent turbulence, the general features of the model are immediately confirmed. (If it produces a high value, then the whole theory of the experiment had better be re-examined!)

Often, at least locally, the simple continuous assumption is valid.

Do you anticipate an exception to this assumption in your experiment? Congratulations! Join the team. Our field, the thermosciences, is replete with discontinuous exceptions to this assumption. Common exceptions are notably at shock wave boundaries, Leidenfrost effect, laminar/turbulent transition, and phase transitions, etc. Even in such exceptional cases, nonetheless, we can often reformulate or remap in order to find a smooth Response Surface map.

9.4 Organize the Data

9.4.1 Keep a Laboratory Notebook

The credibility of the results of your experiment is inextricably tied to the integrity of the data. Diligently determine to guard the integrity of your data with procedures akin to those of your laboratory notebook. You will be daily maintaining your laboratory notebook. A reminder of instructions for the laboratory notebook is discussed in Chapter 14.

9.4.2 Plan for Data Security

These data guidelines are adapted from the rules for laboratory notebooks. Please make careful note of how the refinements protect data integrity.

We recommend these simple, no-nonsense guidelines to ensure the integrity of the data you collect for your experiment. These guidelines transparently display the progress of your experiment, enabling independent confirmation and repeatability, and strengthen the subsequent analysis of the results.

9.4.3 Decide Data Format

This chapter presents these guidelines as if the experimental data are recorded by hand into a single spreadsheet file containing several individual sheets. The same guidelines equally apply to data recorded automatically by computer in ASCII text, unicode, or binary form. Keep a copy of the filenames and summary in your lab notebook.

9.4.4 Overview Data Guidelines

Guidelines:

i) Create a spreadsheet file with at least two tabbed sheets.
ii) Keep a codebook on one sheet of the spreadsheet.
iii) Record the data from a data run in a datatable on a separate sheet of the spreadsheet.
iv) Close the spreadsheet once the data run is completed and all the cases recorded.
v) Change the permission of the spreadsheet file to read-only.
vi) Make a backup of the spreadsheet and store it in a separate, independent location.
vii) Reckon the read-only spreadsheet file as unchanging and incorruptible. Set it apart.
viii) Analyze only derivative files copied and extracted from your original spreadsheet file.

9.4.5 Reasoning Through Data Guidelines

Confirm with your boss and client your procedures for data security and integrity. Expounded guidelines:

i) Create a spreadsheet file with at least two tabbed sheets. One sheet is a codebook that explains the data saved in other sheets of the spreadsheet file. The second sheet contains a datatable containing the data for an individual data run. If there are more data runs, add more sheets, each with its own datatable.
ii) Keep a codebook on one tab of the spreadsheet. The codebook explains each variable that will be recorded. Table 9.1 shows an example spreadsheet codebook.
   a) If a variable is quantitative when recorded, explain in this codebook the physical units (e.g. cm, kg, inch, mpg = miles per gallon, etc....) of the value that is being recorded.
   b) If a variable is categorical (not quantitative), then list and explain the exact possible levels that this variable can take (e.g. T, F, TRUE, FALSE, red, yellow, blue, etc....).
   c) Distinguish and explain levels exactly in the codebook. If a variable may take the level TRUE or T or Y, depending on who writes the data, explain this fact in the codebook. DO NOT assume that T = TRUE without explanation. For the sake of data integrity and future reference, explain overly thoroughly.
   d) If data are recorded by hand by different people, add a variable which identifies the person collecting the data. Explain the identifier in the codebook.
iii) Record the data from a data run in a datatable on a separate sheet of the spreadsheet. The datatable contains your original record of all the data for all cases in this data run.
   a) Each column of the datatable corresponds to one single variable, whether quantitative or categorical.
   b) Each row contains the values of all variables that correspond to a single observation.
   c) The first row may contain labels as titles for the variable in each column.
   d) Design datatables to compactly contain the variables that vary during the data run. Apply Ockham's razor. Tables 8.1, 8.2, and 8.3 show related but distinct datatables.
iv) Close the datatable once the data run is completed and all the cases recorded.
v) Change the permission of the spreadsheet file to read-only.
vi) Guard data integrity.
   a) Reckon the read-only spreadsheet file as holy and untouchable.
   b) DO NOT touch or alter data in the datatable or in the codebook of the read-only spreadsheet file.
   c) DO NOT add do later analysis in the original spreadsheet. Any change to the original spreadsheet exposes the data to being compromised and corrupted.
vii) Analyze only derivative files extracted from your original spreadsheet file.

Design datatables to contain all of the variables (factors) that the operator is to measure and record during the data run. Apply Ockham's razor to keep the datatable as compact as possible.

Table 9.1 Codebook for life jacket safety spreadsheet (total versus partial relationships).

Factor	Level	Units	Explanation
Alive	L		Survived, alive after the incident
	D		Drowned, died from the incident
Vest	V		Vest, wore vest before incident
	N		None, did not wear vest before incident
Weather	Fair		Fair weather, low wave height + easy to access and rescue
	Bad		Foul weather, high wave height + difficult to rescue
Temp	Continuous	C	Temperature of water at time of rescue; Celsius
Time	Continuous	Hour	Time from incident to rescue; decimal hour
Entry	Dry		Never entered water, dry clothes in vessel at rescue
	Wet		Entered water after incident, in vessel at rescue
	Over		Overboard since incident, in water at rescue

Related datatables with pertinent variables, but which do not vary during the data run, can be linked together for later analysis. For example, grade data for a university term may include several related datatables: one datatable recorded the performance of all students in a particular class; a second datatable recorded the professor name, classroom time and location, and class size for each section of a particular course; a third datatable recorded all the courses taken by all the students; a fourth datatable relates the point value of the grade earned for subsequent calculation of the grade point average. The R language allows two or more datatables to be linked during analysis without altering original data.

Data are expensive. High-quality data require many hours of careful planning and collection. During and after collection, guard the data diligently.

9.5 Strategies to Select Next Data Points

Once the first data point has been established, the next points will be sampled at corners of the Sampling Volume as we saw in statistical experiment design (Chapter 8). After completing samples of the corners, where do we sample next? We present two strategies as options to choose subsequent sample points in the operating volume. Depending on your class of experiment (Chapter 5) you may prefer option (i) or option (ii) or a combination of the strategies. The strategies are:

i) Follow a default strategy that relies heavily on experimenter manual involvement and is easy to understand. If your class of experiment is (a) demonstrate system performance (Section 5.2.3) or (b) develop predetermined models (Section 5.2.5), this strategy is well suited. This strategy depends on already having an idea of a Response Surface. If your experiment arises from physics or one of the physics-intensive engineerings (mechanical, electrical, chemical, or aerospace), this strategy will likely prevail.

ii) Get assistance from a computer program designed to optimally select sample points. If your experiment arises from industry, agriculture, economics, medicine, or human society, this option is more helpful. If your class of experiment is exploratory (Section 5.2.1), certification and custody transfer (Section 5.2.6), quality assurance (Section 5.2.7), or if you cannot yet predict the response, we recommend this option. In particular, we recommend the experiment-planning computer program developed at AT&T Bell Labs called "Gosset."

Gosset is a powerful program for optimally selecting data points, offering many features that complement the first strategy. Gosset is a free open-source program, recently released by the authors to the public domain.

Understand the first strategy; then Gosset can help you more.

9.5.1 Overview of Option 1: Default Strategy with Intensive Experimenter Involvement

Since we have focused on the factors that impact the response, we seek locations where there is most variation of the response. Inspect the model built from the factorial experiment data and locate where on the perimeter the response varies most. We can then home in on the steepest variation by bisection search or "interval-halving": the next data point should always be placed halfway between the two points showing the maximum difference. Once the total boundary has been mapped with reasonable density, the interior can be examined, again following the interval-halving approach.

9.5.1.1 Choosing the Data Trajectory

Once the operating volume of an experiment has been sketched out, it can be used in planning the sequence of test points. A sequence of test points defines a path across the operable domain that will be followed by the experiment, step by step. Such a path is called a "data-taking trajectory." A set of sequences of trajectories defines the test program.

The next question is, "In what order shall we take the data?" There are several issues to deal with here. Some of the more important issues are: defining a default strategy, modifications that may be needed to minimize test time, modifications that may be necessary when the test conditions stress the apparatus, and considerations when you need accurate data near the margins of the domain.

9.5.1.2 The Default Strategy: Be Bold

The objective of experiment planning, particularly with a new experiment, is to get the most information you can about the process for the least amount of effort. The most fruitful strategy to accomplish this has two aspects: (1) be bold – go quickly to the limits of the operable domain, and (2) proceed by interval-halving toward where the action is. These two steps comprise what I recommend as the default strategy.

For one thing, you must quickly determine whether or not you really can cover the whole domain you think you can and find out if there are any surprises in store for you. Going immediately to the corners of the domain tells you that.

9.5.1.2.1 *Halve Rather than Nibble*

Then, having seen the results at the corners, you can "interval-halve" and cut back along the boundaries, moving always in the direction of the maximum change. Above all, do not "munch" or "nibble" your way across the domain. That usually results in a lot of wasted effort. The process of interval-halving is more efficient than "nibbling" across a domain following a sequence of small, predetermined steps. Interval-halving allows the experimenter to remain in active control of the data-taking process, to position each data point where it will do the most good. This requires that the data be reduced and plotted as they are acquired, so the experimenter can interact with the process.

9.5.1.3 Anticipate, Check, Course Correct

If, based on the prior art, we expect the data to lie along a straight line in log coordinates, then as soon as we have three data points, we can check whether or not that expectation has been realized. It is important to check expectations early and often, since there are many things that can go wrong with an experiment.

If the first few data points confirm the expected behavior, then, providing there is no reason to suspect any pathology in the physics, you may be finished! If they do not confirm expectations, or if this is an entirely new problem, with no guidance as to how the data should correlate, then the data points could be plotted on linear, semi-log, and log scales to see how they vary. These plots will show the general features of the process, even with only three data points. The main feature to look for is, "Where is the region of maximum change?" In general, it is more effective to put a high concentration of data points in the regions of maximum curvature of the result function.

The general recommendations are: First go to the limits, then interval-halve to fill the interior.

Put more data points in the region of maximum curvature, and try to accomplish two objectives with one data point.

9.5.1.4 Other Aspects to Keep in Mind

There are other aspects to planning the data trajectories that are also important. For example, choosing the data trajectory to minimize the expected scatter in the results.

Consider an experiment that involves taking data for various F-values at constant x-Reynolds number, given the Response Surface shown in Figure 9.3. That corresponds to moving along a line of constant x-Reynolds number in the F, Re(x) plane shown as the dashed line. Such a path would only be acceptable for very high x-Reynolds numbers above 7×10^5, or for very low, below about 2×10^4, but not in between.

The intervening range covers the transition saddle – a region of very rapid change. In that range, data taken along a path of constant Reynolds number, varying F, would likely show a high degree of scatter. Small, accidental variations in the x-Reynolds number will shift the operating point up or down the transition slope. Those small variations are inevitable from day to day and cause a bigger change in Stanton number than would a small change in F.

This problem is especially severe when one attempts to compare data from several different runs which were each at nominally the same Reynolds number. Unless the test rig was exceptionally stable and repeatable in its set point, successive realizations of the

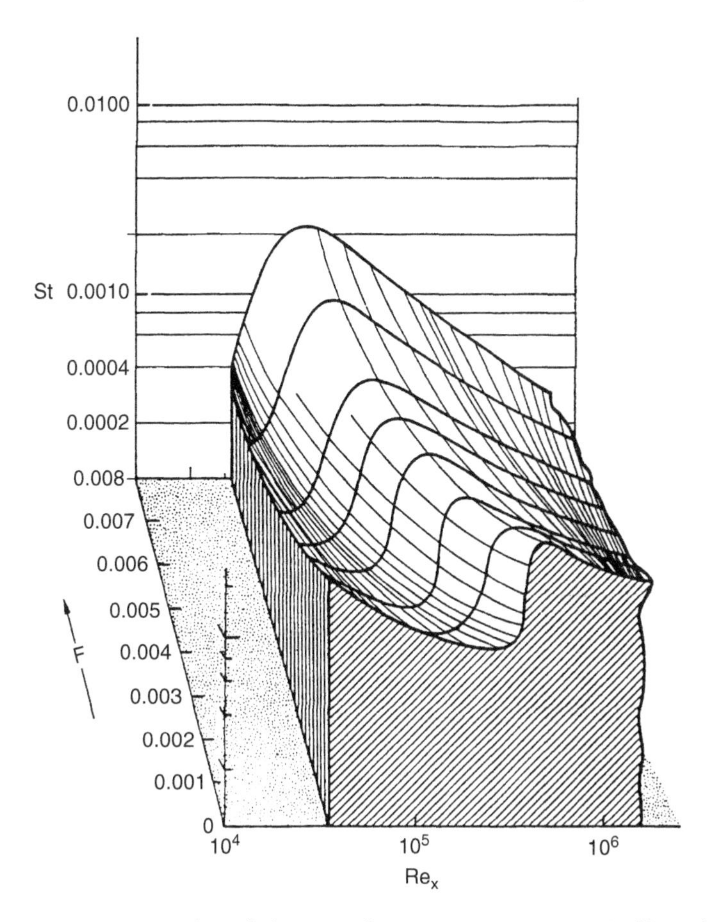

Figure 9.3 The Response Surface of a heat transfer experiment on a transpired boundary layer.

same nominal set point would trace out quite different paths along the steep slope of the transition rise. Testing along such a horizontal trajectory is simply asking for trouble. Stepping a smidgen to one side, you have fallen off the cliff; moving a smidgen to the other side is beyond reach.

Data trajectories should be planned, as much as possible, to lie along paths of steepest descent. This choice will minimize the effects of small failures to hold or repeat the parametric position of the path.

9.5.1.5 Endpoints

Sometimes your Motivating Question requires that you get accurate data concerning both level and derivative near the end of a data trajectory. Even though the experiment may be conceived as a single-sample experiment, it may be necessary to replicate several points near the end of each path. Suppose, for example, that you need to get the partial of Z with respect to x_1 accurately within ± 5% near the end of the data trajectory. The definition of the partial derivative can be expressed numerically and used to describe the Uncertainty in the measurement of the derivative (see Section 10.6 on "Single-Sample Uncertainty Analysis" for the rationale for this development):

$$\frac{\partial Z}{\partial x_1} = \frac{Z_{x_1+\Delta x_1} - Z_{x_1}}{\Delta x_1} \tag{9.2}$$

$$\delta\left(\frac{\partial Z}{\partial x_1}\right) = \sqrt{2}\,\frac{\delta Z}{\Delta\, x_1} \quad \text{if there is no error in } \Delta\, x_1 \tag{9.3}$$

$$\frac{\delta\left(\frac{\partial Z}{\partial x_1}\right)}{\left(\frac{\partial Z}{\partial x_1}\right)} = \sqrt{2}\left[\left(\frac{\delta Z}{\Delta Z}\right)^2 + \left(\frac{\delta x_1}{\Delta x_1}\right)^2\right]^{1/2} \tag{9.4}$$

From Eq. 9.4 we see that the fractional Uncertainty in the derivative is related to the Uncertainties in the measurements of Z and x and on the difference between the two values of Z and the two values of x used to form the derivative. The farther apart are the two points, x_1 and x_2 used to form the difference, in general, the better will be the estimate of the average derivative, but the less "local" the resultant value is. If you are not free to spread apart the two x-values, then you must decrease the uncertainties in the measurements of Z and of x.

If the dominant contributors to the uncertainties are the residual fixed errors in the measurements, you may not have a real problem. Unless the system you are dealing with is more pathological than systems I have dealt with, the fixed measurement errors do not change much for small changes in the variables. That means the fixed errors will not much affect the estimate of a derivative, and they certainly will not introduce scatter.

If the dominant contributors to the uncertainties are randomized errors due to system instability, that can be handled by averaging. Presuming you have already done your best to reduce those Uncertainties, if the Uncertainty is still too high, where does that leave you? You must then use a multiple-sample approach and take several repeated measurements at each of the points you wish to use in getting the derivative. Take, for example, four readings at each point, calculate the average value of Z at each point and the average value of x. The derivative, calculated using those averages, will have a lower Uncertainty, by a factor of 2, than using only one observation. The general rule to reduce randomized error by averaging obeys uncertainty of the mean (Section 7.1.9).

9.5.2 Reintroducing Gosset

At AT&T Bell Labs, N.J.A. Sloane and R.H. Hardin developed an experiment-planning computer program which they titled "Gosset." Gosset is a powerful program for optimally selecting data points, which is the purpose for which we recommend it. Gosset was first introduced to the scientific community in the 1980s. In 2018, Sloane and Hardin agreed to release Gosset to the public domain as a free open-source program.

In the accompanying CD, we have included a release of Gosset, which includes the source code, its own compiler, documentation, and example problems.

Gosset shares a statistical grammar with the statistical language R, which we have already introduced and demonstrated in Chapter 7, and the statistical language S, which is a commercial equivalent of R. Thus the reader has already learned how to read and write model equations for Gosset. The dialect for Gosset is more sophisticated, however, in order to make its features accessible.

Gosset selects future combinations of factors based on previous data and an interactively refined Response Surface. The powerful features which Gosset offers make its learning curve a worthwhile investment for experimentalists.

Gosset can plan experiments beyond the multidimensional cubes typical of most factorial designs. Gosset can also handle sampling volumes with truncated and sliced shapes and with closed or open boundaries. Gosset can handle sampling volumes which are multidimensional spheres.

Gosset has additional features and capabilities which enable it to solve a variety of optimization problems beyond experiment planning.

We urge you to work through examples provided in the Gosset User Manual. We provide a shortened list of examples in the next section.

Later in this chapter, we demonstrate how to use Gosset to select sampling points for an example industrial process. The target results are dependent on a combination of six controllable factors. Three of the factors were discrete: use Machine 1 or Machine 2; process once or twice; manual control or automatic control. Three of the factors could be finely adjusted within operating ranges. We learned in Chapter 8 to choose the initial samples using the Plackett–Burman (PB) screening design. In this chapter, we demonstrate how to use Gosset to select next samples for this industrial example.

Incidentally, we are not the only experts using Gosset for experiment design. Some companies have already employed Gosset to boost their product exploration and bottom line. However, more often than not, these company results are proprietary or trade secrets, so reports detailing the use of Gosset rarely become public. We hope that our advocacy of Gosset promotes its use to benefit industry and science worldwide.

As of this printing, as far as we are aware, Gosset only operates on Linux and Unix operating systems. Thus the experimentalist will require a dual-boot computer or dedicated Linux computer to install and use Gosset.

9.5.3 Practice Gosset Examples (from Gosset User Manual)

We highly recommend that you work through some of the simpler Gosset examples before we use Gosset to design an experiment. In the list below, the number on the left-hand side refers to the section number in the Gosset User Manual.

7.1 Three continuous variables in a cube (similar to our M&H Chapter 8, Example 1)
7.2 Same, but search the built-in library
7.3 Four continuous variables in a sphere
7.4 Same but make three runs at center point
7.5 Six two-level discrete factors
7.6 Three four-level and two two-level discrete numerical variables
7.7 Same, but run as a batch job
7.8 A simple mixture design
7.9 A constrained mixture
7.10 Linear (or PB-type) designs
  7.10.1 Continuous cube
  7.10.2 Two-level variables
  7.10.3 PB design with 24 runs

7.11 A D-optimal example, correcting an error in a published design
7.12 D-optimality: a maximal determinant problem of Galil
7.13 Measuring region discrete, modeling region continuous
7.14 Estimating a response in a room from measurements made only in one half
7.15 Estimating a response on the moon from measurements made on Earth
7.16 A design that guards against loss of one run
7.17 An eye-measurement experiment, illustrating a design with correlated errors
7.18 Block designs
 7.18.1 A second-order Response Surface design in two blocks
 7.18.2 A design with four blocks of size 10
 7.18.3 The central composite design in three blocks
7.19 A balanced incomplete block design
7.20 Packings
 7.20.1 Packing 12 points in an irregular region
 7.20.2 Packing 20 points in a cube
7.21 Qualitative variables with several levels
 7.21.1 One quantitative and three qualitative variables
 7.21.3 A more complicated consumer-response design
7.22 Orthogonal arrays
7.23 Maximizing or minimizing a function of several variables

9.6 Demonstrate Gosset for Selecting Data

Here we begin to demonstrate how to use Gosset for an industrial process described in Example 2 of Chapter 8. We use Gosset to select supplementary output data. In order to explain Gosset features, we review (in Section 9.6.1) the status quo of the experiment and elaborate details overlooked in Chapter 8. Then we begin to employ the Gosset program in Section 9.6.2.

9.6.1 Status Quo of Experiment Planning and Execution (Prior to Selecting More Samples)

In Chapter 8, Example 2, the goal at this particular factory is to produce a certain material with predictable high strength. The class of experiment was initially exploratory; then the class of experiment transformed to type certification or quality assurance, with predictive model.

9.6.1.1 Specified Motivating Question

The Motivating Questions:

i) What combination of six controllable factors produces the material with highest strength?
ii) Create a credible modeling equation of these factors to predict the strength of the material with 95% certainty.
iii) Bonus: Can we plan how to best use machinery at the factory for production?

9.6.1.2 Identified Pertinent Candidate Factors

Candidate controllable factors comprised three continuous factors and three discrete factors. Discrete factors are like a toggle switch to select between two more options: A or B or more. One of the discrete factors was to produce on Machine 1 or Machine 2. Another discrete factor was to operate the tension using manual control or automatic control. The three continuous factors could be finely adjusted within operating ranges.

The six factors are tabulated in Table 9.2, previously shown in Table 8.5.

9.6.1.3 Selected Initial Sample Points Using Plackett–Burman

In Chapter 8, we used the Plackett–Burman screening design to select the initial 12 data points for stage one. The PB technique generates a family of designs (based on number of runs) such as provided in Appendix B. The combinations of six factors for each of the initial samples are tabulated in Table 9.3, previously shown in Table 8.6.

Table 9.2 List of factors for industrial test plan example, plus material property sought in Response.

Label	Factor	–	+	Units	Factor Type
X1	Tension control	Manual □	Automatic ○		Discrete
X2	Temperature	200	250	°C	Continuous
X3	Throughput	10	20	kg/min	Continuous
X4	Mixing	Single ○	Double ⊕		Discrete
X5	Machine used	1	2		Discrete
X6	Moisture	20	80	%	Continuous
Y	**Response**	Strength (etc.)		MPa	

Table 9.3 PB-guided initial 12 runs of industrial example.

	Coded						Natural						
Trial	**X1**	**X2**	**X3**	**X4**	**X5**	**X6**	**X1**	**X2**	**X3**	**X4**	**X5**	**X6**	**Y**
1	+	+	–	+	+	+	Auto	250	10	Dbl	2	80	85
2	+	–	+	+	+	–	Auto	200	20	Dbl	2	20	114
3	–	+	+	+	–	–	Man	250	20	Dbl	1	20	67
4	+	+	+	–	–	–	Auto	250	20	Sgl	1	20	64
5	+	+	–	–	–	+	Auto	250	10	Sgl	1	80	56
6	+	–	–	–	+	–	Auto	200	10	Sgl	2	20	68
7	–	–	–	+	–	+	Man	200	10	Dbl	1	80	13
8	–	–	+	–	+	+	Man	200	20	Sgl	2	80	108
9	–	+	–	+	+	–	Man	250	10	Dbl	2	20	90
10	+	–	+	+	–	+	Auto	200	20	Dbl	1	80	22
11	–	+	+	–	+	+	Man	250	20	Sgl	2	80	130
12	–	–	–	–	–	–	Man	200	10	Sgl	1	20	23

The Sampling Volume for this experiment, from which we choose sample points, is a six-dimensional Factorial Cube. Although it is impossible to clearly show a six-dimensional hypercube on a flat two-dimensional page, we will attempt to display the sample points for all six factors in Figure 9.4a and b. Figure 9.4a shows all the samples run on Machine 1; Figure 9.4b shows all the samples run on Machine 2. By showing two figures (Figure 9.4a and b), one discrete dimension (X5) is covered. The three continuous quantitative factors are shown as the three axes of the oblique projected cube in each figure (X2, X3, X6). Thus far a sum of four dimensions have been covered. For the remaining two discrete factors, we will show by the shape and shading of the symbols at the sample points. If the tension control (X1) was manual, the symbol is square; if automatic control, the symbol is a circle. If the mixing (X4) was single, the symbol is open (no shading); if the mixing was double, the symbol has a criss-crosshatch.

Try to find all 12 samples, six each on Figure 9.4a and b. Hints: The 12-run PB screening design placed the 12 samples on the eight vertices of the spatial cube, therefore four of the eight vertices see two samples. The duplicate sample may be on the same cube or the other cube. A duplicate sample is not a perfect duplicate of sampling conditions, however, because of mismatched shape, square or circle, or mismatched shading, open or crosshatched, or both.

We have provided Figure 9.4a and b in this chapter (rather than Chapter 8) in order to illustrate in Section 9.7 how Gosset guides us to choose subsequent samples, samples which optimally fill the sampling volume. Perhaps such a figure would be useful to your client when you provide a progress report. Perhaps your experiment has many more candidate factors, making it even more challenging to graphically display the multiple dimensions of your sample space.

9.6.1.4 Executed the First 12 Runs at the PB Sample Conditions

We performed all 12 runs in stage one of this industrial experiment to produce material samples. The material strength results we obtained are shown in column Y of Table 9.3. We plotted these results on Figure 9.5, which is a three-dimensional plot. Note the

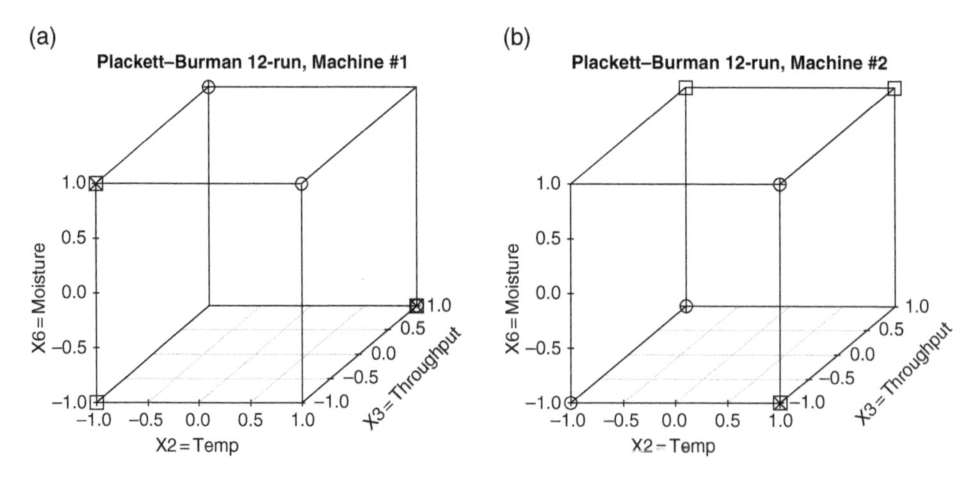

Figure 9.4 (a) Six PB samples on Machine 1. (b) Six PB samples on Machine 2.

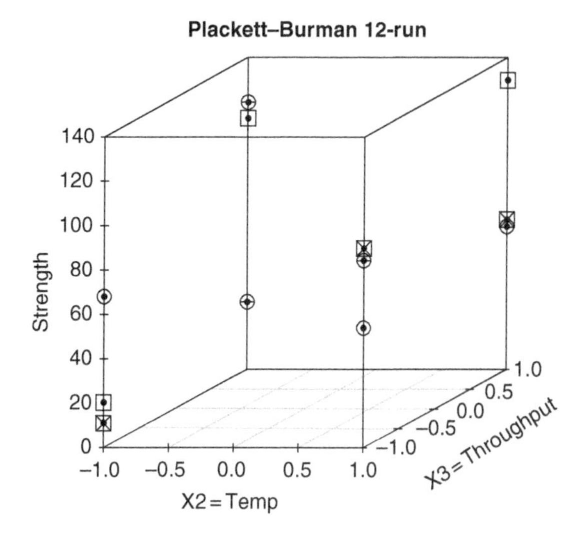

Figure 9.5 Strength results from the PB 12-run screening.

differences from Figure 9.4. The vertical axis now represents strength in (MPa) instead of moisture (X6). Results for both machines are given on the single plot.

Try to find all 12 sample results. Hints: Three distinct runs appear along each vertical edge, equivalent at the four vertices at the base of the cube. Strength results for Machine 2 are notably stronger than for Machine 1. The shape and shading of symbols are exactly the same as in Figure 9.4. If the tension control (X1) was manual, the symbol is square; if tension control was automatic, the symbol is a circle. If the mixing (X4) was single then the symbol is open; if the mixing was double, the symbol has a criss-crosshatch. There are 3 open circles and 3 open squares. As an alternative plot (not shown), each of the 12 runs was plotted with a distinct symbol at its value of strength.

In case you are interested is creating similar plots, these three-dimensional plots were constructed using the R language contributor package scatterplot3D (used with gratitude). Our script for making the plot is included in the accompanying textbook resources.

9.6.1.5 Analyzed Results. Identified Dominant First-Order Factors. Estimated First-Order Uncertainties of Factors

We evaluated the results of the screening design using the statistical language R. We obtained first-order coefficients and Standard Errors for each of the factors from a first-order linear model. Then we performed Analysis of Variance on the data, shown in Table 9.4.

Analysis of Variance quantifies how each X factor contributes to the variance in response Y. The F-value is a ratio of the benefit of a factor divided by the cost of including the factor. According to the F-values, this Analysis of Variance reveals three dominant terms. In order of dominance, the contributing factors are X5 (machine), X3 (throughput), and X2 (temperature). The least contributing factor is X6 (moisture). We conclude from the F-values that the most defensible first-order predictive model includes only factors X5, X3, and X2.

Table 9.4 Analysis of Variance of factors for the full first-order linear model using PB 12-run screening data.

Label	Df	SumSq	MeanSq	F-value	p-Value
X1	1	40	40	0.216	0.66181
X2	1	1728	1728	9.244	0.02874
X3	1	2408	2408	12.883	0.01572
X4	1	280	280	1.500	0.27527
X5	1	10208	10208	54.610	0.00071
X6	1	12	12	0.064	0.81007
Residuals	5	935	187		

By a more familiar measure, the p-value of the Null Hypothesis, we have good reason to "not reject" the Null Hypothesis for factors X6, X1, and X4. In other words, we have good reason to omit factors X6, X1, and X4. This leaves us with the same most defensible first-order predictive model including only factors X5, X3, and X2.

Therefore, we again analyze with R the data of the PB 12-run screening including only factors X5 (machine), X3 (throughput) and X2 (temperature). The reanalysis with fewer factors provides improved first-order coefficients and Uncertainties (Standard Errors). These values are used in the first draft of a predictive equation.

The PB 12-run samples do not allow us to find curvature effects or second-order effects in the factors. Interaction terms between dominant terms are not advised from mere screening data.

In order to find second-order or curvature effects of factors, more data points on in the operating volume are needed. The mathematics requires it. We recommend Gosset for selecting the next samples.

The status quo of the experiment includes one more achievement, a draft predictive equation.

9.6.1.6 Generated Draft Predictive Equation

Using the reanalysis, we have improved coefficients for the draft predictive equation. In Chapter 8 we determined a draft first-order predictive equation to be

$$Y[strength] = 40.83[MPa] + \begin{bmatrix} 0.00, \text{if M1} \\ 58.33, \text{if M2} \end{bmatrix}[MPa] + 2.833\left[\frac{MPa}{\text{kg/min}}\right](flow - 15.)[\text{kg/min}] + 0.480\left[\frac{MPa}{°C}\right](Temp - 225.)\,[°C]. \tag{9.5}$$

At first glance, it is apparent that Machine 2 produced material with more than twice the strength of that by Machine 1. Why consider Machine 1 for future data? That is a decision for your client to consider and make. Your responsibility is to answer the Motivating Questions by providing your client with credible experimental results and analysis.

The reader has already been tasked, as an exercise, to rewrite the draft first-order predictive equation with 95% confidence intervals for each coefficient.

Can the predictive equation be improved with existing data? Yes. How? By using the lessons learned in Sections 7.5 through 7.7.

Having reviewed the status of our screened experiment, we are now motivated to use Gosset to help us pick future points.

9.6.2 Use Gosset to Select Additional Data Samples

We assume you have installed Gosset on your computer and have worked through the practice examples provided in the Gosset documentation. These examples were listed in Section 9.5.3.

9.6.2.1 Example Gosset Session: User Input to Select Next Points

We have Gosset installed in the directory /home/sloane. Computer input and output are printed here in Courier font. At the Linux prompt /home/sloane>, we call the executable program gosset while suppressing nuisance compiler messages. We indicate Gosset prompts requesting our input with the prompt sign % at the left column. Example output from Gosset omits the prompt sign; Gosset probably outputs more lines than shown below.

```
/home/sloane> ./gosset 'cflags -w'
please type 'cd something' to name a local directory for your work
% cd pb12z2e8

% 10 range x2 x3 -1 1
% 20 discrete x5 -1 1

% 30  use x2=1  x3=-1 x5=1
% 40  use x2=-1 x3=1  x5=1
% 50  use x2=1  x3=1  x5=-1
% 60  use x2=1  x3=1  x5=-1
% 70  use x2=1  x3=-1 x5=-1
% 80  use x2=-1 x3=-1 x5=1
% 90  use x2=-1 x3=-1 x5=-1
% 100 use x2=-1 x3=1  x5=1
% 110 use x2=1  x3=-1 x5=1
% 120 use x2=-1 x3=1  x5=-1
% 130 use x2=1  x3=1  x5=1
% 140 use x2=-1 x3=-1 x5=-1

% 150 use x2=4/7 x3=5/9 x5=1
% 160 use x2=2/9 x3=2/7 x5=1

% 170 model  (1+x2+x3+x5)^2-x5^2
% compile
% moments n=1000000
% design runs=22 n=20
% interp
% iv
% quit
```

9.6.2.2 Example Gosset Session: How We Chose User Input

Our reasoning and explanations are in Times font. Computer input and output are printed here in `Courier` font. We indicate Gosset prompts requesting our input with the prompt sign % at the left column.

As a reminder, Linux calls an executable file by appending the executable name to the directory in which the executable resides. Since the executable resides in the current directory `./`, Gosset is called by the first term `./gosset` in the statement below.

Due to the evolution of compilers along with the warnings generated by newer compilers, we chose to suppress compiler warnings with the second term `'cflags -w'` below.

```
/home/sloane> ./gosset 'cflags -w'
```

Gosset prompts us for the name of a local working directory. We already created this directory, choosing a name that reminds us of the proposed design: PB chose 12 points; the user chose 2 points according to strategy one; Gosset will choose 8 points. The total experiment at the conclusion of this stage will be 12 + 2 + 8 = 22 runs.

```
please type 'cd something' to name a local directory for your work

% cd pb12z2e8
```

Line 10 assigns two continuous factors, x2 and x3, each with a range from [−**1**, 1]. We chose to assign factors with names corresponding to our PB example in Chapter 8. We saw that only two continuous factors survived the screening stage.

Line 20 assigns one discrete factor, x5, that can only take the integer values −**1** or 1. Only one discrete factor survived the screening stage.

```
% 10 range x2 x3 -1 1
% 20 discrete x5 -1 1
```

Lines 30 through 140 tell Gosset that 12 points were already chosen during the PB screening. Hence Gosset will choose to optimize other points in the Sampling Volume taking into account these preselected points. The execution of the experiment plan has already provided observed results for these 12 conditions.

```
% 30  use x2=1  x3=-1 x5=1
% 40  use x2=-1 x3=1  x5=1
% 50  use x2=1  x3=1  x5=-1
% 60  use x2=1  x3=1  x5=-1
% 70  use x2=1  x3=-1 x5=-1
% 80  use x2=-1 x3=-1 x5=1
% 90  use x2=-1 x3=-1 x5=-1
% 100 use x2=-1 x3=1  x5=1
% 110 use x2=1  x3=-1 x5=1
% 120 use x2=-1 x3=1  x5=-1
% 130 use x2=1  x3=1  x5=1
% 140 use x2=-1 x3=-1 x5=-1
```

Lines 150 through 160 tell Gosset that the user has adopted strategy one, with approximately interval-halving, to select two points. The values we supplied were fractions, but decimal values within range are also allowed. Hence Gosset will choose to optimize other points in the Sampling Volume taking into account these preselected points. These two points do not yet have observed experimental results.

```
% 150 use x2=4/7 x3=5/9 x5=1
% 160 use x2=2/9 x3=2/7 x5=1
```

Line 170 assigns the model of the Response Surface for which Gosset will optimize points in the Sampling Volume. Notice that the grammar is similar to the R language. The intercept is specified by 1. Individual terms, cross terms, and second-order terms are concisely specified by squaring the bracketed simple linear model. Since the discrete factor x5 cannot logically have a second-order term, we must subtract this parasitic term (due to squaring) from the final model. If you are familiar with the grammar for models in the R language, we think the Gosset dialect provides a clever way to express sophisticated models cogently.

```
% 170 model (1+x2+x3+x5)^2-x5^2
```

We command Gosset to compile a version of itself dedicated to our specified problem in the working directory. This compiled version and all results are inserted in the working directory.

Incidentally, we created a batch script to input all lines from `cd directory` to `compile`.

DO NOT include the prompt symbol in the script, if you create a batch script.

The following Gosset commands must be typed one at a time. Each command requires a few seconds to process so it must not be included in the batch script.

```
% compile
```

We command Gosset to construct one million moments for its optimization. This value n = 1,000,000 was recommended in the Gosset examples. Gosset consumes only a second to calculate all iterations. Gosset outputs several lines (not shown here) to inform the user that moments are calculated, ready and completed.

```
% moments n=1000000
```

We command Gosset to design an experiment with a total of `runs = 22`. We command Gosset to take the best of `n = 20` designs. If the user omits `run=?`, then Gosset chooses the minimum number of runs necessary for the given model. After creating all of the competitive designs, Gosset selects the best design which corresponds to the minimum value of `iv`.

```
design runs=22 n=20
```

We command Gosset to list the results of the design and reveal the experiment plan.

```
% interp
```

Gosset reports the achieved minimum value of iv.

```
% iv
```

We command Gosset to quit. We know, you did not need us. Our wise editor did.

```
% quit
```

9.6.2.3 Example Gosset Session: User Input Along with Gosset Output

We indicate Gosset prompts requesting our input with the prompt sign % at the left column. Computer input and output are printed here in Courier font. Example output from Gosset omits the prompt sign.

```
/home/sloane> ./gosset 'cflags -w'

% please type 'cd something' to name a local directory for your work

% cd pb12z2e8

% 10 range x2 x3 -1 1
% 20 discrete x5 -1 1

% 30  use x2=1  x3=-1 x5=1
% 40  use x2=-1 x3=1  x5=1
% 50  use x2=1  x3=1  x5=-1
% 60  use x2=1  x3=1  x5=-1
% 70  use x2=1  x3=-1 x5=-1
% 80  use x2=-1 x3=-1 x5=1
% 90  use x2=-1 x3=-1 x5=-1
% 100 use x2=-1 x3=1  x5=1
% 110 use x2=1  x3=-1 x5=1
% 120 use x2=-1 x3=1  x5=-1
% 130 use x2=1  x3=1  x5=1
% 140 use x2=-1 x3=-1 x5=-1

% 150 use x2=4/7 x3=5/9 x5=1
% 160 use x2=2/9 x3=2/7 x5=1

% 170 model (1+x2+x3+x5)^2-x5^2

% compile

using 3 variables on -1,1: A(0) A(1) A(2)
  A(0)     A(1)     A(2)     constant
  1.000000 0.000000 0.000000 0.000000 x5
  0.000000 1.000000 0.000000 0.000000 x2
  0.000000 0.000000 1.000000 0.000000 x3
```

The Gosset compiler informs the user how the user names for factors are translated into Gosset-style variables. From the base Gosset-style variables, model variables are generated.

```
model on these variables:
    1
    A(0)
    A(1)
    A(2)
    A(0)*A(1)
    A(0)*A(2)
    A(1)^2
    A(1)*A(2)
    A(2)^2
9 functions in model

creating pb12z2e8/program.log
```

The Gosset compiler informs the user of all Gosset-style terms in the model: the intercept 1, the first-order terms, the cross terms, and the second-order terms. The compiler saves its progress.

```
% moments n=1000000
```

Then Gosset outputs several lines (not shown here) indicating moments are calculated, ready, and completed.

```
% design runs=22 n=20

20 designs for 22 runs

pb12z2e8/v.22.best has IV=0.288158456243
    check pb12z2e8/lib.a
    ar x pb12z2e8/lib.a pb12z2e8/v.3.22.aaaa 2>/dev/null nothing found
    saving pb12z2e8/v.3.22.aaaa
    ar q pb12z2e8/lib.a pb12z2e8/v.3.22.aaaa

ar: creating pb12z2e8/lib.a

local best is pb12z2e8/v.22.best
```

Gosset creates an experiment design with a total of `runs = 22`. Gosset automatically repeats making competitive experiment designs until it has attempted `n = 20` designs. Gosset then takes the best of `n = 20` designs. If the user had omitted `run=?`, then by default Gosset chooses the least number of runs necessary for the given model. The best design corresponds to the minimum value of `iv`.

```
% interp

pb12z2e8/v.22.best
```

```
   x5        x2        x3
 1.0000   1.0000  -1.0000
 1.0000  -1.0000   1.0000
-1.0000   1.0000   1.0000
-1.0000   1.0000   1.0000
-1.0000   1.0000  -1.0000
 1.0000  -1.0000  -1.0000
-1.0000  -1.0000  -1.0000
 1.0000  -1.0000   1.0000
 1.0000   1.0000  -1.0000
-1.0000  -1.0000   1.0000
 1.0000   1.0000   1.0000
-1.0000  -1.0000  -1.0000
 1.0000   0.5714   0.5556
 1.0000   0.2222   0.2857
 1.0000  -0.1425  -1.0000
-1.0000  -0.1087   1.0000
 1.0000  -1.0000  -0.1439
-1.0000   1.0000  -0.1069
-1.0000  -1.0000   0.1069
-1.0000   0.1087  -1.0000
 1.0000  -0.0059  -0.0136
-1.0000   0.0000   0.0000
```

Gosset lists the end result of the competing designs, revealing all sampling points in the best experiment plan. Notice that preset sampling points are returned in the exact same order specified before compilation. In contrast, the samples chosen and optimized by Gosset are randomized according to good experimental practice.

Figure 9.6a and b provide an informative two-dimensional perspective of the total experiment design, stage one with 12 points selected by PB, and stage two with 2 user

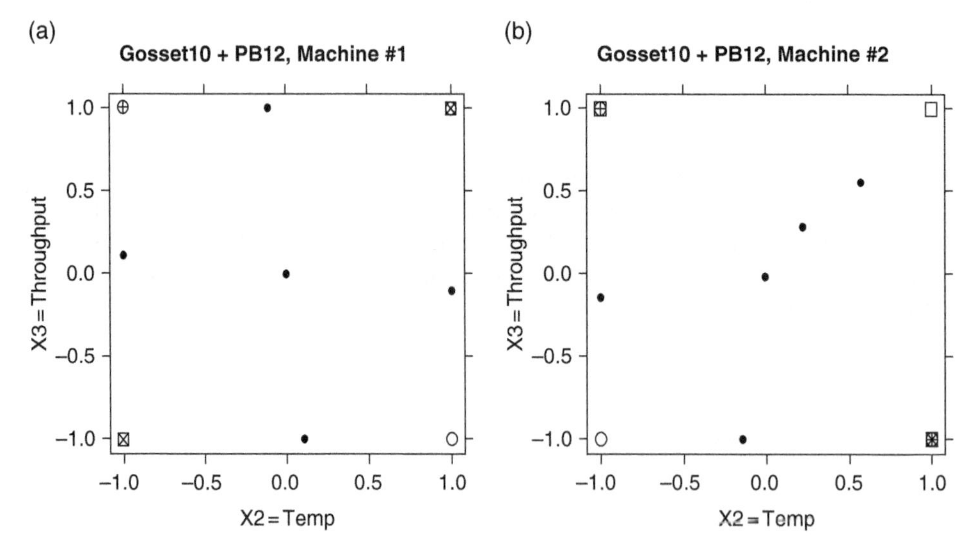

Figure 9.6 (a) Gosset-style samples on Machine 1. (b) Gosset-style samples on Machine 2.

points and 8 points optimized by Gosset. The same six PB points in Figure 9.4a appear with identical symbols in Figure 9.6a. The same six PB points in Figure 9.4b appear with identical symbols in Figure 9.6b. Figure 9.6a obtains five more samples (solid circles) from Gosset that are scheduled on Machine 1. The skewed symmetry of the Gosset-selected points in Figure 9.6a enable a strong predictive model for Machine 1. Figure 9.6b obtains five more samples (solid circles), two user-selected in the upper right quadrant and three optimized by Gosset, that are scheduled on Machine 2. Frankly, being the user who selected the two points, I confess that Gosset selected better sampling points than I did. I hypothesize that Gosset selects better points than I could in exploratory-type experiments.

Congratulations! You have now completed a Gosset design that was not contained within original Gosset documentation. You have harnessed Gosset, a powerful yet nimble generator of optimized experiment designs, for an industrial client.

To implement the Gosset design in our experiment, Gosset variables must next be converted to those values on the experiment's operating map.

9.6.2.4 Example Gosset Session: Convert the Gosset Design to Operator Values

We recall the original six factors for this industrial experiment were listed in Table 9.2. The PB screening design employed a conversion code for all six factors as listed in Table 9.3. After executing the initial experiment plan according to PB, only three factors had a dominant first-order impact on resultant material strength. Two of the dominant factors were continuous, one was discrete.

Our Gosset design in Section 9.6.2.2 used the exact same coding names and ranges as did PB but for only the dominant three factors out of the original six factors. It is a simple matter to convert these three Gosset factors back to their corresponding operating factors. Merely create an inverse of the conversion Table 9.3.

What about the factors which were neglected in the Gosset design? The factors were neglected as a consequence of initial data analysis, Section 9.6.1.4, which suggested little impact due to them. One of the excluded factors, moisture, was continuous. The other two were discrete: tension control and mixing.

Nevertheless, the experiment operator needs guidance on what values to use for all controllable factors for all sample points. Ask the client: what operating conditions does the factory prefer?

Our choices for stage two of the experiment plan? For the moisture factor, X6, we will choose the central value of the operating range. For tension control, we choose automatic. We choose single mixing.

Finally, we have the operator's values for stage two of the experiment plan.

During stage two of the experiment plan, the operator will record the actual operating values and results for each additional observation in stage two. Record achieved values.

Your client may also elect to record additional operating factors for each observation in stage two. May we suggest additional factors such as operator name, time of day, date, raw material supplier, raw material inventory number, weather, etc.? If a factor is recorded, it is available for analysis.

Be aware! When Gosset listed the competing designs (via `interp`), it selected the best experiment plan and revealed Gosset values for all sampling points, old and new. *Beware!* Gosset has reorganized the columns of user factors.

Notice that preset sampling points are returned in the exact same order specified before compilation. The first 12 points already have results due to the PB screening in the first stage of the experiment plan. These points are not scheduled to be resampled.

The next two points in the Gosset list are "user points." These two Gosset points were selected by the experimentalist but not yet sampled. We reviewed the results of stage one and identified which sample had maximum strength. Employing strategy one (not Gosset), we selected two points by interval-halving toward the sample conditions with maximum strength. These two user points must be inserted into the experiment plan schedule for stage two. Inserted how? The two user points ought to be randomly shuffled into the Gosset points.

In contrast to preset points, the sample points chosen and optimized by Gosset are randomized according to good experimental practice. Thus the experimentalist need not make extra effort to randomize a Gosset-generated design.

9.6.2.5 Results of Example Gosset Session: Operator Plots of Total Experiment Plan

In Figures 9.7a and 9.7b, we rework Figures 9.4a and 9.4b to show the operator's values for each of the runs in the total experiment plan, stages one and two. The meanings of the open square and circle symbols for stage one are exactly the same as the previous figures. The Gosset samples for stage two are shown as bullets with vertical trunks. Automatic control tension and single mixing were decided previously.

9.6.2.6 Execute Stage Two of the Experiment Plan: User Plus Gosset Sample Points

With operator conditions in hand for stage two samples, the experiment can continue. During the experiment, the operator will record the actual operating values and results for each additional observation. Record actual achieved values according to the guidelines in Section 9.4 for organizing data collection. Record actual values even if not exactly the same as planned for stage two. If the client requires the record of additional factors, add to each observation.

The data for the complete experiment, all stages, is saved, stored, and archived. These original data record files will never change. Only copies of the original files will be used for analysis.

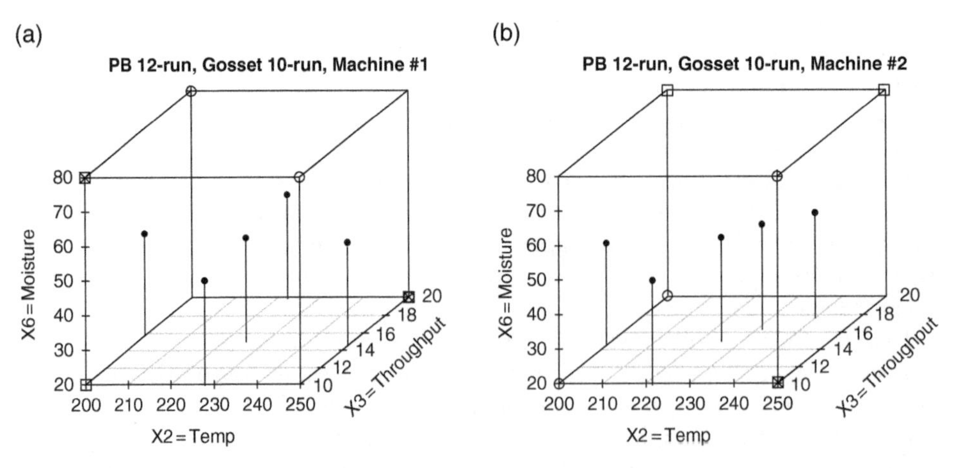

Figure 9.7 (a) Operator conditions on Machine 1 for total experiment plan. (b) Operator conditions on Machine 2 for total experiment plan.

Then the experimentalist analyzes copies of the complete data record, all stages. Apply the chosen model in the grammar of the R language. Evaluate Analysis of Variance. Retain dominant terms of the model. Omit terms in refined models that do not survive the Null Hypothesis, per Ockham's razor.

Reanalyze data with the refined final model, just as we did in Section 8.6.7. Write the modeling equation. Then rewrite the modeling equation with 95% confidence levels for the Uncertainties.

Reanalyze data in two blocks, one for each machine. Write the modeling equations for each machine. Rewrite the modeling equations with 95% confidence of uncertainties. As we saw in Chapter 7, this pair of modeling equations ought to be much improved over your all-in-one equation in the previous paragraph.

9.7 Use Gosset to Analyze Results

Find local extrema in Response Surface. If the Response Surface is a simple continuous two-dimensional surface, many mathematical strategies exist, beginning with Steepest Ascent, which can be employed to search for local extrema.

Gosset can find multiple local extrema for a Response Surface with multiple discrete factors as well as multiple continuous factors. Try that with Steepest Ascent. This capability is part of Gosset's optimization algorithm invented by Sloane and Hardin and was the feature that first caught our attention.

We will demonstrate, in Section 9.9, Gosset's ability to locate local extrema using an example problem from our colleague's book *Design and Analysis of Experiments*, by Douglas C. Montgomery (Wiley, 2017), example 11.2.

The local extrema identified by Gosset become additional points of interest to select and sample for an experiment. The local extrema can guide us to regions or rapid variation where we can engage our expertise through strategy one in Section 9.5.

9.8 Other Options and Features of Gosset

By default, Gosset creates type I-optimal experiment designs (the letter for indigo).

Gosset can create other types of optimal designs (as described in the Gosset User Manual):

A-optimal, D-optimal, or E-optimal designs: type = A, type = D, or type = E
Designs that protect against a missing run: type = B, C, F, or J
Designs with correlated errors
Block designs
Finding packings in arbitrary regions: type = P
Determining minimum number of runs.
Optimizing experiment designs without a model function.
Maximizing or minimizing a function of several variables. (We demonstrate in next section.) Gosset works where steepest descent stumbles. Gosset even works for functions with discrete variables.
Did you notice that Gosset has an option to create experiment designs that robustly overcome lost data?
Finally, did you notice that Gosset can replace PB for creating screening designs with fewer runs?

9.9 Using Gosset to Find Local Extrema in a Function of Several Variables

Gosset can do much more than merely generate optimal experiment designs.

Here we use the R language and Gosset to find local extrema in experimental results. The experimental results are extracted from our colleague's text (Montgomery 11.2). Using R, we find a few models to fit the data and take the best model. This best model turns out to be a quadratic function of several variables. Then we use Gosset to estimate the location of the local maximum. We also show how to use Gosset to find multiple local minima and to find a constrained extremum.

Inspect the datatable from Montgomery 11.2. Note that the datatable has operator variables, *Nl* and *N2*, as well as Gosset-style factors, *X1* and *X2*, and response variables Y1yi, Y2vs, and Y3mw. We will analyze this experiment only for the response variable Y1yi.

```
> mnt=read.csv('mont11_2.csv')
> head(mnt)
  Nl  N2   X1  X2  Y1yi  Y2vs  Y3mw
1 80  170  -1  -1  76.5  62    2940
2 80  180  -1   1  77.0  60    3470
3 90  170   1  -1  78.0  66    3680
4 90  180   1   1  79.5  59    3890
5 85  175   0   0  79.9  72    3480
6 85  175   0   0  80.3  69    3200
```

Use R to fit a first-order linear model to the data. Why is the fit not satisfactory?

```
> mod0=lm(Y1yi~1+X1*X2,data=mnt)
> summary(mod0)
Call:
lm(formula = Y1yi ~ 1 + X1 * X2, data = mnt)

Coefficients:
              Estimate  Std.Error  t-value   Pr(>|t|)
(Intercept)   78.4769   0.3971     197.622   <2e-16 ***
X1             0.9951   0.5063       1.966   0.0809 .
X2             0.5152   0.5063       1.018   0.3354
X1:X2          0.2500   0.7159       0.349   0.7350

Residual Standard Error: 1.432 on 9 Degrees of Freedom
Multiple R-squared: 0.3581,    Adjusted R-squared: 0.1441
F-statistic: 1.674 on 3 and 9 DF, p-value: 0.2413
```

Add second-order terms to the R datatable; note this action did not alter original data.

```
> mnt$X12=mnt$X1^2
> mnt$X22=mnt$X2^2
> head(mnt)
```

```
  Nl  N2  X1  X2  Y1yi  Y2vs  Y3mw  X22  X12
1 80  170  -1  -1  76.5  62    2940  1    1
2 80  180  -1   1  77.0  60    3470  1    1
3 90  170   1  -1  78.0  66    3680  1    1
4 90  180   1   1  79.5  59    3890  1    1
5 85  175   0   0  79.9  72    3480  0    0
6 85  175   0   0  80.3  69    3200  0    0
```

Use R to fit a second-order linear model to the data. The fit is much improved. We also perform Analysis of Variance and find the interaction term has a p-value of 10%. What does the Null Hypothesis say? Shall we keep this term and continue the next step of the analysis (we do) or drop the term and reanalyze with a new model? We kept the interaction term because it appears more instructive for our readers. Typically, we would reckon the interaction term as Null.

```
> mod2=lm(Y1yi~1+X1*X2+X12+X22,data=mnt)
> summary(mod2)
Call :
lm(formula = Y1yi ~ 1 + X1 * X2 + X12 + X22, data = mnt)

Coefficients:
             Estimate Std.Error t-value  Pr(>|t|)
(Intercept) 79.93995 0.11909   671.264  < 2e-16    ***
X1           0.99505 0.09415    10.568  1.48e-05   ***
X2           0.51520 0.09415     5.472  0.000934   ***
X12         -1.37645 0.10098   -13.630  2.69e-06   ***
X22         -1.00134 0.10098    -9.916  2.26e-05   ***
X1:X2        0.25000 0.13315     1.878  0.102519
---

Residual Standard Error: 0.2663 on 7 Degrees of Freedom
Multiple R-squared: 0.9827,     Adjusted R-squared: 0.9704
F-statistic: 79.67 on 5 and 7 DF, p-value: 5.147e-06

> anova(mod2)
Analysis of Variance Table
Response: Y1yi
           Df  Sum Sq   Mean Sq  F-value   Pr(>F)
X1          1   7.9198   7.9198  111.6873  1.484e-05 ***
X2          1   2.1232   2.1232   29.9413  0.000934  ***
X12         1  10.9816  10.9816  154.8663  4.979e-06 ***
X22         1   6.9721   6.9721   98.3225  2.262e-05 ***
X1:X2       1   0.2500   0.2500    3.5256  0.102519
Residuals   7   0.4964   0.0709
```

Having found an acceptable model in R, we construst a Gosset-style model from the coefficients. We call Gosset and follow the typical steps.

```
####### Search for maximum extrema with Gosset

/home/sloane$ ./gosset 'cflags -w'
> please type 'cd something' to name a local directory for your work
% cd gosMontMh
```

```
% 10 range X1 X2 -1 1
% 20 model 1(79.940+0.995*X1+0.515*X2-1.376*X1^2-1.001*X2^2
  +0.25*X1*X2)

% compile

model on these variables:

   1(79.940+0.995*X1+0.515*X2-1.376*X1^2-1.001*X2^2+0.25*X1*X2)

1 functions in model
compile done

% moments n=1000000
% design type=I runs=10
start designs
design time=100000 runs=10 n=1 processors=1
% interp
```

The Gosset optimization results in the location of the maximum extrema of the function in Gosset-style factors.

```
gosMont/v.10.best
X1      X2
0.3893  0.3059
```

We then convert the values of the Gosset-style factors to operator values for the location of the maximum.

```
N1Max =  85+ 0.3893 * 5
N2Max = 175+ 0.3059 * 5
(N1,N2)Max = (86.95,176.53)
```

Had we imposed constraints on the location of the maximum, the Gosset statements would be:

```
% 10 range X1 X2 -1 1
% 20 constraint X2>0.99
% 30 model 1(79.940+0.995*X1+0.515*X2-1.376*X1^2-1.001*X2^2
  +0.25*X1*X2)

# find constrained maximum near edge of sampling volume
```

Had we used Gosset to find all the local minima, the statements would be:

```
# find local minimum by recasting the function as 1+1/f

> please type 'cd something' to name a local directory for your work
cd gosMontMh

10 range X1 X2 -1 1
20 model 1(1+1/(79.940+0.995*X1+0.515*X2-1.376*X1^2-1.001*X2^2
   +0.25*X1*X2))
```

9.10 Summary

We began this chapter insisting that the experimentalist decide and emphasize the absolute must-have data for the experiment. Make sure this is the first data collected. Then we insisted on guarding the integrity of the data.

With determination to perform a credible experiment, we reminded how to organize the data for ease of data analysis.

In order to select the next data points beyond the must-have data, we looked at two strategies.

The default strategy depends heavily on the expertise of the experimentalist for concentrating high-impact data. We discussed the default strategy for identifying and collecting data that will have the most impact. Seek regions of high variation. Interval-halve. The default strategy works well for physics and physics-intensive engineering experiments.

The second strategy employs the computer program Gosset for optimally selecting next data samples. Gosset works well for exploratory experiments and for experiments in fields with many candidate factors. Gosset features go beyond generating optimal experiment designs. We saw that the features of Gosset make it a powerful tool for experimentalists in every field. We think you will agree that Gosset is worthy of the effort required to overcome its learning curve.

The R language combined with Gosset equips the experimentalist with ample computer tools for designing credible experiments. Both the R language and Gosset are free.

Further Reading

The following documents are included in the codemart.cpio file which also includes the source for the Gosset Experimental Design Program.

Sloane, N.J.A. and Hardin, R.H. Computer-Generated Minimal (and Larger) Response-Surface Designs: (I) The Sphere, August 2001a.

Sloane, N.J.A. and Hardin, R.H. Computer-Generated Minimal (and Larger) Response-Surface Designs: (II) The Cube, August 2001b.

Sloane, N.J.A. and Hardin, R.H. Operating Manual for Gosset: A General-Purpose Program for Constructing Experimental Designs (Second Edition), Revised May 2003.

Homework

9.1 Prepare a lab computer for dual boot: Windows and the Linux operating system. Or ....

9.2 Prepare a dedicated lab computer for the Linux operating system. Or Exercise 9.1

9.3 Following the guide in Appendix D.3, download and install Gosset, public domain, open-source, and free.

Exercises 9.4 through 9.23 provide hands-on practice of Gosset to create experiment design, introducing basic features of the program. Type commands individually, waiting for Gosset to complete processing before the subsequent command. Begin execution via `./gosset 'cflags -w'`

9.4 In Gosset Manual pages 5–8, execute the example "Getting started." Be sure to wait. The discrete factor "d" has three levels, which is a capability beyond the two levels demonstrated in Chapter 8, PB Example 2. Factors have mutually dependent constraints, including an inequality constraint. What terms appear in the expanded model equation? The model has interaction and quadratic terms for all variables. Compare and contrast how specification of a quadratic model in Gosset differs from the R language protocol. Compare your design to the results in the manual.

9.5 In Gosset User Manual pages 8–9, execute the example in Gosset section 3.1. Be sure to wait. The discrete factor "d" has two levels. What terms appear in the expanded model equation? The model has interaction and quadratic terms for most variables but omits d^2, since two discrete levels disallow resolution of a quadratic. Remark on the model statement. Present your design.

9.6 In Gosset User Manual pages 19–20, execute the example in Gosset section 3.13 in order to practice finding the location of the absolute maximum of a general function within the prescribed domain. Present the location of the extremum. Evaluate the function to provide the value at the extremum. Identify and evaluate other local extrema on the domain.

9.7 In Gosset User Manual pages 48–49, execute the example in Gosset section 7.1. Did you notice that this exercise has similar factors and model as did Chapter 8, Example 1? Compare and contrast. Be sure to wait. What terms appear in the expanded model equation? Remark on the minimal design having only 10 runs (how would you have determined this is so?). This example specified 14 runs, for which Gosset selected 3 runs near the center. What advantages does Gosset's design have over our Chapter 8, Example 1?

9.8 In Gosset User Manual pages 49–50, execute the example in Gosset section 7.2 which duplicates Exercise 9.7 but employs the Gosset design from Gosset's built-in design library. Be sure to wait for Gosset prompt. How did this design compare to the design you just generated in section 7.1?

9.9 In Gosset User Manual pages 50–51, execute the example in Gosset section 7.3. Be sure to wait. Rather than a Factorial Cube, which is our experience up to this point, these factors continuously fill a four- dimensional sphere. Describe how the values in the Gosset design for a sphere notably differ from cubic designs. What terms appear in the expanded model equation?

9.10 In Gosset User Manual page 51, execute the example in section 7.4. Be sure to wait. This exercise duplicates section 7.3 (Exercise 9.9) but presets three runs at the center of the sphere. How does this Gosset design differ from Exercise 9.9?

9.11 In Gosset User Manual pages 52–53, execute the example in Gosset section 7.5. Be sure to wait. Rather than a Factorial Cube, which is our experience up to this point, these factors continuously fill a four-dimensional sphere. What terms appear in the expanded model equation? Remark on how the results of the "`interp`" step differ from previous exercises. How else do the values in the Gosset design for a sphere notably differ from cubic designs?

9.12 In Gosset User Manual pages 53–54, execute the example in Gosset section 7.6. Be sure to wait. What terms appear in the expanded model equation? Remark on how the results of the "`interp`" step differ from Exercise 9.11.

9.13 In Gosset User Manual page 54, execute the example in Gosset section 7.7 as a batch job. Note how the batch procedure accommodates the wait times you endured in previous exercises. How much time did each wait step consume? How does the batch procedure streamline your use of Gosset?

9.14 In Gosset User Manual pages 54–55, execute the example in Gosset section 7.8. Since this exercise concerns a mixture, the total of the different components must sum to 100%. This requires a constraint that the sum of factors equals 1. This exercise requires extra work from Gosset to construct the moment matrix. Be sure to wait. The results of the "`interp`" step show an additional column not part of the experiment design but related to the constraint.

9.15 In Gosset User Manual page 56, execute the example in Gosset section 7.9. Since this exercise concerns a mixture, the total of the different components must sum to 100%. This requires a constraint that the sum of factors equals 1. This mixture example imposes additional constraints. The results of the "`interp`" step show an additional column related to the constraint, which is not part of the design.

Exercises 9.16 through 9.23 entail the use of Gosset to create experiment designs akin to the PB screening design highlighted in the text. Recall that PB designs required some multiple of four runs: 4n = 4, 8, 12, or 16, etc. The maximum number of factors in a PB design is 4n–1 candidate factors; that is, the number of factors is 3, 7, 11, or 15, etc. For Chapter 8, we used a PB 12 run to screen six candidate factors. Begin execution via `./gosset 'cflags -w'`

9.16 In Gosset User Manual page 57, execute the example in Gosset section 7.10.1. This exercise designs an experiment for five continuous variables on a Factorial Cube. Describe how the design notably differs from the PB designs given in Appendix B.2. How many dimensions does the Factorial Cube have? How many runs did Gosset choose?

9.17 In Gosset User Manual pages 57–58, adjust the Gosset statements in order to execute the last case in Gosset section 7.10.1. This exercise designs an experiment for six continuous variables on a Factorial Cube. Describe how the design notably differs from the PB designs given in Appendix B.2. How many dimensions does the Factorial Cube have? How many runs did Gosset choose?

9.18 In Gosset User Manual pages 58–59, execute the example in section 7.10.2. This exercise designs an experiment for five discrete two-level variables on a Factorial Cube. Describe how the design notably differs from Exercise 9.16. How many dimensions does the Factorial Cube have? How many runs did Gosset choose?

9.19 Following the example in Exercise 9.16, use Gosset to create a design for seven continuous variables on a Factorial Cube. Describe if and how the design differs from the PB design given in Appendix B.2, Table B.4. How many dimensions does the Factorial Cube have? How many runs did Gosset choose?

9.20 Following the example in Exercise 9.18, use Gosset to create a design for four continuous variables and three discrete two-level variables on a Factorial Cube. Describe if and how the design differs from the design in Exercise 9.19. Describe if and how the design differs from PB design given in Appendix B.2, Table B.4. How many dimensions does the Factorial Cube have? How many runs did Gosset choose?

9.21 Following the example in Exercise 9.16, use Gosset to create a design for 11 continuous variables on a Factorial Cube. Describe if and how the design differs from the PB design given in Appendix B.2, Table B.3. How many dimensions does the Factorial Cube have? How many runs did Gosset choose?

9.22 Following the example in Exercise 9.18, use Gosset to create a design for seven continuous variables and four discrete two-level variables on a Factorial Cube. Describe if and how the design differs from the design in Exercise 9.21. Describe if and how the design differs from the PB design given in Appendix B.2, Table B.3. How many dimensions does the Factorial Cube have? How many runs did Gosset choose?

9.23 In Gosset User Manual pages 59–60, execute the example in Gosset section 7.10.3. This exercise designs an experiment for 23 discrete two-level variables on a Factorial Cube. Describe the limits of Gosset. In light of these Gosset limitations, how would you propose to create a screening design for 19 or more variables (continous, discrete, or mixed)? Note whether there are any limitations for PB designs. If not, as experimentalist, what are the advantages of choosing a standard PB design with 4n runs, where 4n exceeds the number of variables? What are disadvantages of excess PB runs? Hence the text, in Chapter 8, introduced PB with a 12-run design.

Exercises 9.24 through 9.27 practice the examples Chapter 9, Section 6. Begin execution via `./gosset 'cflags -w'`

9.24 Refer to this book (M&H) Section 9.6.2.1 to command Gosset to optimally select additional runs after an initial PB screening. Where do the initial PB screening runs appear among the commands? What is the purpose of the two commands which immediately follow? How many additional runs remain for Gosset to design? Explain why some of the original factors do not appear as variables as you execute Gosset.

9.25 Refer to this book (M&H) Section 9.6.2.3 and compare your Exercise 9.24 design results with the design listed in the textbook. Remark on how the total experiment design (including initial PB screening) compares to commands you supplied to Gosset. Rescale and convert your Gosset design to actual (operational) experimental parameters in realistic units.

9.26 Refer to textbook website http://wiley.com/??/chp9resources and download *.csv, *.R and *.Rhistory files.

Each *.csv file contains data for analysis and plotting using the R language.

The scatterplot3d.R file contains functions and subroutines for plotting (x, y, z) data in a three-dimensional graph on two-dimensional paper. The scatterplot3d.R file was created by a user of R and generously provided for free in the contributors folder on the r-cran site. The most recent updated version of scatterplot3d.R is available on the r-cran site. From the R-command window, select `File`, then select `Source R Code`, then locate and open the file scatterplot3d.R. This must be loaded before accessing any of its functions and routines.

Each *.Rhistory file lists a sequence of R language commands. The *.Rhistory file is conveniently read as a pure text file, for example using `gedit` or `notepad++`. The Rhistory commands can be copied and pasted into the R command window, individually or as a block. From the R-graphics window, select `History,` then select `Recording`, in order to retain in memory all the plots generated during the R session. A checkmark appears. Before closing the R session, use `pageUp` and `pageDown` in order to select and save journal-ready graphs in your preferred format. As a final step before closing the R session, select `File`, then select `Save history`. Devise a new dated name to save all the commands issued during the R session into a new *.Rhistory file. From this new *.Rhistory file, you can quickly analyze data and regenerate any graphs.

9.27 Open a new R session and load the file `scatterplot3d.R` as instructed in Exercise 9.26. Using the files `ch9plot3dCmd180915.Rhistory, ch94a_plot3d1.csv`, and `ch97a_plot3Dgos1.csv`, practice generating three-dimensional plots. With these commands you can duplicate Figures 9.4a, 9.6a, and 9.7a. Starting with the file ch9_pb.csv and the results of Exercise 9.24, practice generating the other three-dimensional figures.

Exercises 9.28 through 9.32 introduce more advanced features Gosset for creating experiment designs. Begin execution via `./gosset 'cflags -w'`

9.28 In Gosset User Manual pages 62–63, execute the example in section 7.13. The measurement domain is discrete, yet the modeling region is continuous. Primed variables match the unprimed variables. Compare to the experiment design of Exercise 9.7, which is fully continuous for both measurement and modeling regions. What advantages does the discrete measurement domain confer?

9.29 In Gosset User Manual pages 63–64, execute the example in section 7.14. The measurement domain is half of the modeling region. Continuous primed variables match the unprimed variables. Compare to the experiment design of

Exercise 9.7, which measured and modeled the whole region. What warnings would you give the client about extrapolating outside the measured region?

9.30 In Gosset User Manual pages 64–65, execute the examples in Gosset section 7.15. Here the measurement region is disconnected from the modeling region. One example measures from the volume of the Earth to model the response of the moon's volume. The second example measures from the volume of the Earth to model the response of the sun's volume.

9.31 In Gosset User Manual pages 69–77, consider the block designs in Gosset sections 7.18 and 7.19. Blocks can be balanced or imbalanced, complete or not.

9.32 In Gosset User Manual pages 77–80, data points can be optimally packed throughout odd-shaped regions with constraints. If there is no known model for the response, select the packing option.

9.33 In Gosset User Manual pages 84–86, execute the example in Gosset section 7.21.3, regarding a complicated consumer response. The idea of this experiment is to estimate consumer response to a new service with a variety of features. Features are contained in two-level or three-level discrete variables. This is a most complex experiment for business and economics.

10

Analyzing Measurement Uncertainty

10.1 Clarifying Uncertainty Analysis

There are a lot of misconceptions about Uncertainty Analysis and what it purports to do.

Results from a new experiment often fail to agree with results in the literature and also often show more scatter on repeated trials than was expected. There is more to Uncertainty Analysis than just drawing error bars on a plot of the data or calculating the root-mean-square (RMS) deviation of data relative to a curve fit. Those are simply techniques for describing the scatter in the results and provide no way of judging whether or not the observed scatter was "reasonable" or not – that is where Uncertainty Analysis comes in.

As used here, the term "Uncertainty Analysis" refers to a formal mathematical procedure for estimating the Uncertainty in a result that has been calculated using a known set of equations to process input data whose individual Uncertainties are known.

> Uncertainty Analysis cannot tell you how much scatter you *will* get in your experiment, but it will tell you how much of the scatter you *do* get can be explained by what you already know about your data.
>
> If your results scatter more than the Uncertainty Analysis predicts they should, that usually means that your experiment has problems that need to be resolved.

The approach to Uncertainty Analysis described here begins with Single-Sample Uncertainty Analysis. Although soundly based in statistical methods, it does not require a large number of trials: one observation at each test point is enough. The method requires very little work to implement in a spreadsheet or online data reduction program (DRP). It is not statistically driven. It is mainly judgment driven: what should I use for the Uncertainties?

This kind of Uncertainty Analysis is a powerful tool for experiments. We will demonstrate its use for debugging and analyzing heat-transfer experiments. It can be used in the early planning stages of an experiment to choose the best of several experimental approaches and to assess the suitability of the instrumentation. During the shakedown and qualification period, it can guide the search for heat losses and measurement errors and can provide the closure criterion for energy balance and mass balance tests. During the data production phase, it can be used to monitor the stability

Planning and Executing Credible Experiments: A Guidebook for Engineering, Science, Industrial Processes, Agriculture, and Business, First Edition. Robert J. Moffat and Roy W. Henk.

Companion website: www.wiley.com/go/moffat/planning

of the experiment. Finally, a well-documented estimate of the Uncertainty in the results allows other users to compare their results with yours, in a rational manner.

Uncertainty Analysis is often done "after the fact," mainly to satisfy contractual or editorial requirements. What a waste! It is far more useful when it is done first and used in all phases of the experiment: planning, executing, and reporting. In the planning stage, an *a priori* Uncertainty Analysis can assess whether or not the instrumentation is accurate enough to use for the intended purpose and whether the method is appropriate or not. During execution, online Uncertainty Analysis can be used to monitor the scatter on repeated trials, providing a warning that the results are drifting. Accurately reporting the Uncertainty in the result allows meaningful comparison with other data in the literature.

10.1.1 Distinguish Error and Uncertainty

The error in a measurement is defined as the difference between the True Value and its measured value. This definition is clear and absolute. If we already knew the True Value, however, we need not perform an experiment. Since we don't know the exact True Value, we cannot know the error for certain. This leaves us experimentalists with the task of estimating the error; we plan to estimate an Uncertainty band that encompasses the error 95% of the time.

As an aside: when do we ever know an absolute True Value? We do know that the speed of light in a vacuum is 299,792,458 m/s. We know absolute zero temperature and that heat cannot be drawn from something at absolute zero. However, there is no pure vacuum in this universe; neither have we achieved absolute zero temperature. A third instance truly occurs when we know the exact values for each member of the total population. A fourth instance may occur in everyday life or engineering; we assume that the True Value of a property is the value:

i) tabulated by the National Technical Information Service (NTIS), formerly the National Bureau of Standards,
ii) provided as baseline data, or
iii) referenced to one of the basic conservation laws of physics or engineering.

Other examples of the fourth instance: we assume the gasoline pump provides exactly one gallon of fuel at the advertised price; we assume the liter water bottle contains 1000 ml of fluid; we assume that when electrical probes contact, the resistance is zero. When we calibrate an instrument or qualify an experiment, keep alert to note your assumptions.

In most situations, we cannot talk confidently about what the error in a measurement is, we can only talk about what it *might be* – about the limits which bound the possible error.

The term "Uncertainty" is used to refer to "a possible value that an error may have." Kline and McClintock (1953) attribute this definition to Airy (1879), and it seems still an appropriate and valuable concept. The terms "Uncertainty interval" and "Uncertainty" are commonly used interchangeably and will be so used in this discussion, both referring to the interval around the measured value within which the True Value is believed to lie. The term "Uncertainty Analysis" refers to the process of estimating how much effect the Uncertainties in the individual measurements have on the calculated result.

10.1.1.1 Single-Sample vs. Multiple-Sample

The generic term "Uncertainty Analysis" includes two branches, single-sample and multiple-sample, each appropriate for a different type of experiment.

The terms "single-sample" and "multiple-sample" were used by Kline and McClintock (1953) to distinguish between experiments in which each variable is measured once per observation (raw data) and those in which each measurement is taken many times and averaged before the results are processed. The distinction is important, as this aspect of the experiment dominates the handling of Uncertainties from the beginning (when the component Uncertainties are measured in the test cell) to the end, when the results of the analysis are interpreted.

The distinction hinges on whether or not a "large" or a "small" number of independent data points are taken at each test point. In this era of high-speed digital data acquisition, the issue of independence takes on more subtle overtones than it had in the early 1950s when the term was coined. Consider, for example, a measuring system capable of acquiring data at 100 kHz, a system readily available under current technology. A 100 kHz sample rate would produce 100 readings in a 1 ms period. Applied to an experiment whose output varied at about 10 MHz (the system's own time constant), these 100 readings can each be regarded as independent measures of the process. The time between consecutive readings is large compared with the autocorrelation time of the signal. This set of 100 readings would be, then, a multiple-sample set of observations. This same equipment, applied to a system whose output changed only at 1 Hz, would produce a single-sample measure of the process, tainted, perhaps, by a multiple-sample contribution from the high-frequency random errors present in the measuring system.

Classification of a given experiment as single-sample or multiple-sample must take into account not only the number of observations made at each test point but the relationship between the data sampling rate and the frequencies in the process being studied.

Single-Sample Uncertainty Analysis has been described in the engineering literature by the works of Kline and McClintock (1953) and Moffat (1982, 1985). The techniques of Multiple-Sample Analysis are described by Abernethy and Thompson (1973) and by ANSI/ASME PTC 19.1-1985. We provide an abbreviated list of references and further reading at the end of this chapter.

Both Single-Sample and Multiple-Sample Uncertainty Analysis have the same final objective (to estimate the effect of the accumulated measurement Uncertainties on the accuracy of the result), even though somewhat different procedures are required. Either the Nth-Order Uncertainty interval of Single-Sample Analysis or the U_{95} interval of Multiple-Sample Analysis answers the same question: "How close to this result does the True Value probably lie, assuming that the right equations were used, that all the important variables have been included, and that the test situation is representative?"

In addition to this final result, each method produces auxiliary information about the experiment, mainly useful as diagnostics, during the developmental phase of an experiment or in monitoring its "health" during a long series of runs. The auxiliary outputs of the two types of Uncertainty Analysis are similar but not identical. For example, either the First-Order Uncertainty interval of a Single-Sample Analysis or the precision index of a Multiple-Sample Analysis provides an answer to the question: "How much scatter should be expected from repeated trials with this experiment, using the same equipment and the same procedures?" The answer is, "with 95% confidence, additional results are expected to lie within ± 2σ of the quoted results." For Single-Sample Analysis, σ is taken as the Standard Deviation of a set of at least 30 observations and is a property of the population of possible observations. For Multiple-Sample Analysis, σ is interpreted as the Standard Deviation of the mean of the set. The "result" referred to is the next "single observation" in a Single-Sample Analysis and the next "set of N observations" for a Multiple-Sample Analysis.

An Uncertainty estimate of either type is only as good as the equations on which it is based. If those equations are incomplete and don't acknowledge all of the significant factors which affect the result, or if falsely low values are used for the component Uncertainties, then the analysis will underestimate the Uncertainty in the result. On the other hand, if the component Uncertainties are exaggerated, then the analysis will overestimate the Uncertainty.

10.1.2 Uncertainty as a Diagnostic Tool

The most common, and most visible, use of Uncertainty Analysis is in reporting results to the technical community through publications, but it must be noted here that it has far broader uses. During the shakedown period of an experiment, it is a powerful diagnostic tool in seeking out the sources of residual errors. In the early stages of an experiment, for example, when comparing the first results from a new test rig with those from an existing baseline set, the most frequent question is: "Is the difference I see significant or just a consequence of the Uncertainties in my measurements?" In this phase of an experiment, Uncertainty Analysis provides clear, unambiguous guidance. If the observed difference exceeds the expected Uncertainty interval, then the observed difference is probably significant. Even earlier in an experiment, Uncertainty Analysis can be used to help choose the most reliable technique for a given measurement or to identify which are the critically important instruments in a system (and thereby determine where expensive instruments are needed!).

10.1.2.1 What Can Uncertainty Analysis Tell You?

Uncertainty Analysis cannot tell you how much scatter you will get from your experiment, but it will tell you how much of the scatter you do get can be explained by what you already know about the raw data.

It is important to keep in mind that the scatter in results is not all caused by measurement Uncertainty – there are many other factors.

For example: Figure 10.1 is a classic textbook image showing the Nusselt number as a function of Reynolds number for a very simple situation: heat transfer from a round cylinder perpendicular to an air flow. Note that in order to present the broad range of Reynolds numbers, the horizontal scale is logarithmic; likewise, the vertical scale is logarithmic for the Nusselt number. Most of the experiments that contributed to this collection probably claimed Uncertainties of ±5–10% (if they mentioned the issue at all!). By close inspection, noting the logarithmic scales, the figure shows ±50% Uncertainty at the higher end and ±20% Uncertainty in the lower range. What does this discrepancy mean?

It means that at least some of those experiments were influenced by factors not considered: tunnel blockage, turbulence intensity, nonuniformity in the flow, etc.

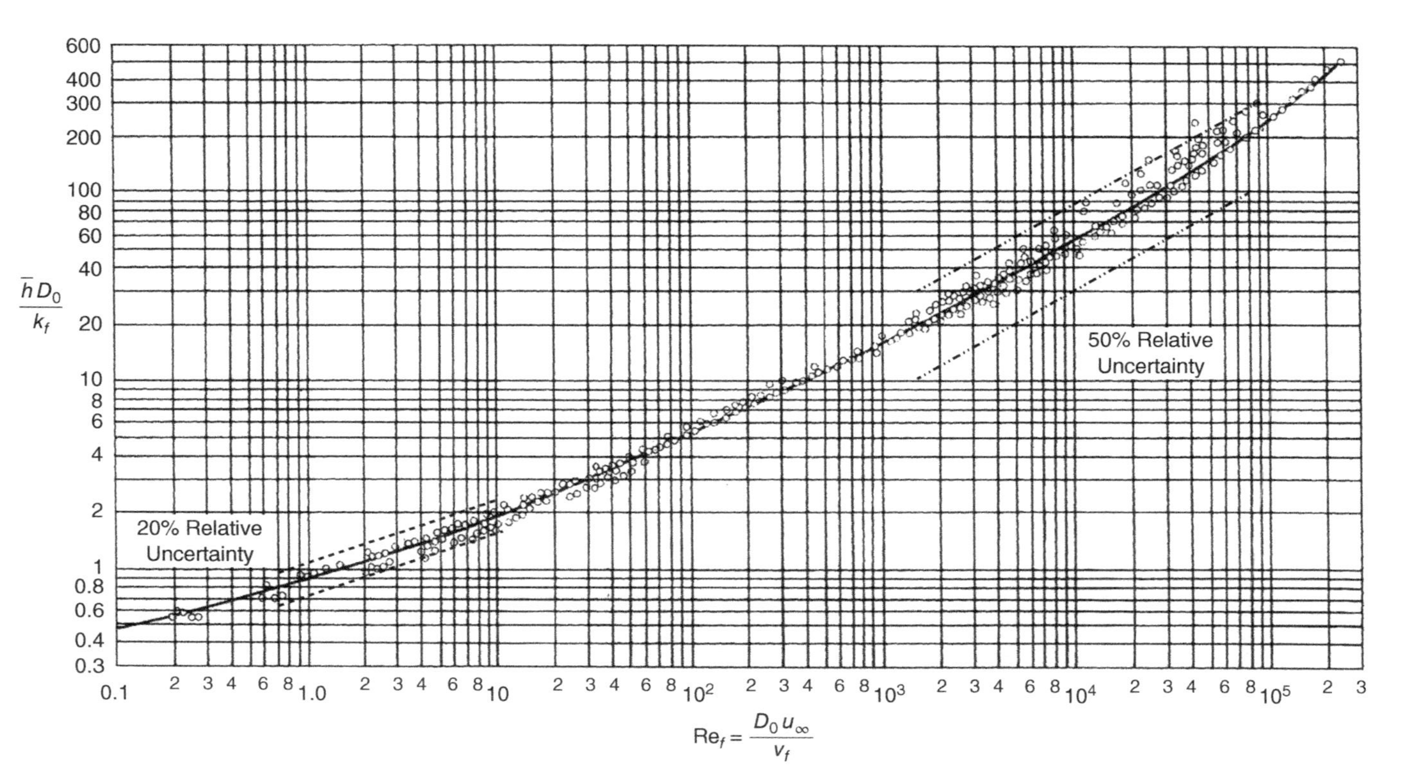

Figure 10.1 Typical heat-transfer data from a simple situation: a round cylinder perpendicular to an air flow.

Imagine that you have done an Uncertainty Analysis on your experiment and have repeated the experiment several times, and you are now trying to reconcile how the scatter in your results exceeds the predicted Uncertainty interval. NOTE: When I say "exceeds," I mean exceeds in a statistically significant sense – not merely splitting hairs, but significantly bigger.

There are three possible outcomes:

- The observed scatter is significantly less than the predicted Uncertainty interval. (You probably were padding your estimates of the individual Uncertainties in the input data. You should re-examine your input descriptions.)
- The observed scatter is about equal to the predicted Uncertainty interval. (Your experiment is behaving about as you thought it would – there are no surprises.)
- The observed scatter significantly exceeds your predicted Uncertainty interval. (There is something going on in your experiment that you don't know about: you have an opportunity to learn something about the system you are studying.)

10.1.2.2 What Is Uncertainty Analysis Good For?

Uncertainty Analysis is a powerful tool for an experimenter (recall Figure 1.1):

- Used in the planning stage of an experiment, it can reveal whether or not a proposed experiment plan will yield usefully accurate results with the proposed instruments.
- Used during the shakedown period (the initial phases of running the experiment), Uncertainty Analysis can be used to check the stability of the apparatus and the test procedure: is the repeatability of the data as good as predicted by the Uncertainty Analysis?
- At the "qualification stage," Uncertainty Analysis can be used to quantitatively answer the question: do the present data agree with the expected values within the expected Uncertainty interval?
- During data production, Uncertainty Analysis can be used to check for drift in the experiment by looking at the scatter in repeated trials at a "baseline" test point.
- Finally, at the reporting stage, Uncertainty Analysis can be used to estimate the overall accuracy of the results, for presentation to others.

10.1.2.3 Uncertainty Analysis Can Redirect a Poorly Conceived Experiment

Now that Uncertainty Analysis is widely known, people expect it, and many contracts require it. If you plan a test series and don't assess the Uncertainties before you run the experiment, you run the risk of looking a bit foolish if your results show too much scatter to be useful. Uncertainty Analysis cannot make scatter go away, but by the end of the shakedown period, you will know enough about the system to predict the expected Uncertainty. You then have an opportunity to modify the test rig before you run the production data.

10.1.2.4 Uncertainty Analysis Improves the Quality of Your Work

Simply being aware of the importance of Uncertainties will focus your attention on the quality of your instruments and the care with which you run your experiments.

Combining Uncertainty Analysis with the idea of checking against baseline data gives you a powerful pair of tools. Uncertainty Analysis tells you how close you should expect to be; comparing against a baseline tells you how close you are.

When you are as close as you have a right to expect, you are doing a good job.

Having baseline data to check against is half the battle. (When you don't know what to expect, anything looks OK.) The other half is being able to recognize the significance of an observed difference. Suppose you know that the answer should be 5.00 and you get 5.15; is that close enough? That depends on the Uncertainty prediction. It may be that you cannot expect to be closer than ±0.15 using the instruments you have, and you knew this from the start (due to your planning Uncertainty Analysis). Conversely, you might have expected to see agreement within ±0.0015 based on the original Uncertainty Analysis. In that case, you know you must do more work on the experiment to clear up sources of error.

10.1.2.5 Slow Sampling and "Randomness"

Often the appearance of random scatter in data is caused by sampling too slowly. We touched upon this in Sections 2.4.3 and 5.4 but revisit it from the viewpoint of Uncertainty. For example, consider two functions: $Z = \sin(\omega t)$ and $Z = \sin(\omega t) + 0.1 \sin(10\omega t)$. Can you tell these apart? Perhaps, and perhaps not! Consider the situation shown in Figure 10.2a. If you expect to see a simple sine function, and yet have the data shown in Figure 10.2a, it would be easy to exclaim, "There is the sine!" and claim the scatter is simply noise or "experimental error."

Now consider, your experiment may require determining the impact of high frequency scatter on a steady (or slow) process. How do you adjust sampling to quantify scatter? There are a host of tools and techniques already developed based on the Fast Fourier Transform (FFT) and on long time-series of raw data. Choose the sampling rate to be at least twice the highest frequency of interest. The Nyquist frequency, equal to half the sampling rate, is the highest detectable frequency. Choose the length of each time-series to be multiple ($N \geq 16$) periods of the longest time-constant of your process. The FFT calculates each frequency's contribution to the signal. Sampling at a higher rate revealed the high frequency component, as Figure 10.2b shows.

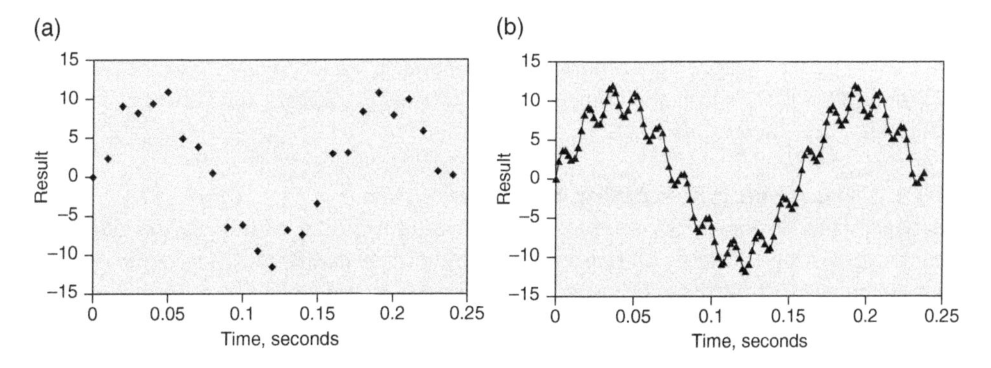

Figure 10.2 (a) Slow sampling on a rapidly changing signal gives the appearance of scatter in the data. (b) Rapid sampling on the same signal shows that the data are smooth and continuous.

Uncertainty analysis applies to frequency results. When a FFT is applied to the raw data, the Uncertainty of the amplitude at each frequency is 100%. Be not dismayed. Forming a power spectrum, to average FFTs, provides the reason for sampling N multiple periods. The amplitude Uncertainty at each frequency improves by $1/\sqrt{N}$, just like uncertainty of the mean in Section 7.1.9.

Perhaps the Motivating Question for your experiment focuses only on steady-state or stationary conditions. We advise against blaming randomness. We rather advise that you view the results of your experiment as completely deterministic – that the processes obey natural laws for all factors that impact your experiment. If your experiment deals with animated creatures, human nature, economics, information-based decisions, your experiment likely does transcend physical determinism. Placebo and nocebo effects may pertain. We urge heeding the cautions and warnings of Ioannidis (cf. Chapters 1 and 2).

For the rest of us, who can expect processes to be naturalistically deterministic, we advise against ignoring "scatter". For example, in heat transfer at the continuum level there might be contamination by dust or insects, but there are no random processes. Rather, we urge a diligent search for the source of higher-frequency components and its removal or quantification. Otherwise, the ignored factor inflates the Uncertainty associated with the experiment.

10.1.2.6 Uncertainty Analysis Makes Results Believable

When you present your results, you have an obligation to describe them as accurately as you can – not by increasing the number of digits you present, but by describing the range within which your result might have fallen, by chance alone, if you had repeated the experiment many times. For example, quoting the heat-transfer coefficient as $50.0\,W/m^2C \pm 10.0\,W/m^2C$ at 20:1 odds, means a great deal more than simply quoting $h = 50.0\,W/m^2C$. An Uncertainty estimate is an absolutely necessary part of a quoted result.

10.1.3 Uncertainty Analysis Aids Management Decision-Making

Uncertainty Analysis plays an important role in setting the performance goals for products sold with warranties. Consider a company manufacturing a product that is sold with a performance warranty. Customers expect that the "guaranteed" performance will be realized in the field. Unfortunately, even though the customer's field testing may have higher Uncertainty than the manufacturer's test-cell testing, customers tend to trust their own results and complain when the product appears to not meet its guaranteed performance, based on their own tests. In part, that problem is political, not technical, but it is real, in any case.

Different entities within the company have different interests and different roles in addressing the Uncertainty issue.

10.1.3.1 Management's Task: Dealing with Warranty Issues

Management's responsibility is to balance the marketing and warranty issues against the realities of testing and settle on reasonable, economically justifiable Uncertainty goals. If a performance warranty is to be issued, the guaranteed value must not be set so high that the risk of not meeting that goal is unacceptably large. The task of choosing the warranty value involves the accuracy and repeatability of both field tests and factory tests. Field tests cannot be controlled as well as factory tests, so the variance in field test results is larger, yet it is likely to be the field tests that the customer relies upon most heavily.

The situation is illustrated in Figure 10.3, showing the cumulative probability distributions for a set of hypothetical field tests and factory tests conducted on a single compressor whose efficiency truly is 80%. The distributions shown are entirely due to experimental variances introduced by the testing procedures (uncertain calibrations, unsteadiness in the system, uncontrolled variations in test conditions, etc.). The distribution of values from a population of compressors would show even more dispersion due to manufacturing variability.

Note that if the warranty efficiency was set at 0.795, just 97.5% of the time would this unit pass the factory test based on company experience. This same unit would pass the field test only 84.5% of the time, even though the unit had a true efficiency of 0.80. In order to statistically ensure that 97.5% of units pass the field test, a warranty value of 0.79 must then be specified.

The point of this discussion is that the experimental Uncertainties of both the factory tests and the field tests are important in assessing the risk involved in choosing the warranty value. The smaller the variance in both testing situations, the more accurately warranty judgments can be made. The variance due to testing can be reduced either by development work on the test procedures or improving the data interpretation program. In the factory test setting, test procedures and equipment can be refined so the test results are more repeatable and more accurate. In the field, where test conditions are difficult to control, the data interpretation program can correct for variations that cannot be controlled. In both of these situations, Uncertainty Analysis is key in questions of assessing warranty compliance.

Note: Figure 10.3 should not be taken too seriously. The concept is valid, but the distribution of values shown in that figure is based upon a theoretical normal distribution which may or may not apply to real hardware. Values in a normal distribution are symmetric about the mean value of the population, but the hardware performance measurements may not be symmetric. Development of a realistic statistical base upon which to base warranty judgments may take some time (and a good experimentalist and statistician!).

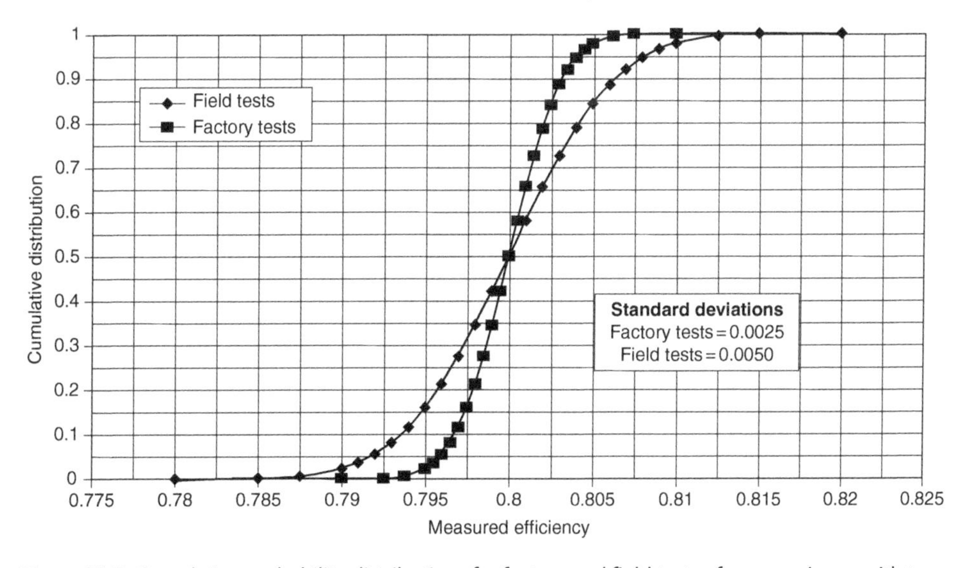

Figure 10.3 Cumulative probability distributions for factory and field tests of one specimen, with true 80% efficiency. Variances of Factory and Field tests differ due to testing conditions.

10.1.4 The Design Group's Task: Setting Tolerances on Performance Test Repeatability

The group responsible for advancing the hardware design must rely upon factory test results to accept or reject proposed design changes. To the design group, the most important aspect of test performance is repeatability. In order to recognize an improvement by comparing the results of two tests, the difference in the results must be larger than 1.5 times the long-term repeatability in the individual tests. For example, if the long-term repeatability of the tests were ± 0.1% (at 95% confidence), differences smaller than 0.15% cannot be regarded as significant based on a single pair of tests.

The design group and the performance test group must negotiate an allowable value for the long-term repeatability of the test (a sort of "Long-Term First-Order Uncertainty," as will be discussed later). This goal can be expressed in two ways: (i) a tolerance on the calculated parameter (e.g. compressor efficiency ± XX%), or (ii) a tolerance on the individual measurements from which the parameter will be calculated (e.g. T ± XX, etc.). The "hidden issue" here is: Which group does the Uncertainty Analysis? The Uncertainty Analysis links the Uncertainty in individual measurement to the Uncertainty in the calculated parameter using the data interpretation equations.

The doctrine of "Separation of Powers" seems to suggest that the design group should ask for what it needs (an accurate measure of efficiency) and let the performance test group decide how to get it. The design group has the right to ask for proof that this has been successful. In some cases, there may be more than one way to make the same measurement, and the performance test group is in the best position to know the alternatives. Of course, selection of the test method can be part of the negotiation between the design group and the performance test group.

In either case, regardless of which way the tolerance is described or who does the Uncertainty Analysis, both parties must agree that the goal is reasonable and necessary. Setting the repeatability criteria needs to be an iterative procedure. The design group will undoubtedly want zero tolerance, and the performance test group cannot deliver that. Reducing Uncertainty costs both time and money. It is not uncommon to find that the cost of a test doubles each time one cuts its Uncertainty in half. Improving from ±2% Uncertainty to ±1% should not be viewed as a 1% improvement in accuracy – it is a 50% reduction in every error source! Tentative goals must be costed out by the performance test group, and the proposed tolerance gradually relaxed until an acceptable compromise has been reached.

Each test request issued by the design group to the performance test group should record the (agreed-upon) allowable Uncertainty intervals and confidence levels (generally 95% confidence) expected, whether on individual measurements or on calculated parameters. It then becomes the responsibility of the performance test group to achieve this goal.

10.1.5 The Performance Test Group's Task: Setting the Tolerances on Measurements

The test engineer's responsibility is to achieve the agreed-upon allowable Uncertainty. If the test request is in terms of individual measurements, the test group's responsibility is to document the residual fixed errors in the calibrations used, and the First-Order Uncertainty observed during the tests. If the test request is in terms of the allowable

Uncertainty in a calculated output parameter (such as efficiency), more is required. If this is a new type of test request, the test group will have to have access to the data interpretation code that will be used in calculating the output parameters and freedom to run that code to assess the Uncertainties and sensitivity coefficients. Before any tests are run, representative data values should be processed, accompanied by tentative tolerances, and Single-Sample Uncertainty Analysis procedures used to predict the consequent Uncertainty interval in the result.

If existing instrumentation on the test rig cannot achieve the required Uncertainty allowance, then the test cannot be run. Changes must be made. It may be necessary to purchase more accurate instruments, modify the test rig, or modify the data interpretation equations by adding corrections for some of the error sources.

10.2 Definitions

Some of the terms used in discussing Uncertainty Analysis must be rather strictly defined to avoid confusion. The following definitions will be used.

10.2.1 True Value

The objective of the measurement. There are two aspects of this definition: (i) the conceptual definition – the full name of the thing you are trying to measure, and (ii) the value itself. The "value" is easy to describe, the full name is not. For example, when the phrase "coolant temperature" is used, the implied full name (usually) is "mixed mean temperature of the coolant at this location after correction for recognized heat losses." The accurate identification of the True Value is vitally important. Until you acknowledge the full name, you cannot list all of the possible errors there might be in its measurement, and you cannot estimate the Uncertainty in that measurement. (See Section 10.3.3, "Specifying the True Value.")

10.2.2 Corrected Value

Your best estimate of the numerical value of the True Value, after you have applied all the corrections you know about. Applying corrections for all the errors that you know about is not "cooking the data," it is good engineering: it is part of the test engineer's job.

10.2.3 Data Reduction Program

The complete set of equations by which the Corrected Value is calculated from the raw data (the indicated values). The DRP includes all the equations needed to make your best effort at correcting for all the error sources you know about: system disturbance errors, system/sensor interaction errors, calibration errors, and conceptual errors. (See Section 10.3.2, "Sources of Errors: The Measurement Chain.")

10.2.4 Accuracy

The absence of error.

10.2.5 Error

The difference between the True Value and the reported value.

10.2.6 XXXX Error

The component of the total error attributed to the cause named XXXX.

10.2.7 Fixed Error

An error that does not change on repeated readings at the same set point, over the entire period of the experiment. Fixed errors can be detected only by calibration against a reference value.

10.2.8 Residual Fixed Error

An estimate of the possible range of the remaining fixed error after you have corrected the indicated value by your best estimate of its "average fixed error," determined by the calibration. The best estimate of the residual fixed error is equal to the Uncertainty in the calibration correction, at 20:1 odds. In many cases, there is no statistical evidence regarding the Uncertainty in the calibration correction, and it must be estimated based on whatever information is available.

The estimated residual fixed error will later be used in combination with statistically determined random error estimates, made at the 95% confidence level (20:1 odds), but the estimated residual fixed error is not, itself, a statistically justifiable value: it may only be a "best estimate." The value you quote for the estimated residual error should be such that you would be willing to bet $20.00 of your money against $1.00 of mine that the residual fixed error lies within the interval you describe. The interval should be zero centered and symmetric: i.e. the interval should be centered about zero error. We strongly advise against "one-sided" estimates like "plus nothing, minus 0.5 units." The probability of a positive residual should be just as likely as the probability of a negative residual.

In Uncertainty Analysis, we work only with the residual fixed error. If no calibration correction is applied, then the whole calibration error becomes residual. For example, if a calibration report simply says that a particular voltmeter always reads within ±0.005 V of the true electromagnetic force and does not give a curve or equation by which its readings can be corrected, then the entire ±0.005 must be treated as the residual fixed error, at every point in its range. If a calibration curve is provided, and claimed to be accurate within ±0.0005 V, then we have two choices: (i) live with the overall error tolerance of ±0.005, or (ii) apply the correction suggested by the curve and use the lower residual fixed-error tolerance of ±0.0005 at every point in its range.

10.2.9 Random Error

An error is called "random" if the Observed Value changes randomly on repeated observations, even when sampled as rapidly as possible with the equipment at hand. Errors that must be considered "random" for a given instrumentation suite can be identified by taking a set of repeated readings at the same set point and plotting the data as a function of time. If the readings scatter randomly about a fixed mean value, there are random errors in the system. It is this random component of error that we deal with in Uncertainty Analysis.

To reiterate Section 10.1.2.5, don't expect or excuse really random processes in fluid mechanics or heat transfer at the continuum level. Rather, expect deterministic causes. Expect that "apparently random" fluctuations in data are the result of sampling too slowly, are contamination in the flow or impurities on the surface, are generated by the electronics, or were picked up as electrical noise from the environment. If the sampling frequency is high, compared with the time constant of the apparatus, and there still appears to be "random scatter" in the data, that suggests electrical or instrumentation problems.

10.2.10 Variable (but Deterministic) Error

An error which changes systematically, not randomly, on repeated observations at the same set point. Variable errors can be detected by taking a set of repeated readings at the same set point and plotting the data as a function of time. If the data show a smooth trend with time, this is evidence of a variable error. The cause of a variable-but-deterministic error can often be identified by consideration of the processes involved (i.e. radiation error or velocity error in a gas temperature measurement). A zero-centered, symmetric correction should be applied to the Observed Value (a "best estimate" of the possible error) and the estimated Uncertainty in the correction used as the residual Uncertainty.

10.2.11 Uncertainty

A possible value that the total error may have, including the effects of fixed and random errors.

10.2.12 Odds

One way of describing the confidence that you have in your estimate of the Uncertainty. If you describe a measured temperature as 24.5±0.5 °C, at 20:1 odds, this means that you are willing to bet $20.00 of your money against $1.00 of mine that if you made another 20 observations at the same set point using the same instruments, 19 of the 20 new values would lie between 24.0 and 25.0 °C.

10.2.13 Absolute Uncertainty

The Uncertainty expressed in measurand units, i.e. δZ = XX degrees, W/m^2-K, CFM etc.

10.2.14 Relative Uncertainty

The Uncertainty expressed as a fraction of the value of the measured value, $\delta Z/Z$ = XX%.

10.3 The Sources and Types of Errors

The Uncertainty attributed to a measurement is an estimate of the possible error in that measurement. We cannot begin to estimate the Uncertainty until we know what error sources should be considered. In this section, we will discuss the sources of errors and the types of errors using a classification system which will allow a systematic examination of any proposed experiment.

There are two main issues to deal with in trying to classify errors: (i) what is the source of the error (i.e. what sort of mechanism causes the error), and (ii) what type of error is it (fixed, random, or variable but not random)?

10.3.1 Types of Errors: Fixed, Random, and Variable

An error source is categorized as "fixed," "random," or "variable" depending on whether the error it introduces is steady, changes randomly during the time of one complete experiment, or changes smoothly during the time of one complete experiment.

Calling an error "random" implies that each observation of the error is independent of the preceding observation, except, perhaps in a statistical sense, when the error in question appears to have no systematic variation. Most errors that are not "fixed" are "variable" but not really "random." Very few processes in nature are truly random; most processes proceed smoothly from one state to another. If, however, we sample any variable process very slowly relative to its autocorrelation time, the resulting data will *appear* to be random. Usually, measurements in fluid mechanics or heat transfer that display apparently random variability come from situations where the interval between observations is longer than the autocorrelation time of some variable that affects the process.

It is useful to recognize a separate category for these variable errors: "variable but deterministic." This class includes all errors which change during the experiment but not randomly. For example, the radiation error of a gas temperature sensor depends on the wall temperature – if the wall temperature changes during the test (for example, due to a change in test-cell ventilation rate), then the error will change – but not randomly.

Classification of an error source as "fixed," "random," or "variable but deterministic" thus depends not only on the characteristics of the source but also on the frequency of the observations. In this discussion, the term "variable error" will include both the random and the deterministic components of unsteady error.

10.3.2 Sources of Errors: The Measurement Chain

To discuss where the errors come from, it is useful to have a picture of the information flow in the measurement process.

Errors are introduced into measurements from the instrumentation and from corrections which are (or should be) applied to the Observed Values to learn the True Values of the measurements.

Any mechanism that affects the result we seek can be considered to be "the source of an error" in our measurements.

Figure 10.4 shows a diagram of the Measurement Chain (Moffat 1973), which outlines the steps information passes through going from the Undisturbed Value (the value the named property would have had in the system if the act of measurement had not altered it)

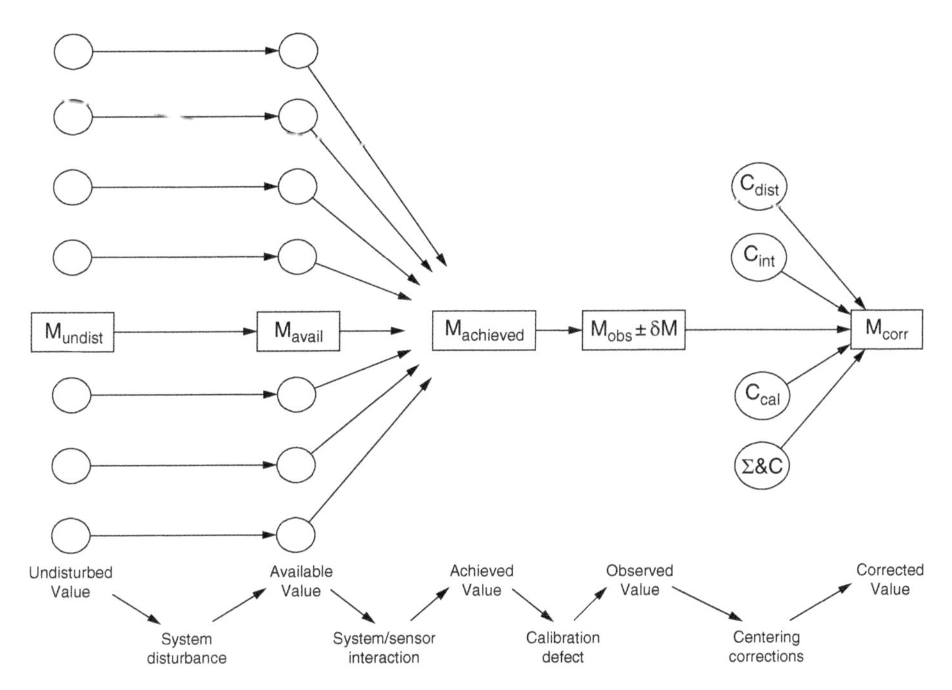

Figure 10.4 The measurement chain.

to the Recorded Value. The Measurement Chain provides the nomenclature with which we can classify the sources of error in a measurement.

The horizontal center line of the Measurement Chain diagram represents the sequence of apparent values of one measurand, the "considered measurand." Each step in the sequence (from left to right) increases the amount of "reality" allowed. The vertical columns of circles represent other measurands (properties of the system), some of which might affect the measurement being considered. Some of these might be observed and recorded during the experiment, but many are not even recognized as possibly having an effect.

Make the considered measurand clear.

For clarity in discussing this figure, let's assume the considered measurand is a gas temperature (the horizontal center line). The columns of circles represent all of the descriptors of the system whose values might affect the measured gas temperature (e.g. the wall temperature, gas velocity, probe emissivity, probe recovery factor, etc.).

The stations along the Measurement Chain are (from left to right): the Undisturbed Value, the Available Value, the Achieved Value, the Observed Value, and the Recorded Value.

Notice that the True Value does not appear in the Measurement Chain. The True Value is in a class by itself. The error is defined against the True Value.

Alert: Within statistical analysis there exists a procedure entitled "Hypothesis Testing" which distinguishes "type 1 errors" and "type 2 errors." Type 1 and type 2 errors are distinct from the Uncertainty Analysis concepts in this chapter. Whole books are dedicated to Hypothesis Testing; in this text we describe and employ it in a later chapter.

Often engineering designers adopt (by convention) a True Value which is not a directly measurable value. Such a True Value exists as a conceptual value but not as a physical, observable value. For example, we note the mixed mean temperature of a gas flow in a duct. The mixed mean temperature is useful (and standardized) for engineering design, but the temperature itself is not measurable at any single location.

Let's elaborate the stations of the Measurement Chain.

10.3.2.1 The Undisturbed Value

The Undisturbed Value of the gas temperature is the value which would have existed at the location of the sensor, if the instrumentation had not been installed (i.e. no sensors, anywhere, except control sensors required to operate the system). The Undisturbed Value is not available to the sensors and it is not necessarily equal to the True Value. The Undisturbed Value is simply that: the value of the considered measurand, at the location of the sensor, in the undisturbed state.

10.3.2.2 The Available Value

The Available Value is the value of the measurand at the sensor location when the sensors are present – not just the considered sensor, but all of them. This is the value the considered sensor is exposed to.

There is often a difference between the Undisturbed Value and the Available Value because putting sensors into a system may alter the temperature distributions, the flow distribution, and the pressure level.

If we have defined the objective of the measurement to be "to measure the Undisturbed Value of the gas temperature," the amount that the temperature changed due to the presence of the sensors would have to be considered as an error in the measurement.

The difference between the Undisturbed Value and the Available Value is called a "system disturbance error" for the measurement in question. System disturbance errors can be estimated analytically if enough is known about the system.

The system disturbance error should be treated as a fixed error, since it is simply a calculation based on a known equation with a set of estimated inputs: there will be no scatter in repeated estimates if the inputs are kept the same.

10.3.2.3 The Achieved Value

A sensor exposed to the Available Value of gas temperature will generally not be exactly at the same temperature as the gas but will be either higher or lower than the Available Value, due to radiation error, velocity error, or conduction error. The temperature the sensor does reach is called the "Achieved Value."

The difference between the Available Value and the Achieved Value is due to interactions between the system and the sensor and is called a "system/sensor interaction error." System/sensor interaction errors are governed by the laws of heat transfer and fluid mechanics, and their magnitudes can be estimated quite accurately in most cases. The system/sensor interaction error is a variable but deterministic error. It is a fixed error if and only if the peripheral conditions that cause the error (wall temperature, gas velocity, etc.) are constant.

The Available Value (and hence the Achieved Value) may not be steady with time, hence the signal passed down to the instrumentation system may not be steady. Sticking to the gas temperature example, a correction should be applied to account for the time constant of the sensor. That correction will rescale the amplitude of the variation and phase shift its peaks with respect to time. That constitutes a correction for a system/sensor interaction (the time constant). With only one sensor, this unsteadiness introduces a conceptual problem: how to define "the" gas temperature. With two or more sensors, the problem becomes more complex.

In general, with several probes, each will have a different time constant. A set of observations acquired at the same instant of time instantaneously (e.g. with a sample-and-hold data acquisition system [DAS]) will represent observations from different real times (because of the difference in time-shifts for the sensors). The relationship between the temperatures of the different sensors should not be examined in the "raw" state but only after each reading has been corrected for the time-shift due to its response lag.

10.3.2.4 The Observed Value

The output of the sensor depends on the Achieved Value of temperature in the sensor. That output is transmitted to the instrumentation system and interpreted using the calibration curve assigned to the sensor. The resulting estimate of temperature is called the "Observed Value." If there are defects in the calibration of the sensor, or if electrical noise is picked up by the system, or if the system itself is out of calibration, the Observed Value may not reflect the Achieved Value.

The difference between the Achieved Value and the Observed Value is called an "instrumentation error" (or calibration defect). The instrumentation error may include both fixed and random error components, depending on the type of nuisance encountered. It might be possible to correct for the fixed-error portion, if its cause were known, but the random portion will simply become part of the variance in the Observed Value, indistinguishable from unsteadiness in the process.

10.3.2.5 The Corrected Value

The engineer responsible for the accuracy of the measurements should make corrections for all of the error sources for which the mechanisms are known and for which the values can be reasonably estimated.

The corrections themselves are not perfectly accurate, and applying them leaves some residual error. These residual errors are fixed errors, because they don't cause scatter on repeated observations (all conditions remaining constant). The corrections arise from equations used to estimate the various error sources, and those equations don't yield different results each time they are considered, assuming their input data remain constant.

The Uncertainty which should be reported, along with Corrected Value, is the best estimate of the probable difference between the Corrected Value and the True Value. This Uncertainty estimate has two components: (i) the accumulation of all the residual fixed errors left over from all the applied corrections, and (ii) the random component introduced by unsteadiness in the system and its instruments. Those components are combined using the Root-Sum-Square (RSS) method to be described later.

It is worth repeating, however, that the True Value does not even appear in the Measurement Chain. The True Value may not exist as a measurable entity: it may be a calculated construct, like the "average velocity" or the "mixed mean temperature."

If the True Value is not a directly measurable parameter, there may be large errors in the assumptions used in calculating the True Value from the measurements. Those errors are conceptual errors, not physical ones, and there is no limit to how large they can be.

For example, engineers frequently use the "bulk mean temperature" of a fluid stream in energy balance equations relating to heat transfer or engine performance. The "bulk mean temperature" is defined as the product of local density, velocity, specific heat, and temperature, integrated across the flow path. Most measurements that are used as "bulk mean temperature" are actually the average of only a few data points from the temperature field and a few samples of the velocity field, along with average density and average specific heat. The point measurements might individually be very accurate, but if the distribution of temperature or velocity is not uniform, the estimate of "bulk mean temperature" can be seriously in error.

It is vital to the success of any Uncertainty Analysis that the True Value be clearly defined, so the investigator knows how many of the error modes described above must be accounted for in estimating the total Uncertainty in the reported value. This is the topic of the next section.

10.3.3 Specifying the True Value

The error in a measurement is defined as the difference between the Observed Value and the True Value of the intended measurand. Generally, the error has several components, each introduced by one aspect of the situation. The problem is to identify all of the possible error sources.

The Observed Value is easy to identify – it is the value returned by the measuring system modified by the appropriate corrections: it exists.

The True Value can only be identified by the end user of the data: only the end user knows what the data will be used for. There is no realistic way to assign an Uncertainty to a measured temperature, for example, without knowing what role that temperature will play in subsequent calculations.

To illustrate, consider a thermocouple probe installed downstream of a gas turbine combustor, in a duct with cold walls, Figure 10.6. There are (at least) four options which could be chosen for the definition of the True Value of that temperature. The classification and nomenclature used here follows that used in the Measurement Chain

i) The temperature of the thermocouple junction (the Achieved Value). This value might be entirely appropriate for a controls person.
ii) The gas temperature at the junction location (the Available Value).
iii) The gas temperature that would have existed at the junction location if the instrumentation system had not disturbed the distribution (the Undisturbed Value).
iv) The mass-flow-weighted average temperature the gas in the duct would have had, at the axial location of the thermocouple probe, had the instrumentation system not disturbed either the temperature or the flow distribution (the True Value). This is the value needed for an energy balance or a cycle analysis.

The number of effects which must be counted as possible error sources depends on which of these options is adopted as the definition of the "True Value." Different choices lead to different lists of errors which need to be handled.

10.3.3.1 If the Achieved Value Is Taken as the True Value

If the Achieved Value is elected as the "True Value," then only the instrumentation errors need be considered. This includes all of the fixed and variable errors introduced by every component of the measuring system due to such effects as errors in gain (fixed error), ripple in power supplies (random), and drift due to temperature changes in the instrument (variable but deterministic). Keeping track of this collection of sources is no small problem and has been the main focus of the measurement literature for many years.

Errors introduced by the measurement system can be estimated by either experimental or analytical methods. Experimentally, one can do an "end-to-end" calibration by providing a known and constant input to each channel of the measuring system and observing that channel's output as a function of time – fixed errors are evidenced by an offset of the mean value and variable errors by variations in the output. Typically, if the fixed errors exceed some acceptable level they are "zero'd out" by pretest adjustments to the system gain using a two-level check: "zero input" and "full scale for the test." *The tolerance used to accept imperfect performance establishes the Uncertainty in the measuring system fixed error* – this tolerance value should be recorded for each channel of the measuring system and used as one of the inputs to the Uncertainty Analysis. Good practice requires both pre- and post-test checks. When a change is found, the initial tolerance plus one-half of the change can be used as the estimate of fixed error.

The variance of a set of repeated pretest trials can be used to measure the variable error introduced by the measuring system, but the trials must cover a time interval representative of a true test, and the environmental conditions must be representative as well, else the "variable but deterministic" components of variable error will not be picked up. A set of readings taken over a short time interval can detect only the variability which occurred during that interval. The variable error is usually not measured except specifically as a check on measuring system electrical performance, since both Single-Sample and Multiple-Sample Uncertainty Analyses pick up evidence of variable errors during their routine data acquisition.

The analytical approach to estimating measuring system errors requires an estimate of the fixed and random components of error from each component of the system, usually based on the manufacturer's specifications. Since few of those are described in terms of statistical properties of the population of instruments, this involves a great deal of interpretation. Taylor (1986) presents a systematic approach to the identification and estimation of measuring system errors based on manufacturer's specifications. Figure 10.5 shows one of the worksheets used by Taylor in organizing the error analysis of a thermocouple channel.

A separate worksheet must be generated for each type of measurement channel used. There must be a column for every component in the channel (including one for lead wires, one for connectors, etc.). Each column must contain as complete as possible a documentation of the characteristics of that measurement channel.

Data to be entered on these worksheets must be converted from the form given in the manufacturer's specifications into an estimate of the Standard Deviation, σ, which might describe the data. This is not straightforward. For example, if a manufacturer might claim that a certain component is "accurate within 0.1% of full scale" but there is no way to know whether 0.1% represents a 2σ or a 3σ excursion: it is a judgment call, based on experience. Once an interpretation has been chosen, the error can be described in terms of the Standard Deviation, σ, which is seen to be either 0.05% or 0.033%. The estimated values of σ, for the fixed and variable errors of each component, must be entered on the worksheet in consistent terms, as well as that can be done, and the original decision about how the specification was interpreted (and why) should be noted for the record. The estimates for the fixed (and variable) errors of the different components in each measurement channel are then found as RSS combinations of the fixed (and

Sensor	Temperature reference junction	Differential amplifier	Active filter	MPLXR	PGA	S/H	ADC
Type ____ Range ____	Ref. temp.= ____ •o Range ____	CMV CMV •o δ_{CMV} Gain= ____ •o ____ V CMV ____ V	•o Gain= ____ •o = ____ V		•o Gain= ____ •o = ____ V		F.S.= ____ V No bits= ____
Mfg. ____ Model ____	Mfg. ____ Model ____	Mfg. ____ Model ____	Mfg. ____ Model ____	Mfg. ____ Model ____	Mfg. ____ Model ____	Mfg. ____ Model ____	Mfg. ____ Model ____
	Ambient junction RTD meas. error b7.1= ____ S7.1= ____ Stability S7.2= ____ Constant temp. junction Junction accuracy b7.1= ____ Temp. uniformity S7.1= ____ Stability S7.2= ____	Gain accuracy b6.1= ____ Linearity b6.2= ____ Offset b6.3= ____ δ_{CMV} b6.4= ____ Noise S6.1= ____ Gain stability S6.2= ____ Zero stability S6.3= ____ Normal mode S6.4= ____	Gain accuracy b5.1= ____ Linearity b5.2= ____ Offset b5.3= ____ Noise S5.1= ____ Gain stability S5.2= ____ Zero stability S5.3= ____	Gain accuracy b4.1= ____ Linearity b4.2= ____ Offset b4.3= ____ Noise S4.1= ____ Gain stability S4.2= ____ Zero stability S4.3= ____ Crosstalk S4.4= ____	Gain accuracy b3.1= ____ Linearity b3.2= ____ Offset b3.3= ____ Noise S3.1= ____ Gain stability S3.2= ____ Zero stability S3.3= ____	Gain accuracy b2.1= ____ Linearity b2.2= ____ Offset b2.3= ____ Noise S2.1= ____ Gain stability S2.2= ____ Zero stability S2.3= ____	Gain accuracy b1.1= ____ Linearity b1.2= ____ Offset b1.3= ____ Noise S1.1= ____ Gain stability S1.2= ____ Zero stability S1.3= ____ Quantization S1.4= ____
B8= S8=	B7= S7=	B6= S6=	B5= S5=	B4= S4=	B3= S3=	B2= S2=	B1= S1=

Figure 10.5 Worksheet for estimating fixed and random errors in a thermocouple measurement from manufacturer's specifications. After Taylor (1986).

variable) errors of the components. This same procedure is followed for both Single-Sample and Multiple-Sample Uncertainty Analysis.

The experimental and analytical estimates of channel performance should agree, but if they don't, then the experimental values should be used.

10.3.3.2 If the Available Value Is Taken as the True Value

The Available Value differs from the Achieved Value because of system–sensor interactions.

In heat-transfer experiments, errors arising from system–sensor interactions are usually larger than the measuring system errors. In gas temperature measurement, for example, the Achieved Value (the temperature of the sensor) is never equal to the Available Value (the total temperature of the gas [stagnation temperature] at the probe location) unless the surrounding walls and support structure are exactly at the same temperature as the gas, and the gas velocity is low.

Those conditions never exist in engines and compressors. In the usual situation, the temperature sensor probe comes to thermal equilibrium (reaches a steady state) when the convective heat-transfer rate from the gas to the probe equals the rate at which heat is being lost to the surrounding walls by radiation and conduction. If, for example, the convective thermal resistance ($1/hA$) between the probe and the gas is equal to the radiative thermal resistance between the probe and the surrounding walls, then the Achieved Value (the probe temperature) will be halfway between the wall temperature and the gas temperature (ignoring conduction loss, which would usually make matters worse!).

With those facts in mind then, if the Available Value is taken as the definition of the True Value, then all of the applicable system/sensor interaction errors must be added to the list of possible errors.

For example, consider a probe used to measure the temperature of a high-velocity flow of a hot gas in a duct with cold walls. Figure 10.6a illustrates a wind tunnel duct with probe inserted in the center of the flow. Figure 10.6b shows a close-up of flow near the front tip of the probe.

Our target is the total temperature of the flow in the duct. The total temperature, T^0, includes the static temperature of the flow, T_1, adjusted by addition of the kinetic energy in the flow.

For a measurement of gas stream total temperature, T^0, the possible errors include radiation, velocity, and conduction errors of the probe in addition to all of the

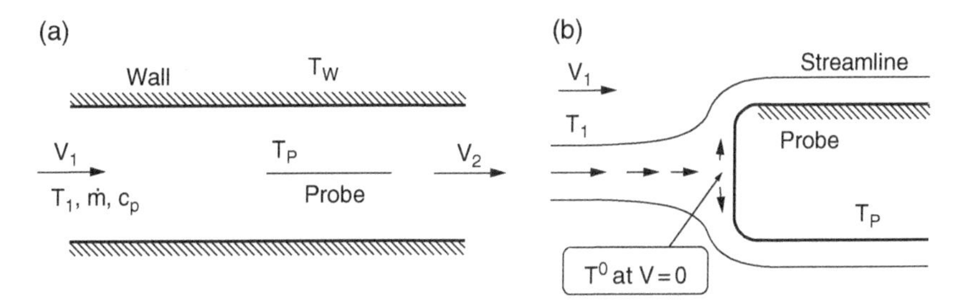

Figure 10.6 (a) Schematic of high-velocity flow in a duct with probe. (b) Schematic of flow near the front tip of the temperature probe.

instrumentation errors (which include the calibration of the sensor). Note: total temperature is stagnation temperature.

The first estimate of the total temperature of the gas is the probe temperature itself. The probe temperature is the value we measure, that is, the Achieved Value.

The best estimate of the total temperature (the Available Value) would be "probe temperature corrected for radiation, conduction, and velocity errors and for the residual fixed errors in the instrumentation system."

All three error tendencies apply, but generally the conduction error is small and can be ignored. When short probes are used, and attached to cold walls, the conduction term can be important.

The velocity error must be addressed before radiation or conduction errors can be dealt with.

We begin by imagining a test where heat losses by radiation and conduction are zero. Then the probe temperature, T_P, will be determined by its recovery factor, α, and the flow characteristics as shown in Eq. 10.1. The temperature error of the probe due to velocity is the difference between the stagnation temperature T^0 and the probe temperature T_P.

The velocity contribution to the temperature error is

$$E_V \equiv \text{Velocity Error} \equiv \left(T^0 - T_P\right) = \left(1-\alpha\right)\frac{V^2}{2c_p} \tag{10.1}$$

In this imaginary test, we have stipulated no heat loss; by definition this is the "adiabatic temperature," T_{ad}, of the probe in that flow. The adiabatic temperature is therefore the "effective temperature" of the gas for heat-transfer calculations, since it is the temperature the sensor attains when there is no heat transfer to or from it.

$$T_P = T^0 - \left(1-\alpha\right)\frac{V^2}{2c_p} \equiv T_{ad} \tag{10.2}$$

Now we include probe heat loss due to radiation and thermal gain due to convection. When the probe is in a duct with cold walls, it loses heat to the wall by radiation heat transfer. The gas, now warmer than the probe, seeks to equilibrate the probe temperature by convection heat transfer.

The radiation error to temperature, E_R, can be estimated by Eq. 10.3. The radiation term is the numerator of the right-hand side. The convective coefficient is the denominator.

$$E_R \equiv \left(T_{ad} - T_P\right)_{RAD} = \frac{\sigma\varepsilon A\left(T_P^4 - T_W^4\right)}{hA} \tag{10.3}$$

Combining Eqs. 10.2 and 10.3 yields the final equation for dealing with the system–sensor interactions when estimating T^0 from T_P:

$$T^0 = T_P + \left(1-\alpha\right)\frac{V^2}{2c_p} + \frac{\sigma\varepsilon\left(T_P^4 - T_W^4\right)}{h} \tag{10.4}$$

where the terms are defined as:

α = Recovery factor of the probe, [ ]
σ = Stefan–Boltzmann constant, $W/(m^2K^4)$
ε = Emissivity of the probe, [ ]
c_P = Specific heat of gas, J/kg K
h = Convective heat-transfer coefficient, W/m^2K
T_P = Probe temperature, K
T^0 = Gas stream total temperature, K
T_W = Wall temperature, K
V = Velocity of the gas, m/s

Equation 10.4 is the equation that should be used to correct the indicated temperature of the probe for velocity error and radiation error. The need for these corrections is well established in the literature. Making these corrections does not constitute "cooking the data" as long as reasonable (and verifiable) values are used for the terms used in the corrections.

This "Available Value" is an improved candidate for the True Value of the total temperature.

Corrections for these errors can be estimated from well-known models but, in order to make these corrections, the experiment must be instrumented to provide the necessary peripheral data.

Such extra instrumentation requires your foresight from your expertise!

These corrections will always be uncertain, due to Uncertainties in the modeling of peripheral data (the heat-transfer coefficient and the recovery factor) used in estimating them. One needs a source for values of h and α, and their Uncertainties, as a function of Reynolds number and Mach number, appropriate for the gas to be measured. The recovery factor is a function of the Prandtl number of the gas and, to some extent, the surface texture on the probe (especially for small diameter probes, perpendicular to the flow).

Recovery factors range from 0.67 to 0.95, depending on geometry, and often require a tolerance of ±0.05.

The heat-transfer coefficient is a function of probe geometry and Reynolds number as well as turbulence scale and intensity. Rarely can one trust a heat-transfer coefficient value to better than ±25% unless it was measured in situ, in the engine under test.

Heat-transfer coefficients reported in the literature or in textbooks tend to have been measured under idealized conditions – steady flow, low turbulence. These tend to be "low-limit" values. When these values are compared with values measured in engine-type flows, they may be low by 50%. If the source of h is a textbook correlation, it might be safest to increase its value by 25% and assign an Uncertainty equal to ±25% to it. This has the benefit of making the correction "zero centered" with the residual error equally likely to be positive or negative – a requirement for the statistical method we are using.

The Uncertainty in the correction remains as an Uncertainty in the final value of gas temperature, a contribution like any other "fixed error" in the system.

Equation 10.4 may be regarded as a small "data interpretation program" aimed at calculating T^0 from T_P. Its Uncertainty contribution can be calculated separately, or the same equation could be embedded in the main data interpretation program as a correction subroutine and the overall Uncertainty in the final result calculated in one pass.

From the standpoint of keeping the Uncertainty calculation up to date with the data interpretation program, it is better to embed it but to display its output as an independent item, for periodic review. Then the contributions from different sources can be compared and the most likely source of trouble identified when the experiment misbehaves.

10.3.3.3 If the Undisturbed Value Is Taken as the True Value

If the Undisturbed Value is used as the True Value, then the system disturbance caused by the introduction of the measuring system must be corrected for. Furthermore, the Uncertainty in that correction is retained as a residual Uncertainty in the measurement.

The amount by which a measurand changes because of the effect of the instrumentation on the system depends on both the process system and the measuring system. The usual advice is to keep the sensors as small as possible, to minimize the disturbance, and then to estimate the remaining effect with a simple equation.

It is difficult to describe a representative system disturbance error for a gas temperature measurement, since the effect of the probe blockage on gas temperature depends on too many factors – any example would seem contrived. Instead, a simpler situation will be used to illustrate this effect – measurement of the surface temperature of a metallic specimen.

Consider a thermocouple attached to a hot metal part but exposed to a cooler gas flow. The thermocouple will act as a fin, transferring heat to the gas and locally cooling the metal part – just where the thermocouple junction is attached. This is a typical system disturbance effect: the temperature at the sensor location is different when the sensor is present than it would have been without the sensor.

For simplicity, consider an intrinsic thermocouple welded to the surface, with no radiation error or velocity error. The thermocouple junction is exactly at the surface with this construction, and the junction temperature (the Achieved Value) is equal to the temperature of the metal where the junction is attached (the Available Value). There is still an error, however, because the heat loss out the thermocouple wires cooled the point of attachment so the Available Value of surface temperature is lower than the Undisturbed Value. The effect is described by the following equation:

$$\text{System disturbance error} \equiv \left(T_{undist} - T_{avail}\right) = \frac{\phi}{1+\phi}\left(T_{undist} - T_{gas}\right) \tag{10.5}$$

where:

$$\phi = \frac{\sqrt{hD_W k_W}}{2k_S} \tag{10.6}$$

and the individual terms are defined as:

h = Convective heat-transfer coefficient, $W/(m^2\ K)$
D_W = Diameter of the thermocouple wire, m
k_W = Thermal conductivity of the thermocouple wire (average), W/(m-K)
k_s = Thermal conductivity of the substrate, W/(m-K)

Equation 10.5 estimates the difference between the Undisturbed Value (metal temperature with no disturbance) and the Available Value (metal temperature at the thermocouple location, with the disturbance present), using estimates of the heat-transfer coefficient, the wire diameter, and the conductivities of the materials involved. This difference can be used as a correction for the system disturbance effect of installing the thermocouple. The Uncertainty in that correction must be estimated and becomes part of the overall measurement Uncertainty if the Undisturbed Value is to be estimated.

There are six variables in Eq. 10.5, each of which is somewhat uncertain, but most of the Uncertainty will come from the heat-transfer coefficient, *h*, which rarely can be trusted within ±50%, thus if hand calculations are being used, a single-term Uncertainty estimate would be justified. The same procedure recommended in the preceding section is recommended for treatment of the Uncertainty in *h* in this application.

This correction can be done in a subroutine inside the main program, but the output of that subroutine should be presented separately for its diagnostic value.

10.3.3.4 If the Mixed Mean Gas Temperature Is Taken as the True Value

If the mixed mean gas temperature is elected as the desired True Value, then the velocity and temperature distributions must be estimated. The mixed mean temperature is defined as the integral, over the flow area, of the local product of velocity, specific heat, and temperature. When those parameters are not uniform, it is very tedious to measure them. If not measured, the profiles must be estimated, and that leads to Uncertainty. The Uncertainties in the correction factors which account for the maldistributions become part of the overall Uncertainty, when the mixed mean gas temperature is taken as the desired True Value. Fully developed laminar or turbulence velocity profiles are often assumed but seldom realized. The distributions of flow and temperature in a duct following a bend are particularly nasty – often skewed to one side or the other.

10.3.4 The Role of the End User

Faced with several options, as in the preceding situation, how is the experimenter to know what to consider as the True Value definition for a given measurement? Until the True Value is specified, it is impossible to even start the Uncertainty Analysis because the list of possible errors has not been identified.

The definition of the True Value must come not from the experimenter but from the end user of the data. The key questions are:

How will this value finally be used?
What does this value represent in the end-use equations?

10.3.4.1 The End-Use Equations Implicitly Define the True Value

Consider the following examples:

$$Q = \dot{m} c_p \left(T_{out} - T_{in} \right) \tag{10.7}$$

Equation 10.7 represents an energy balance on a flowing fluid at mass-flow rate $\dot{m}$, based on the change in the bulk mean temperature between inlet and outlet. This requires that T_{out} and T_{in} be defined as the mass-flow-weighted average temperatures of the fluid flowing, while c_p must be the average specific heat over the temperature range.

$$Q = hA(T_s - T_g) \tag{10.8}$$

Equation 10.8 represents convective heat transfer from a gas to a surface in terms of the difference of two temperatures. Equation 10.8 requires that T_s be the average temperature of the surface over the area A, while T_g must be defined in the same manner as it was defined when h was measured! The heat-transfer coefficient is sometimes calculated in terms of the mass-flow-weighted average gas temperature, sometimes in terms of the temperature of the gas far from the surface, and sometimes in terms of the adiabatic wall temperature of the surface. There is no way to know what T_g must mean without knowing which definition of h will later be used with it.

$$BSFC = \frac{W/(\theta_1 - \theta_2)}{KTN} \tag{10.9}$$

Equation 10.9 represents the calculation of brake specific fuel consumption (BSFC) from the quantity of fuel consumed over a measured time interval. For calculating the brake specific fuel consumption of an engine, W must be the mass of fuel consumed by the engine over the time interval (θ_1–θ_2) and KTN (a calibration constant K times the torque times the speed) must be the average power produced over that same interval. Someone must decide whether that "power" should be measured before or after allowance for the accessories (fan, air cleaner, generator, air conditioner, etc.). Uncertainties in those allowances constitute Uncertainties in the corrected result. The torque, in this equation, must be the torque produced by the engine. Bearing and windage losses might make the torque different at the torque meter than at the output of the engine.

Only by looking at the end-use equations can the required definition of the "True" Value be identified for a given measurement. Only when the definition of the True Value is clearly understood can all of the possible error sources be identified, and only then can the Uncertainties in their corrections be estimated. For any one measurement, there may be several error sources, each contributing a fixed error and a random error. These can all be dealt with according to the techniques appropriate for Single-Sample or Multiple-Sample Uncertainty Analysis, as discussed in the following sections.

10.3.5 Calibration

The verb "to calibrate" refers to the act of comparing the response of one system (the specimen) to that of a reference system (the standard). The "standard" is presumed to be of significantly higher precision and accuracy than the specimen (often a factor of 3 or 4 is expected). Precision refers to the readability or repeatability of the system; accuracy refers to the extent to which the systems output is correct. Note: a stopped watch is very repeatable! The noun "calibration" can refer to the report that presents the results of that act or to the process itself.

There are many kinds of "calibration." For example, here are seven different ways a calibration could be presented:

i) a "go/no-go" statement confirming or denying that the specimen meets the manufacturer's specifications
ii) a linear calibration with a uniform tolerance expressed as % of full scale
iii) a linear calibration with variable tolerance expressed as % of reading
iv) a nonlinear calibration with uniform tolerance expressed as % of full scale
v) a nonlinear calibration with variable tolerance expressed as % of reading
vi) a "difference calibration" with uniform tolerance expressed as % of full scale
vii) a "difference calibration" with nonuniform tolerance expressed as % of reading.

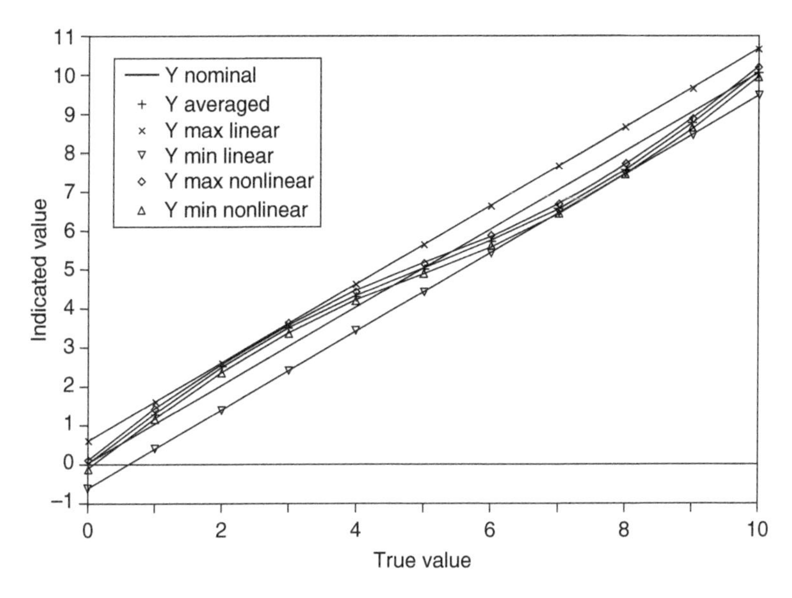

Figure 10.7 Calibration data with its "scatter band" and two ways to describe the calibration: (i) a linear calibration with uniform tolerance, and (ii) a nonlinear calibration with a uniform tolerance band.

Figure 10.7 shows a set of calibration data and two ways of presenting it: (i) a linear calibration with a uniform tolerance band, and (ii) a nonlinear calibration with a uniform tolerance band. The uniform tolerance bands are often presented as "% full scale." An instrument which is considered quite accurate near full scale must be regarded as quite inaccurate near the low end of its range when its calibration tolerance is expressed in terms of full scale. For this reason, instruments with linear calibrations are generally limited to the top half of their rated range and should be selected accordingly.

Linear calibration of sensors was important when readout instruments were designed to respond linearly to their input signal. With the advent of digital signal processing, it is almost as easy to implement a nonlinear calibration as a linear one, and the benefits of improved accuracy can be considerable. It is also possible to implement nonlinear calibrations in which the tolerance is expressed as % of reading, rather than of full scale.

The width of each band represents the smallest value of tolerance which includes 95% of the calibration data, relative to the model function. Since the response of the instrument is nonlinear (as shown), using a linear calibration requires a large tolerance.

The bands cover the total residual Uncertainty in the calibration – the Root-Sum-Squared (RSS) of the fixed and random errors of the calibration process. This residual Uncertainty is treated as a fixed error in assessing the overall Uncertainty when the instrument is used in the field. The process by which the random error in the calibration process becomes a part of the fixed error in field measurements is called "fossilization" and is discussed in more detail in a later section.

The type of calibration to be used depends on the type of data acquisition and interpretation software (DAS) available. If only linear signal interpreters are available in the DAS, then linear calibrations must be used, at least at that stage of data handling. Nonlinear calibrations can then be built into the subsequent data processing software. The net result is just as good as having used nonlinear calibrations in the DAS.

End-to-end calibrations generally yield smaller Uncertainties than assembling the Uncertainties from component calibrations, chiefly because of the accumulated effects of fossilizing the random errors in the several calibrations. For this reason, end-to-end calibrations are preferred when possible.

There is no simple rule for establishing sensible tolerances for calibrations. Naturally, it is easy to focus on calibration accuracy as though it alone determined the end-use accuracy of the field measurements. Partly, this is because the calibration tolerance is so easily identifiable – it comes in its own report! Unfortunately, the residual Uncertainty in a calibration is only one of several factors in determining the end-use Uncertainty and, often, not even the most important component. The Uncertainties in the corrections applied for system disturbance effects, system–sensor interaction errors, and idealization errors are often far larger than the residual Uncertainty in the sensor calibration.

10.4 The Basic Mathematics

When we do an Uncertainty Analysis, we aim to estimate the Uncertainty in a calculated result at the same odds (with the same confidence) that we used in expressing the Uncertainties in the measurements from which it is calculated. This aids the statistics.

Consider, for example, an experimental model constructed from the sum of five variables. Assume that each variable is uncertain by ±1.0 unit. The worst case happens when all variables are simultaneously high or when all are simultaneously low. The novice experimentalist, which we admit to being at one time, will estimate the Uncertainty as the worst-case value. It is highly unlikely, however, that during a measurement that all of the values will simultaneously be high, just as it is unlikely that they will all be low. If the variables are mutually independent, it is more likely that some will be high while others are low. Thus, it is highly unlikely that the sum will be in error by as much as ±5 units. If each Uncertainty had been estimated at 20:1 odds, the probability of the "worst-case" value of ±5 units error is $\left(\frac{1}{20}\right)^5$, that is, only occurring once in 32×10^5 observations. Thus, the maximum worst case is not the kind of Uncertainty estimate we generally want to report. What we want is an estimate of the Uncertainty in the result at the same odds as those chosen for assessing the Uncertainties in the inputs.

10.4.1 The Root-Sum-Squared (RSS) Combination

Kline and McClintock (1953) showed that an Uncertainty estimate *at constant odds* could be found by using an RSS combination of the contributions of the individual measurements, for most equations used in engineering. The "contribution" of a measurement is defined as the change in Z per unit change in the variable "x" times the Uncertainty in "x." There may be some equations for which Kline and McClintock's RSS combination is not statistically accurate, but such exceptional equations were so rare that the issue had not even been addressed in the literature. We can safely assume that the RSS combination of the contributions yields a good estimate of the constant-odds combination. Thus, if we estimate the Uncertainty in the measured variables at 20:1 odds, and use the RSS combination method to combine the contributions, the Uncertainty calculated for the result will be the appropriate interval for 20:1 odds. This is the basic form used in Uncertainty analyses. It is used in both Single-Sample and Multiple-Sample Analyses, for combining Uncertainties or Uncertainty components.

$$\text{RSS}(x_1, x_2, x_3, \ldots\ldots x_N) = \left\{(x_1)^2 + (x_2)^2 + (x_3)^2 \ldots\ldots (x_N)^2\right\}^{1/2} \tag{10.10}$$

We will use the RSS combination method to estimate the most likely total fixed error, random error, or overall Uncertainty in a single measurement subject to several error sources, or to describe the most likely Uncertainty in a result calculated from several measurements.

10.4.2 The Fixed Error in a Measurement

The fixed error in a measurement or a calculated result is often called its "bias" or "bias error" and may represent the sum of several component fixed errors. These components are often denoted by the symbol "B" and subscripted to identify which measurement or which source of error is involved.

There are usually several sources for fixed errors in each measurement: residual fixed errors in the instrument calibration and terms representing possible errors in the estimated corrections applied. The best estimate of the total fixed error in a single measurement is the RSS combination of all of the individual fixed-error components which apply, assuming that each was described as a "zero-centered error" (equally likely to be positive or negative).

$$B_X = \text{RSS}(Bx_1, Bx_2, Bx_3, \ldots\ldots Bx_N) = \left\{(Bx_1)^2 + (Bx_2)^2 + (Bx_3)^2 \ldots\ldots (Bx_N)^2\right\}^{1/2} \tag{10.11}$$

Usually, if we recognize a fixed error, we correct for it. The problem is that the correction is never perfect. What we deal with in Uncertainty Analysis is the residual fixed error, after our best effort at correction. For example, suppose we run an end-to-end calibration of a voltage measurement channel and determine that it reads low by 0.015 V at design point, with an Uncertainty of ±0.001 V. That gives us a correction which can be applied but leaves a possible residual fixed error of ±0.001 V (20:1). The residual error is treated as a "fixed error" of the measurement channel because it does not change from observation to observation when reading that channel, even though it was evaluated from the scatter in the end-to-end calibration tests. The random error in the calibration process is said to have been "fossilized" when it is handed off for use in the field.

When the estimate of a residual fixed error comes from a statistical analysis of a large set of observations, we can calculate the Standard Deviation of the population of data points and use that as a basis for estimating the possible value of the residual fixed error and the probability that the estimate is correct. Before the mean and Standard Deviation are calculated, however, the data should first be examined as a time series (a plot of measured value versus time). If the data display a stationary mean value over the observation period, then the mean and Standard Deviation of the dataset can be calculated in the usual manner and the Standard Deviation used to estimate the random error of the measurement. If there is a trend with time, the system was not steady and the interpretation of the data is more complex. First, a low-order curve fit must be passed through the data (first order or second, at most) and then the Root-Mean-Squared (RMS) deviation of the data points from the curve fit calculated. The best estimate of the random error in the measurement is twice (for 95% confidence interval) the RMS deviation.

When we have no statistical basis for estimating the residual fixed error, we guess. When we guess, however, we must make a best effort to guess a number that has the same meaning as the statistical measure would have had if we had the data. For example, if we are guessing the residual fixed error in an instrument, the value we propose must be equally likely to be positive or negative, at 20:1 odds.

Sometimes the instrumentation components of the fixed errors in a measurement must be estimated from manufacturer's descriptions of the accuracy of their instruments. The most difficult problem here is interpreting the manufacturer's specifications. What is needed for Uncertainty Analysis is an estimate that corresponds to the Standard Deviation which would be observed if a large number of observations were made. Manufacturers rarely provide such a description, preferring instead to "guarantee" the "accuracy" to be ±0.001% of full scale or some such statement. It is up to the investigator to decide whether the quoted interval represents σ, 2σ, or 3σ. Once that decision is made, σ is "known," and the residual fixed error can be estimated as the 2σ value needed for the Uncertainty estimates. Estimating Uncertainties from manufacturer's tolerances can lead to unrealistic estimates of overall Uncertainty. An end-to-end calibration over representative time span is far better.

10.4.3 The Random Error in a Measurement

The random error in a single measurement can only be determined by taking a set of readings with the entire system (instrumentation and process hardware) operating in their normal mode, with the set of observations spread over the usual time period required for production data taking under normal running conditions. The data should be examined as a time series or plot of measured value versus time before it is analyzed. Inspect visually and perform a spectrum analysis as in Section 10.1.2.5. If the data display a stationary mean value over the observation period, then the mean and Standard Deviation of the dataset can be calculated in the usual manner and the Standard Deviation used to estimate the random error of the measurement. If there is a trend with time, the system was not steady and the interpretation of the data is more complex. First, a low-order curve fit must be passed through the data (first order or second, at most), and then the RMS deviation of the data points from the curve fit calculated. The best estimate of the random error in the measurement is twice the RMS deviation.

For each channel, a set of at least 30 observations should be made, and the Standard Deviation of the observations must be estimated. The random Uncertainty in the measurement is estimated as 2σ (for 20:1 odds).

The contribution of the instrumentation system alone to the total random error can be estimated from bench test data on the instrumentation system or from the manufacturer's specifications, but the result is of little value since the process unsteadiness is usually larger than the instrumentation unsteadiness.

In Multiple-Sample Uncertainty Analysis, the random Uncertainty contribution from process instability is suppressed because only the mean values of data are processed, not individual observations, and long runs are averaged. The Standard Deviation of the mean of N samples is smaller than the Standard Deviation of the samples, by a factor of

$1/\sqrt{N}$. Hence, the Uncertainty in the mean of a set of 100 observations is only 1/10 the Uncertainty in a single observation (Section 7.1.9).

Processing only the mean values of sets of data gives the appearance of improving the accuracy of the measurement but does not, in fact, alter the basic process. If the different variables measured in an experiment all display variations, it may be more accurate to process individual datasets and average the results rather than to average the data and process the averages. When the process being studied is nonlinear, and the excursions in the data are significantly large, there will be a difference in the two outcomes.

10.4.4 The Uncertainty in a Measurement

The Uncertainty in a measurement is an estimate that includes both fixed and random components.

Consider a single measurement of the variable x_i. The measured value is the best estimate you have of the value of x_i, but it is not perfect – the measurement has both fixed errors and randomized errors which you don't know. To properly describe what you do know, and how accurately you know it, you must estimate the most likely value of the total error, by whatever tools you can bring to bear. To describe your conclusions in a way which will be clear to other users, you must use a standard format for describing the Uncertainty. The following form has been adopted internationally for representing a measurement and its Uncertainty:

$$x_i = x_i \pm \delta x_i (20/1) \tag{10.12}$$

This statement should be interpreted to mean the following:

- The best estimate of the value of x_i is its Recorded Value, x_i.
- There is an Uncertainty in x_i which may be as large as $\pm\delta x_i$
- The odds are 20:1 against the Uncertainty in x_i being larger than δx_i.

The Uncertainty in a measurement has two components: its fixed error and its random error, which are sometimes referred to as the bias error (B) and the precision error (P) of the measurement. Both the bias error and the precision error are treated as though they came from normally distributed populations of possible values, centered around zero error. The magnitudes assigned to the bias error and the precision error both will correspond to estimates of the 95% probability that the real error will not exceed the estimate.

With this in mind, the Uncertainty in a single measurement can be described as follows:

$$\delta x_i = \left\{ (B_i)^2 + (P_i)^2 \right\}^{1/2} \tag{10.13}$$

Sometimes we deal separately with the bias and precision terms, and sometimes we deal only with the overall Uncertainty.

In Single-Sample Uncertainty Analysis, the word "measurement" is used to refer to a single observation, and in Multiple-Sample Uncertainty Analysis, it often refers to the mean of a set of measurements. The value of x_i could represent the Recorded Value from one observation (in a single-sample experiment), or it could represent the mean of a set of N observations (in a multiple-sample experiment). Equation 10.13 would be used in both cases, but the meaning of the symbol δx_i depends on the meaning of the symbol x_i.

If x_i represents a single observation, then δx_i represents 2σ, where δ is the Standard Deviation of the population of possible measurements from which the single-sample, x_i, was taken. This can be interpreted to mean that if 19 more observations were taken, not more than 1 would lie further than $\pm\delta x_i$ from the present measured value. This is the interpretation used in Single-Sample Uncertainty Analyses. For multiple-sample experiments, δx_i means the Uncertainty in the mean of the N samples. This can be interpreted to mean that if 19 more sets of N samples were taken, and the means of each set were calculated, not more than 1 of the 19 new means would lie further than $\pm\delta x_i$ from the presently Available Value. In Multiple-Sample Uncertainty Analyses, the symbol U_{95} is used instead of $\pm\delta x_i$ but has the same meaning.

10.4.5 The Uncertainty in a Calculated Result

The result, Z, of an experiment is usually calculated from a set of measurements using a group of equations which will be called the "data interpretation program." The calculations might be done by hand or by computer. We can represent the statement that the result Z is a function of several variables as follows:

$$Z = Z(x_1, x_2, x_3 \ldots\ldots\ldots x_N) \tag{10.14}$$

We now wish to estimate the Uncertainty in Z, assuming we know the Uncertainties in each of the x_i.

There are two issues to deal with: first, some of these variables will be more important than others in setting the value of Z, and second, when several variables are involved, it is highly unlikely that all of their Uncertainties would be positive or negative at the same time.

The effect of the Uncertainty in any one measurement if only that one measurement were in error would be

$$\delta Z = \frac{\partial Z}{\partial x_i}\delta x_i \tag{10.15}$$

The partial derivative of Z with respect to x_i is the "sensitivity coefficient" for the result Z, with respect to the measurement x_i.

As an example, consider a case where Z is a simple function of two variables:

$$Z = 3x^2 + 2y \tag{10.16}$$

We seek to estimate the effect on the Uncertainty in Z for data taken near the point $x = 1$, $y = 1$. To illustrate the effects of one variable, first let's assume that one of the two is known exactly.

The sensitivity of the result Z to Uncertainty in x alone is

$$\frac{\partial Z}{\partial x} = 6x = 6 \tag{10.17}$$

On the other hand, the sensitivity of the result Z to Uncertainty in y alone is

$$\frac{\partial Z}{\partial y} = 2 \tag{10.18}$$

Suppose that, when x is uncertain, the following describes its Uncertainty:

$$x = x \pm 0.01(20/1) \tag{10.19}$$

and when y is uncertain:

$$y = y \pm 0.01(20/1) \tag{10.20}$$

An Uncertainty of 0.01 units in x would cause an Uncertainty of 0.06 units in Z, if x were the only source of error, whereas an Uncertainty of 0.01 units in y would cause only 0.02 units Uncertainty in Z, if it were the only source of error.

Using this simple equation as an example, let's consider the Uncertainty in Z when both x and y are uncertain: the question is, how shall we combine the Uncertainty contributions from both x and y? The worst case would be simply to sum up the above results and claim that the Uncertainty in Z could not be larger than ±0.08 units; that is certainly true, but it is very unlikely to occur. Only 1 time in 20 will the error in x be as large as 0.01 units (causing an error of 0.06 units in Z), and only 1 time in 400 will both x and y have their largest values. Thus, Z may occasionally be in error by as much as ±0.08 units, but the odds are that will only occur once in 400 trials.

In a typical Uncertainty Analysis situation, our objective is to express the Uncertainty in the calculated result at the same odds as were used in estimating the Uncertainties in the measurements. This issue was taken up by Kline and McClintock (1953), who showed that the Uncertainty in a computed result could be estimated with good accuracy using a "Root-Sum-Squared" combination of the effects of each of the individual inputs, and that the RSS operation preserved the odds. Thus, if the Uncertainty in each measured variable was described at 20:1 odds, then an RSS combination yielded an estimate of the Uncertainty in Z at 20:1 odds.

In the present example, using the RSS combination yields

$$\delta Z = \left\{ \left(\frac{\partial Z}{\partial x} \delta x \right)^2 + \left(\frac{\partial Z}{\partial y} \delta y \right)^2 \right\}^{1/2} \tag{10.21}$$

or, after substituting from above, $\delta Z = 0.063$.

Note that the Uncertainty is dominated by the larger term. The RSS combination has the effect of suppressing terms which are smaller than one-third of the largest (unless there are many of the small terms).

The general form, to be used when several independent variables are used in the equations which calculate Z, is shown below. The individual Uncertainty contributions are combined by an RSS method.

$$\delta Z = \left\{ \left(\frac{\partial Z}{\partial x_1} \delta x_1 \right)^2 + \left(\frac{\partial Z}{\partial x_2} \delta x_2 \right)^2 + \left(\frac{\partial Z}{\partial x_3} \delta x_3 \right)^2 + \ldots + \left(\frac{\partial Z}{\partial x_N} \delta x_N \right)^2 \right\}^{1/2} \tag{10.22}$$

This Eq. 10.22 is the basic equation of Uncertainty Analysis.

Each term represents the contribution made by the Uncertainty in one variable to the overall Uncertainty in the result, Z. Each term has the same form: the partial derivative of Z with respect to x_i multiplied by the Uncertainty interval for that variable, δx_i The estimated Uncertainty in the result, δZ, has the same probability of occurrence as have the Uncertainties in the individual measurements.

Equation 10.22 applies as long as:

i) *The errors or Uncertainties in the δx_i are independent of one another.*
This does not mean the *values* of the x_i are independent, just their errors or Uncertainties. For example, changing the flow in a system while holding the power constant will change the coolant temperature in a predictable way (the coolant temperature rise depends on the flow, at constant power), but the Uncertainty in the measurement of coolant temperature is not dependent on the Uncertainty in the measurement of flow.
Coleman and Steele have introduced the notion of "correlated bias error" to account for the situation when two or more transducers are calibrated against the same standard (thus embedding the same fixed error in both) and have developed the mathematics with which to handle this case (C&S; see preface).
ii) *The distribution of errors or Uncertainties is approximately Gaussian, for all the x_i.*
This assumption is not very restrictive. Both experimentalists and statisticians assume that the random component of the error in a series of physical measurements has a Gaussian distribution. As a consequence, the residual fixed error after correction for a calibration defect also can be assumed to have a Gaussian distribution.
iii) *The various δx_i are all quoted at the same odds.*
This is also not very restrictive. If you are going to combine Uncertainties at constant odds, then surely you must start with Uncertainty estimates at the same odds.

(adapted from Kline and McClintock [1953]).

The partial derivative of Z with respect to x_i is often referred to as the "sensitivity coefficient," and the product of the sensitivity coefficient and the Uncertainty, δx_i, is "the contribution." The form of Eq. 10.22 is often called the "RSS form."

The data interpretation program may be simple enough that all of the partial derivatives can be evaluated analytically, or it may be too complex for that and may require direct computer analysis – the procedures are the same. A general technique for computerized Uncertainty Analysis is given in the next section.

In most situations, the overall Uncertainty in a given result is dominated by only a few of its terms. Terms in the Uncertainty equation which are smaller than the largest term by a factor of 3 or more can usually be ignored. This is a natural consequence of the RSS combination – small terms have *very* small effects. There are exceptions, of course, where there are many terms of approximately the same size, but in general that is not the case.

10.4.5.1 The Relative Uncertainty in a Result

In many applications, the Uncertainty estimate is wanted in relative terms, as a fraction of reading, $\delta Z/Z$, rather than in absolute terms (engineering units), δZ. While this can always be calculated starting from the results of the general form in Eq. 10.22, it is sometimes possible to do the calculation of Relative Uncertainty directly and very simply.

Whenever the equation describing the result is a pure "product form," such as shown in Eq. 10.23, or can be put into that form, then the Relative Uncertainty can be found directly:

If

$$Z = Cx_1^a x_2^b x_3^c x_4^d \ldots\ldots \tag{10.23}$$

Then:

$$\frac{\delta Z}{Z} = \left\{ \left(a\frac{\delta x_1}{x_1} \right)^2 + \left(b\frac{\delta x_2}{x_2} \right)^2 + \left(c\frac{\delta x_3}{x_3} \right)^2 + \left(d\frac{\delta x_4}{x_4} \right)^2 + \ldots\ldots \right\}^{1/2} \tag{10.24}$$

For example, in calculating the velocity of a fluid flow using the incompressible Bernoulli equation,

$$V = \sqrt{\frac{2\Delta P}{\rho}} \tag{10.25}$$

In the form of Eq. 10.23

$$V = \sqrt{2}\left(\Delta P\right)^{1/2} \rho^{-1/2}$$

The Relative Uncertainty of V is then:

$$\frac{\delta V}{V} = \left\{ \left(\frac{1}{2}\frac{\delta \Delta P}{\Delta P} \right)^2 + \left(\frac{-1}{2}\frac{\delta \rho}{\rho} \right)^2 \right\}^{1/2} \tag{10.26}$$

The Relative Uncertainty is a natural and convenient way to describe the Uncertainty in a result, especially in situations where the Uncertainties of the measurements are known in terms of "percent of reading," and the result is needed in the same terms. The exponent of each variable in the computing equation becomes the sensitivity coefficient for its Relative Uncertainty contribution.

This same form is sometimes derived by showing that the logarithm of the result is the sum of the logarithms of the component factors, each with their appropriate coefficient, and using the RSS combinatorial form on the terms.

The equations given above can be dealt with by hand for simple relationships, and this provides a very quick and easy way to estimate the attainable accuracy in an experiment in the early stages of planning the experiment.

Most engineering data interpretation is done by computer programs which may involve table lookups, curve fits, and approximations. These operations cannot be dealt with by such analytical notions as finding the partial derivatives. In the next section, it will be shown that the main data interpretation program itself can be used to find the Uncertainty in the result, using the principles established above but using numerical means to find the values (Figure 10.8).

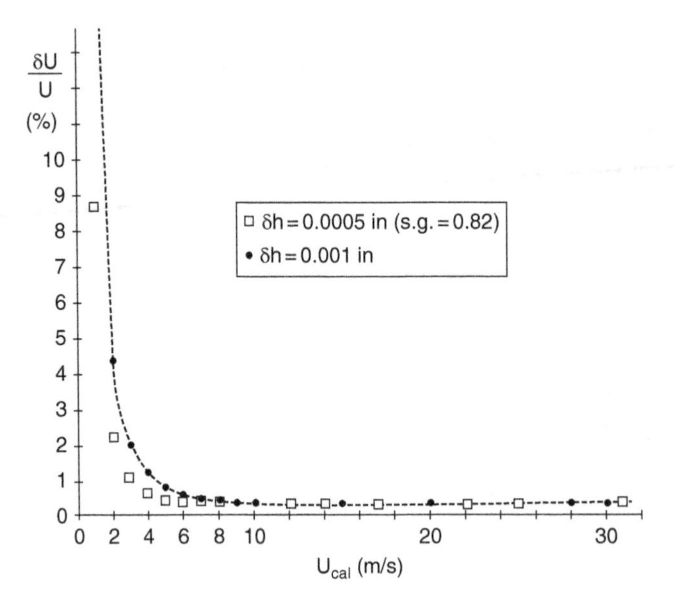

Figure 10.8 Uncertainty in velocity measured with two different micro-manometers.

10.5 Automating the Uncertainty Analysis

When Z is calculated using a spreadsheet or computer program, often the case in practical situations, or when the equations describing Z involve operations which are difficult to differentiate (e.g. involving table lookups or numerical integration), the analytic operations of Uncertainty either cannot or will not be done by hand. It is not practical, from the maintenance standpoint, to write a separate Uncertainty Analysis program: it is too difficult to ensure that the Uncertainty code is updated each time the Data Reduction Program (DRP) is revised. Fortunately, there is an easy solution to this problem.

If a DRP exists by which the result of an experiment can be calculated from its observed data, then that same program can be used to estimate the Uncertainty in the result. This is accomplished numerically, rather than analytically, by sequentially perturbing the input values by their respective Uncertainties and collecting their individual contributions using an RSS combination to find the Uncertainty in the result.

10.5.1 The Mathematical Basis

To see the basis for this process, let's write a typical term in the RSS equation, using a strict mathematical interpretation of the partial derivative operator. We then make the assumption that the Uncertainty interval δx is small enough that its magnitude can be used in estimating the derivative, as opposed to Δx:

$$\frac{\partial Z}{\partial x_1}\delta x_1 = \left(\lim_{\Delta x \to 0} \frac{Z_{x_1+\Delta x_1} - Z_{x_1}}{\Delta x_1}\right)\delta x_1 \cong \left(Z_{x_1+\Delta x_1} - Z_{x_1}\right) \tag{10.27}$$

What this means, in practical terms, is that the contribution of variable x_1 to the Uncertainty in Z can be found by calculating Z twice: once with the Observed Value of

x_1 and once with $x_1 + \delta x_1$ and subtracting the two value of Z. When several variables are involved, the overall Uncertainty in the result can be found by sequentially perturbing the individual variables and then finding the square root of the sum of the squares of the individual terms.

The steps in this direct computer-executed Uncertainty Analysis can be summarized as follows (Moffat 1985). For each variable used in computing Z, establish its Uncertainty interval, δx_i and enter that value as an additional data entry. Calculate the result Z for the recorded data and identify the original value as Z_o and store it.

Then, for i = 1 to N, where N is the number of variables used in calculating Z:

Increase the value of the ith variable, x_i, by its Uncertainty interval, δx_i and calculate Z_i, using the augmented value of the ith variable with all other variables at their Recorded Values.

Find the difference $(Z_i - Z_o)$ and store it as C_i, the contribution to the Uncertainty of Z caused by the ith variable, assuming a positive excursion.

> If the result, Z, is likely to be a strongly nonlinear function of x_i, including consideration of the size of its Uncertainty interval, it may be worthwhile to also calculate C_i using a negative change in x_i. The present recommendation would be to use the average of the absolute values of C_i+ and C_i- as the working value of C_i, but that recommendation must be regarded as tentative, since no definitive analysis has been done to investigate the issue.

Return the value of x_i to its Recorded Value (unperturbed). Call "Next I" and repeat until all variables have been examined. The Uncertainty in the result is the RSS of the C_i.

A primary advantage of this method is that the actual working data interpretation program (DIP) itself is used in the assessment of Uncertainties. Thus, each time the program is modified, the modifications are automatically incorporated into the Uncertainty calculations.

The process flowchart is illustrated in Figure 10.9.

This same operation can be done on a spreadsheet, as illustrated in the following example.

10.5.2 Example of Uncertainty Analysis by Spreadsheet

In this example, the heat-transfer coefficient from a component is calculated after corrections are applied to the raw data to account for several effects. Strictly speaking, there are generally two kinds of equations involved in calculating any result: one which applies corrections to the data for the potential errors you know about, and another that calculates the result from the corrected data. Sometimes these auxiliary equations are embedded into the DRP itself.

In the present example, it is assumed that an electrically heated specimen (one component of many on a circuit board) is exposed to a coolant flow. Electrical power, surface area of the specimen, surface temperature, and flow temperature are measured.

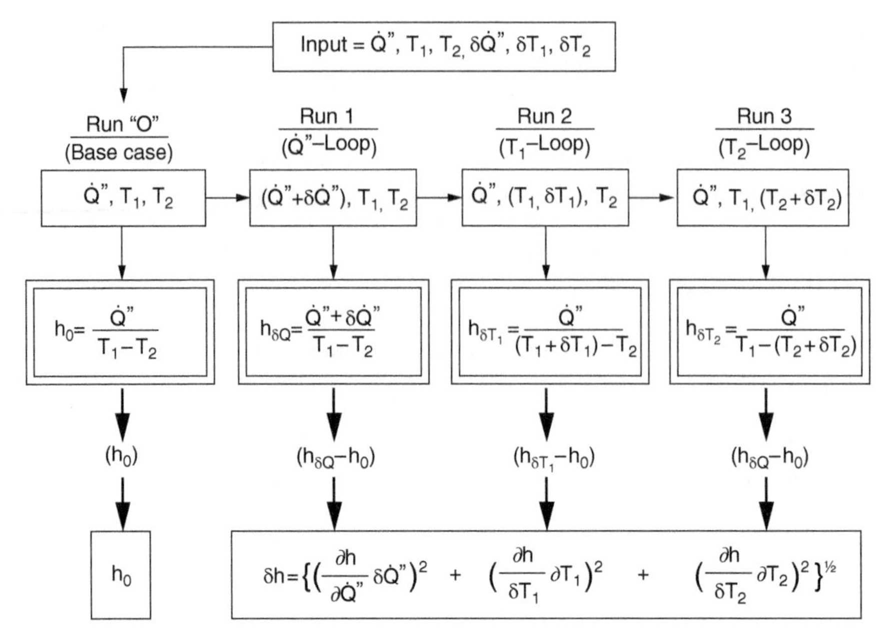

Figure 10.9 Flowchart for estimating the Uncertainty in a result by sequentially perturbing the variables.

The equations by which h is calculated are known as soon as the decision is made to run the experiment:

$$h = \frac{q_{conv}}{A(T_o - T_{cool})} \tag{10.28}$$

The convective heat-transfer rate, q_{conv}, is not directly measurable, however. It must be deduced from the measured electric power provided to the component, W, by subtracting the estimated heat losses by conduction and radiation.

Auxiliary equations:

$$q_{conv} = W - q_{cond} - q_{rad} \tag{10.29}$$

The conduction loss term may be represented by an algebraic equation using an overall conduction shape factor, K. In this example, the component is assumed mounted to a circuit board, whose temperature is known at the point of attachment (T_{board}).

$$q_{cond} = K(T_o - T_{board}) \tag{10.30}$$

The radiation term can be calculated from the Stefan–Boltzmann constant, the emissivity, the area, and the difference between the fourth powers of the two temperatures involved: the component and the surrounding walls.

$$q_{rad} = \sigma \varepsilon A\left(T_o^4 - T_w^4\right) \tag{10.31}$$

The power measurement may come from a wattmeter that is known to read high by 2% (to make this example easy). Then W_{actual} is found from W_{ind} by a simple correction equation:

$$W_{actual} = 0.98 \cdot W_{ind} \tag{10.32}$$

In a more elaborate situation, there may be several corrections to apply for sensor errors. In measuring the temperature of a hot gas stream, for example, corrections are needed for radiation error, conduction error, and velocity error on the sensor. These application "errors" represent unwanted consequences of heat transfer between the sensor and the surroundings. Their magnitude is set by physical mechanisms and can be estimated. There are some types of errors, however, which are not limited by physical mechanisms: these are the conceptual errors. For example, consider trying to "measure" the average (mixed mean) temperature of coolant flow in a channel using only the reading from one sensor at an arbitrary location in the flow. That "measurement" could be very far off, not because the sensor did not measure the local coolant temperature, but because the local temperature was not equal to the mixed mean temperature. If the mixed mean temperature is needed, and only one sensor is used, then the difference between the mixed mean temperature and the sensor temperature must be estimated, and a correction applied to the sensor temperature. That estimation will involve a considerable amount of judgment and will have a large Uncertainty.

Equations (10.28) through (10.32) together constitute a DRP by which h might be calculated from the measured data, for this simple example. Together, they constitute "the" equation for Z as a function of the x_i. These are the equations which would have to be differentiated to find the Uncertainty in h analytically. Luckily, as was mentioned in the earlier section, this can be done numerically.

An example spreadsheet is shown in Figure 10.10a, illustrating the use of Eqs. 10.28 through 10.32 with one set of data to calculate one value of h. The same format could have been used to assess the fixed error or the random error, by providing those estimates instead of the Uncertainty estimates. The codebook is Figure 10.10b.

10.6 Single-Sample Uncertainty Analysis

Single-sample experiments are those in which each test point is run only once or, at most, a very few times. Exploratory tests, development tests, and research experiments are usually single-sample experiments. They usually cover a broad range of one or more parameters, with only a few data points over the range and little, if any, replication of set points. The problem is to find a statistically sound way of assessing the Uncertainties in an experiment which produces only one or two observations at each set point. It is this need that lead to the development of Single-Sample Uncertainty Analysis techniques by Kline and McClintock (1953).

Over the intervening years, Single-Sample Uncertainty Analysis has evolved several new techniques for working with this type of experiment. It can now be used both to provide a diagnostic tool during the developmental phase of the experiment and to provide a formal estimate of the Uncertainty in the reported result.

(a)

Sequentially perturbed parameters: perturbed by Uncertainty																		
Units	W	m^2	C	C	C	C	$W/(m^2K^4)$	[ ]	W/C	W	W	W	W	$W/(m^2C)$				
Perturb	0.5	2.5e-6	1	2	2	2	1.304E-013	0.1	0.01									
Parm	Wind	A	To	Tc	Tb	Tw	StefBoltz	Ems	K	Qcond	Qrad	Wact	Qconv	h(iter)	δh	$(\delta h)^2$	Perturbed	SensCoef
Iterate 0	4.0	0.002	80.0	40.0	55.0	55.0	5.67E-008	0.8	0.06	1.50	0.3591	3.92	2.061	25.76	0	0	none	
1	**4.5**	0.002	80.0	40.0	55.0	55.0	5.67E-008	0.8	0.06	1.50	0.3591	4.41	2.551	31.89	6.13	37.516	**Wind**	12.250
2	4.0	**0.002**	80.0	40.0	55.0	55.0	5.67E-008	0.8	0.06	1.50	0.3596	3.92	2.060	25.72	-0.04	0.001	**A**	-15106.1
3	4.0	0.002	**81.0**	40.0	55.0	55.0	5.67E-008	0.8	0.06	1.56	0.3752	3.92	1.985	24.21	-1.56	2.420	**To**	-1.556
4	4.0	0.002	80.0	**42.0**	55.0	55.0	5.67E-008	0.8	0.06	1.50	0.3591	3.92	2.061	27.12	1.36	1.838	**Tc**	0.678
5	4.0	0.002	80.0	40.0	**57.0**	55.0	5.67E-008	0.8	0.06	1.38	0.3591	3.92	2.181	27.26	1.50	2.250	**Tb**	0.750
6	4.0	0.002	80.0	40.0	55.0	**57.0**	5.67E-008	0.8	0.06	1.50	0.3332	3.92	2.087	26.08	0.32	0.105	**Tw**	0.162
7	4.0	0.002	80.0	40.0	55.0	55.0	**5.67E-008**	0.8	0.06	1.50	0.3591	3.92	2.061	25.76	0.00	0.000	StefBoltz	-8E+007
8	4.0	0.002	80.0	40.0	55.0	55.0	5.67E-008	**0.9**	0.06	1.50	0.4040	3.92	2.016	25.20	-0.56	0.315	**Ems**	-5.611
9	4.0	0.002	80.0	40.0	55.0	55.0	5.67E-008	0.8	**0.07**	1.75	0.3591	3.92	1.811	22.64	-3.13	9.766	**K**	-312.5
								Uncertainty in convection coefficient, h						7.363				
								Relative uncertainty						0.286				

Figure 10.10 (a) A spreadsheet for estimating the Uncertainty in *h* by the perturbation method. (b) Codebook for spreadsheet shown in Figure 10.10a.

(b)

Data to be analyzed is listed in the second tab (sheet) of this spreadsheet:
Save the data (in the second tab) as a csv (comma separated values) file.

Label	Name or description	Units	Elaborate
Iterate	Iteration count from basis		Basis at iteration 0 has unperturbed values
Parm	Parameter or factor		Key terms in data reduction equation
Perturb	Amount factor is perturbed		Value comes from precision or bias uncertainty
Units	Parm dimensions		Physical dimensions particular to this term
Wind	Power indicated	W	Watts Power as measured on equipment
A	Area	m^2	Meter^2
To	Temperature, specimen	C	Degrees Celsius
Tc	Temperature, cooling air	C	Celsius for convection heat transfer
Tb	Temperature, board	C	" for conduction heat transfer
Tw	Temperature, wall	C	" for radiative heat transfer
StefBoltz	Stefan–boltzmann constant	W/ (m^2 K^4)	5.67E-008 1.30E-013
Ems	Emissivity	[ ]	
K	Shape factor	W/C	Conduction shape factor
Qcond	Heat transfer, conduct	W	Watts, attributed to conduction heat transfer
Qrad	Heat transfer, radiative	W	" attributed to radiation heat transfer
Wact	Power, bias removed	W	" actual power, corrected bias in indicated
Qconv	Heat transfer, convect	W	" attributed to convection heat transfer
h(iter)	h coefficient at iteration	W/ (m^2 C)	Convection heat transfer coefficient
δ h	Δ h due to perturbation	W/ (m^2 C)	del_h, change of convection coef
(δ h)^2	del_h^2		Square of del_h
Perturbed	Select parameter perturbed		
SensCoef	Relative sensitivity of delh to this perturbation		
Uncertainty in convection coefficient, h			Square root of (sum of del_h^2)
Relative Uncertainty			h uncertainty / unperturbed value of h

Figure 10.10 (Continued)

In this era of high-speed data acquisition, it is important to be clear about the meaning of single-sample and multiple-sample. Consider a 100 kHz DAS set up to measure the temperatures, pressures, and flows in an engine running at a steady engine condition. A burst of data taken in 0.01 second contains 1000 observations. Is that a single-sample or a multiple-sample dataset?

It could be regarded as either! Those 1000 observations comprise both a multiple-sample of the high-frequency components of the random error in the instrumentation system, and, at the same time, they comprise a single-sample of the behavior of the engine.

Throughout these Single-Sample Uncertainty notes, there will be repeated calls for "30 observations over a representative time interval." The purpose of the words "over a representative interval of time" is to avoid a situation in which a high-speed burst of data provides 30 observations that sample only the instrumentation variability and not the process variability. It is the process time-scale that is important.

10.6.1 Assembling the Necessary Inputs

Before we can assess the overall Uncertainty in a calculated result, we must have estimates of the Uncertainties in each measurement, and we must know the equations by which the measurements will be used to calculate the result. The set of equations is referred to as the DRP, and the generic result will be named "Z."

The Uncertainty in each variable can be described in two different ways:

- By separately stating the fixed error (B_i) and the random error (P_i) in the measurement.
- By stating the overall Uncertainty in the measurement.

If the fixed error and the random error are provided separately, they can be combined to describe the overall Uncertainty using the RSS combination, provided that both error estimates were presented at the same odds (probability of occurrence), usually 20:1 odds or 95% confidence (Equation 10.13):

$$\delta x_i = \left\{ (B_i)^2 + (P_i)^2 \right\}^{1/2}$$

If the overall Uncertainty is provided directly, it should be described in the standard form (Equation 9.1):

$$x_i = x_i \pm \delta x_i (20:1)$$

Single-sample experiments rarely provide more than one or two observations at each test point, not enough to provide a statistically valid estimator of the Standard Deviation based on the production data alone.

To supplement the production data, Single-Sample Uncertainty Analysis uses information about the Uncertainty characteristics of the system deduced from replicated data taken during the rig shakedown. The concept of "pooled variance" can be relied on provided that the test rig behaves the same during the shakedown runs and the production runs. It is this peripheral data, characterizing the experimental system, that justifies use of the statistical approach to the analysis of a single set of observations,

During the "shakedown" and "debugging" phase of the experiment, the Uncertainty properties of the system must be determined and stored for use in the interpretation of the future "production" data.

In particular, two types of information must be gathered before production data taking starts:

- At least 1 set of 30 readings on each data channel during an end-to-end calibration using a stable and accurate source, spanning a representative time interval. From this set we learn the required correction and the residual fixed error for each channel.
- At least 1 set of 30 readings of each data channel during one normal run, at typical test conditions, spanning a representative time interval. From this set we learn the total "random" error in each channel.

If you have reason to suspect that the stability characteristics of the rig might be different in different parts of its operable domain, then the "running" tests should be repeated in different parts of the domain.

These diagnostic datasets will provide all the information needed to assess the Uncertainty in the production datasets. The following sections discuss how this is done.

Collecting these diagnostic data seems like a lot of work, and it is, but it is necessary in order to validate the final results. Research experiments are like icebergs: only about 10% of the total effort (the output data) are ever visible. The other 90% is unseen – relegated to the logbook, but that 90% is what establishes the validity of the visible 10%!

10.6.2 Calculating the Uncertainty in the Result

Once the overall Uncertainty estimates are available for each measurement channel, production data may be taken as single-sample sets, and the Uncertainty in the result calculated using the analytical or numerical method already described (Equation 10.22):

$$\delta Z = \left\{ \left(\frac{\partial Z}{\partial x_1} \delta x_1 \right)^2 + \left(\frac{\partial Z}{\partial x_2} \partial x_2 \right)^2 + \left(\frac{\partial Z}{\partial x_3} \delta x_3 \right)^2 + \ldots\ldots + \left(\frac{\partial Z}{\partial x_N} \delta x_N \right)^2 \right\}^{1/2}$$

There are several different lessons that can be learned from an Uncertainty Analysis, depending on what constituent errors are included in the Uncertainty estimate for each variable. Thus, although the mathematics is straightforward, the values to be used in the equations may change depending on the intent of the analysis: this is where judgment comes in. The experimenter must know what he (or she) wants to learn from the Uncertainty Analysis before assigning Uncertainty intervals to the measurements. The meaning of the outcome, as well as its value, depends on the inputs.

Uncertainty Analysis can be a powerful diagnostic tool. By making special choices of the inputs, the same equations can be made to yield different types of information about the experiment.

Three fundamentally different levels of Uncertainty estimation have been proven useful in experiment planning and development.

These different estimates of Uncertainty can be used to answer the three most important questions which have to answered for any experiment.

i) During the planning of the experiment:
   Is the proposed instrumentation system accurate enough?
ii) During the shakedown and debugging of the experiment:
   How much of the scatter on repeated trials can be explained by what is already known about the data, and how much represents instability in the test rig?
iii) When reporting the results in the literature:
   What is the overall Uncertainty in the reported result?

These questions can be answered unambiguously if one acknowledges that there are different values which should be assigned to the Uncertainty in a measurement, depending on what one wishes to learn from the Uncertainty Analysis. This is the subject of the next section.

10.6.3 The Three Levels of Uncertainty: Zeroth-, First-, and Nth-Order

To answer the diagnostic questions which arise in developing a valid experiment, Single-Sample Uncertainty Analysis generates three descriptors for each result: the Zeroth-, First-, and Nth-Order Uncertainty estimates.

Each of these is a "total Uncertainty" for a particular kind of replication of the experiment – a measure of the scatter which would be encountered if the experiment were repeated subject to a particular set of constraints. The terminology arises from considering the role of replication in setting the levels of Uncertainty in experiments. All of the techniques for estimating Uncertainty either require actual replication (e.g. Multiple-Sample Uncertainty Analysis) or require that one imagine that replications might be made (Single-Sample Uncertainty Analysis). From experience we know that when an experiment is actually repeated several times, the way in which the replications are conducted affects the scatter in the results. The more Degrees of Freedom allowed, the greater will be the variance. There is no single value for "the Uncertainty" in a measurement; there as many possible values as there are possible ways to replicate the experiment.

When discussing hypothetical replications, as we do in dealing with Single-Sample Uncertainty Analysis, we are free to assign different Uncertainty intervals to different levels of replication. While any significant experiment could be "replicated" in hundreds of slightly different ways, experience has shown that three baseline levels of replication are useful to consider, as discussed in the following sections.

10.6.3.1 Zeroth-Order Replication

The intent of the Zeroth-Order Replication is to assess the errors introduced by the instrumentation system alone. The name "Zeroth-Order" was chosen because in this level of replication, nothing about the experiment itself is allowed to change – not even time. Only the instrumentation can introduce variable error. We seek data on the variability introduced by the instrumentation system itself when it is looking at truly steady inputs (the random error), as well as the offset between the signal reported by the instruments and the signal provided to the sensors (the fixed error). An accurately controlled, steady input must be provided to each channel, and the instrumentation system scanned through the channels as though a normal dataset were being taken.

A Zeroth-Order Replication consists of taking several consecutive sets of data with precisely known "dummy" inputs representing each sensor, as is done in "end-to-end" calibrations. The recommended number of sets is 30 or more, and they should be taken over a period of time typical of the time usually needed to acquire data from a regular run and with the usual conditions existing inside the test cell. For each data channel, the set of 30 observations should be processed to estimate the mean error and the Standard Deviation, S_{30}, of the set of 30. Note that S_{30} is very close to σ for the population of possible measurements made using that channel and can be used to represent the Standard Deviation of the data. This is not the Standard Deviation of the mean of the data, but of the data population itself. The mean error of the data provides an estimate of the fixed error of that channel in the instrumentation system, and 2σ constitutes the best estimate of the interval of variable error which would contain 95% of the data.

When precisely known dummy signals are used, there is no variability in the apparent "process" itself, and the True Value of each input is accurately known. This allows the overall instrumentation system errors to be determined, both fixed and random.

If the Zeroth-Order error estimates are used in conjunction with a representative dataset, the Uncertainty interval thus determined is the Zeroth-Order Uncertainty of the result, denoted by δZo. The value of δZo is the smallest Uncertainty interval which can be achieved using the proposed instrumentation system. No real experiment will do as well, because every process introduces some unsteadiness and will increase the apparent random errors in the measurements.

10.6.3.2 First-Order Replication

The First-Order Replication level allows real time to run. Any instability in the test apparatus will cause variations in the parameters, which will show up as variations in the data. The intent of the First-Order Replication is to assess the short-term repeatability of data taken with the entire system running in its normal condition. Note that this does not assess the *accuracy*, just the repeatability.

The data for assessing the First-Order Replication should be acquired from a set of special tests run during the shakedown period with the apparatus running at a real test points – preferably a significant one. Ideally, 30 sets of data should be taken at each of the selected "representative" test points, with the usual conditions existing inside the test cell. The 30 sets should each be taken over the time interval required for a normal test series.

The data from each set of 30 observations should be processed in two ways: first (as datasets), to learn the mean value of the result, $\bar{Z}$, and the Standard Deviation in the population of 30 calculated results, σZ_1; and second (channel by channel), to estimate the mean and Standard Deviation, $\bar{x}_i$ and σx_i, of the data from each channel.

The First-Order Uncertainty in the result is $\delta Z_1 = 2\sigma Z_1$. This is the interval within which 95% of measured results will be expected to lie when that same set point is repeated in the future. If more than 1 out of 20 future trials at that same set point differs from the observed mean by more than $\pm\delta Z_1$, that should be taken as a warning that some problem is developing within the rig: something is changing.

The First-Order Uncertainty in the measurement of x is $\delta x_1 = 2\sigma x_1$. This is the interval within which 95% of measured results will be expected to lie when that same set point is repeated in the future. If more than 1 out of 20 future trials at that same set point begin to drift away from the observed mean by more than $\pm 2\delta x_1$, that is another warning that some problem is developing within the rig: something is changing.

The Standard Deviations of the x_i, the σx_1 at these "representative" test points, will be used to estimate the First-Order Uncertainty. That same Standard Deviation can be expected when running other "nearby" set points. This is another application of the "pooled variance" assumption. We assume that although there will be different data values associated with different set points, the unsteadiness in the observations will be about the same, independent of which set point we are running, as long as they are not too far from a set point for which we have some data on the variance. Note that this produces a different kind of estimate of repeatability than collecting the results from repeated trials gathered as add-ons during data production. The First-Order Uncertainty is estimated by 30 repeated datasets taken one after another with the rig running steadily at a single set point. Collecting 30 datasets from tests taken on repeated starts, or on different days, introduces several new possibilities for variability (e.g. How well was the same set point achieved? What was the effect of changes in the barometric pressure, relative humidity, or time of day, etc.?). Knowing the First-Order Uncertainty allows one to learn whether or not results on different days are being affected by some hidden change in conditions: if results on different days display the same scatter as repeated runs on the same day, then there are no additional sources of variation. That means the test is very well controlled.

The First-Order Uncertainty in the results, δZ_1, describes the total random Uncertainty in the results: the combination of random errors from the instrumentation and unsteadiness in the process. No fixed-error components are sampled in the First-Order Uncertainty. If more than 1 out of 20 future trials at that same set point begin to drift away from the observed mean by more than $\pm 2\delta x_1$, that is a warning that some problem is developing within the rig: something is changing.

10.6.3.3 Nth-Order Replication

This is a hypothetical replication, not a real one. The intent is to assess the overall Uncertainty in the reported results, considering all sources, fixed and variable. To make that estimate, we must include an estimate of the fixed errors in the instrumentation. To cast this in a statistical guise, we "imagine" a set of experiments in which the serial numbers of all the instruments were changed from run to run, keeping with the same type of instrument. In that case, the fixed errors in the instrumentation contribute scatter to the data from this hypothetical experiment. To assess the fixed errors, we must rely either on a large sample of calibration results from different instruments or on the manufacturer's data.

The Nth-Order Uncertainty estimates the total Uncertainty in each measurement or in the calculated result. It includes the effects of random instrumentation errors and process unsteadiness, and also the possible fixed errors made in correcting for system/sensor interaction errors, system disturbance errors, and errors due to improperly acknowledging the idealizations and assumptions linking the True Value to the Measured Value.

The Nth-Order Uncertainty of each measurement is calculated by combining all of the residual fixed-error estimates into one value, called the "bias estimate" for that measurement, B_N, and combining the bias estimate with the measured First-Order Uncertainty for that channel. All of these combinations are made using the RSS combination method. If the fixed-error estimate is called the bias estimate, B, then:

$$\delta X_N = \left\{ (B_X)^2 + (\delta X_1)^2 \right\}^{1/2} \tag{10.33}$$

10.6.4 Fractional-Order Replication for Special Cases

The three fundamental replication levels (Zeroth-, First-, and Nth-Order) described in the preceding sections are sufficient for most experimental development work. There are some situations in which intermediate replication levels are useful, especially when looking for the source of scatter in experimental work. Every increase in the versatility of replications (the number of things that might change) will increase the scatter on repeated trials. Sometimes, it is useful to "turn the world on a bit at a time" and see where the trouble starts!

As an example, consider the problem of identifying the source of Uncertainties in the data from performance tests of commercial silicon solar cells under "receiving inspection." In one situation, it appeared that a shipment of commercial cells would have to be rejected, since they showed too much variability in resistance. Taking the cell resistance as the result of the test, a diagnostic test program was put together to determine whether the variance was attributable to the purchased solar cells or to the inspection process.

The program involved the following steps:

i) *Find the Zeroth-Order Uncertainty* by repeated observations of the readings when the measuring instrument was left attached to a reference resistor and read repeatedly

(30 times) over a representative interval of time. Twice the Standard Deviation of a set of 30 observations is a good estimator of the variable error of the measuring system, at 20:1 odds (95% confidence). If S_o, the sample Standard Deviation, is 0.5 Ω then the best estimate of the Zeroth-Order Uncertainty (20:1) is 1.0 Ω. Next, we calculate the variance, S_o^2. In this example, $S_o^2 = 0.25$.

ii) *Find the First-Order Uncertainty* by repeated observations of the measuring instrument, left attached to a single, real solar cell. In the present example, we assume $S_1 = 1.5\,\Omega$, based on a sample of 30 readings. Twice the Standard Deviation of a set of 30 observations is a good estimator of the First-Order Uncertainty in the measurement – the total random error at 95% confidence under test conditions. The square of the Standard Deviation, S_1^2, is the variance under test conditions. The difference $S_1^2 - S_o^2$ represents the variance introduced by unsteadiness in the test conditions, at nominally steady conditions. For the present example, this difference is 2.0. Thus, there is measurable unsteadiness in the test system, beyond that introduced by the instruments.

iii) *Find the First-Order Uncertainty* – a fractional increase in the order of replication by repeatedly attaching and detaching the same specimen cell to the same instrument, making one measurement for each attachment. This experiment introduces an additional Degree of Freedom to the replications: the attachment process. The Standard Deviation of the resulting population will be denoted by $S_{1.1}$. Once again, we estimate the variance at this level from the Standard Deviation and compare the variance at order 1.1 to the Variance at order 1.0. The difference of the variances of these two datasets measures the variance introduced by the act of attaching and detaching the instrument. The square root of the difference of the variances is the Uncertainty contributed by the act of attaching the specimen to the fixture. In the present example, $S_{1.1} = 4.0$. Thus, the 95% confidence Uncertainty interval at order 1.1 is 8.0 Ω, the variance of the data is 16, and the contribution to that total variance that came out of the attachment process is 13.75 units.

iv) *Find the Nth-Order Uncertainty*, the overall Uncertainty of the inspection process, by another set of trials in which a new cell is attached for each measurement (as would normally be done in receiving inspection). Denote the Standard Deviation of a set of 30 trials as S_N, and let $S_N = 4.5\,\Omega$. The Nth-Order variance is 20.25 units, of which 13.75 came from the attachment variance and only 4.25 from variance in the cells themselves!

When fractional replication-level experiments are used, it is important to clearly describe the increase in complexity which accompanies each level and clearly define the operations involved. Since fractional replication-level experiments usually involve only one set of instruments, fixed errors may not enter the picture.

10.6.5 Summary of Single-Sample Uncertainty Levels

10.6.5.1 Zeroth-Order

The Zeroth-Order Uncertainty for a result can be estimated by processing a "typical" dataset through the Uncertainty program using the Zeroth-Order estimates for the Uncertainty of each measurement.

The Zeroth-Order Uncertainty interval for a measurement is the RSS of the residual fixed error of its calibration (estimated at 20:1 odds either from calibration experience or from the manufacturer's specifications) and its variable errors (measured by repeated observations with a steady, known input). The Zeroth-Order Uncertainty estimate includes only contributions arising from the instrumentation system itself. It is used to assess the suitability of a proposed instrumentation system. This is the smallest Uncertainty interval which could be achieved with the proposed instrumentation. The only situation under which this could describe the overall Uncertainty of the experiment would be if the experiment could be replicated with no variations in the process and no Uncertainty in any error corrections.

10.6.5.2 First-Order

The First-Order Uncertainty interval estimates the scatter in the results of repeated trials using the same instrumentation, but with the process running.

The First-Order Uncertainty interval accounts for all variable errors, including the effects of process unsteadiness, but does not include any fixed errors. It differs from the variable error component of the Zeroth-Order Uncertainty because it includes the process variability.

The First-Order Uncertainty is used during the debugging phase of an experiment to assess the significance of scatter in the output and during production data taking to monitor the condition of the test (if results from repeated trials at a reference set point begin to drift outside the allowable band, this means the rig is changing.)

10.6.5.3 Nth-Order

The Nth-Order Uncertainty interval estimates the overall Uncertainty in the result or an individual measurement. It acknowledges all sources of fixed errors, including Uncertainties in the estimates of the corrections applied, and all variable errors, including process unsteadiness.

This is the value which must be reported when results are published, since it includes the fixed errors of the measuring system as well as the total variable error.

References

Abernethy, R.B. and Thompson, J.W. Jr (1973). "Handbook Uncertainty in Gas Turbine Measurements," Arnold Engineering Development Center, AEDC-TR-73-5, AD 755356, Feb. 1973.

Airy, S.G.B. (1879). *Theory of Errors of Observation.* London, England: Macmillan and Company.

ANSI/ASME (1985). *Measurement Uncertainty. Supplement to ASME Performance Test Codes, PTC 19.1-1985.* New York: The American Society of Mechanical Engineers.

Kline, S.J. and McClintock, F.A. (1953). Describing the Uncertainties in single-sample experiments. *Mech. Eng.* February: 3–8.

Moffat, R.J. (1973). The Measurement Chain and Validation of Experimental Measurements. *Acta IMEKO*: 45.

Moffat, R.J. (1982). Contributions to the theory of single-sample Uncertainty Analysis. *ASME J. Fluids Eng.* 104 (June): 250–260.

Moffat, R.J. (1985). Using Uncertainty Analysis in the planning of an experiment. *J. Fluids Eng.* 107 (2): 173–178. https://doi.org/10.1115/1.3242452161-164.

Taylor, J.L. (1986). *Computer-Based Data Acquisition Systems: Design Techniques.* Instrument Society of America.

Further Reading

Dieck, R.H. (1992). *Measurement Uncertainty*. Instrument Society of America.
Abernethy, R.B., Benedict, R.P., and Dowdell, R.B. (1985). ASME measurement Uncertainty. *J. Fluids Eng.* 107 (2): 161–164. https://doi.org/10.1115/1.3242450153-160.
Burns, G.W. and Scroger, M.G. (1989). *NIST Measurement Services: The Calibration of Thermocouples and Thermocouple Materials*, vol. April, 250–235. NIST Special Publication.
Coleman, H.W. and Steele, W.G. (2018). *Experimentation, Validation, and Uncertainty Analysis for Engineers*, 4e. Wiley.
Diamond, W.J. (2001). *Practical Experiment Design for Engineers and Scientists*, 3e. Wiley.
Gülham, A. "Heat Flux Measurements in High Enthalpy Flows," RTO AVT Measurement Techniques for High Enthalpy and Plasma Flows, 25–29 Oct 1999.
International Organization for Standardization. Fluid Flow Measurement Uncertainty. International Organization for Standardization, Technical Committee TC30 SC9. Draft Revision of ISO/DIS 5168. May 1987.
Kline, S.J. (1985). The purposes of Uncertainty Analysis. *J. Fluids Eng.* 107 (2): 153–160. https://doi.org/10.1115/1.3242449.
Lindgren, B.W. (1976). *Statistical Theory*, 3e, 575. MacMillan Publishing Company.
Possolo, A. "Simple Guide for Evaluating and Expressing the Uncertainty of NIST Measurement Results," NIST Technical Note 1900. United States Department of Commerce Technology Administration, National Institute of Standards and Technology, October 2015. This publication is available free of charge from: http://dx.doi.org/10.6028/NIST.TN.1900
Press, W.M., Teukolsky, S.A., and Vetterling, W.T. (2007). Numerical Recipes, 3e. Cambridge University Press.
Range Commanders Council, Telemetry Group. "Uncertainty Analysis Principles and Methods," RCC Document 122–07, September 2007. http://www.dtic.mil/dtic/tr/fulltext/u2/a619551.pdf.
Taylor, B.N. and Kuyatt, C.E. "Guidelines for Evaluating and Expressing the Uncertainty of NIST Measurement Results," NIST Technical Note 1297, 1994. United States Department of Commerce Technology Administration, National Institute of Standards and Technology.
Wells, C.V. "Principles and Applications of Measurement Uncertainty Analysis in Research and Calibration," National Renewable Energy Laboratory. NREL/TP-411-5165, UC Category: 270, DE93000034. https://www.nrel.gov/docs/legosti/old/5165.pdf

Homework

We hope that you will take a look at Exercises 10.7 and 10.8, which highlight current research areas.

Exercises 10.1 through 10.3 require the ideal gas law, written as $\frac{P}{\rho} = RT$, where

P is Pressure [Pa = N/m^2] or [psi]
ρ is Density [kg/m^3]
R is the Specific gas constant, specific to each gas
T is Temperature [K]

Gas	Specific Gas Constant, R [J/kg K]	Molecular Weight [g/mol]
Air	287.05	28.965
Carbon dioxide	188.92	44.01
Helium	2077.1	4.003
Hydrogen	4124.7	2.016
Universal R	8.31446 [J/mol K]	

10.1 Your laboratory has a tank equipped with a thermometer which can read from −30 °C to 150 °C, marked in intervals of 2 °C. The tank absolute pressure gauge can read from 50 to 500 kPa in intervals of 10 kPa. The tank contains 370 kPa air at room temperature of 22°C. What is the density inside the tank? Quantify the Uncertainty of your density estimate. Justify your values. What are your answers if the tank contains helium? Carbon dioxide?

10.2 Your laboratory has a tank equipped with a thermometer which can read from −30 °C to 150 °C, marked in intervals of 2 °C. The tank pressure gauge can read from 0 to 3000 kPa in intervals of 50 kPa. A barometer in the room is reading 29.8 inHg and a thermometer is reading 77 °F. The tank contains 550 kPa air at room temperature. What is the density inside the tank? Quantify the Uncertainty of your density estimate. Justify your values. What are your answers if the tank contains helium? Carbon dioxide?

10.3 Your laboratory has a test section with ample optical access. All pressure and temperature probes are located on the wall. Laser diagnostics and schlieren are used to measure absolute temperature, density, and velocities within the three-dimensional flow field. The fluid is carbon dioxide gas at a nominal temperature of 47 °C and nominal absolute pressure 110 kPa. Based just on the ideal gas law, write an overall Uncertainty equation for pressure in terms of the measured quantities.

Exercises 10.4 through 10.6 require the following definitions.

h	Convective heat-transfer coefficient, [W/m^{2}K]
k	Thermal conductivity, [W/m K]
A	Frontal area = length · diameter = $L \cdot D$, [m^2]
T	Surface temperature, [°C]
T_∞	Fluid stream temperature, [°C]
M	Mass of cylindrical rod specimen, [kg]
$\overline{c}$	Specific heat capacity of cylindrical rod material, [J/kg °C]
t	Time, [s]
τ	Characteristic time, [s]
P	Power added to heat the system, [W]

10.4 The convective heat-transfer coefficient h is a convenient model for engineering designs. When the temperature of a solid surface and flowing fluid are both known, as well as the exposed surface area, then h allows the designer to estimate the amount of heat transferred from the surface to the fluid. Your job, using an experiment designed to mimic the situation, is to provide the value h to the engineer. Along with the model value, you must also provide its 95% Uncertainty. The equation to estimate h is given as

$$h = \frac{P}{A\left(T - T_{\infty}\right)} \tag{10.34}$$

The nominal values measured are

Power	P = 200.0 W
Diameter	D = 0.03 m
Length	L = 0.3 m

For the instruments available, the Uncertainties are

δP =	0.5 W
δD =	0.02 mm
δL =	0.1 mm
$\delta \Delta T$ =	0.2 °C

10.5 When a hot object is suddenly immersed in a fluid (for example, quenching or diving), a transient technique is used to experimentally estimate the convective heat-transfer coefficient h. Your job, using an experiment designed to mimic the actual situation, is to provide the value h to the design engineer. Along with the model value, you must also provide its 95% Uncertainty. The equation to estimate h is given as

$$\tau = \frac{t_2 - t_1}{\ln\left(\dfrac{T_1 - T_{\infty}}{T_2 - T_{\infty}}\right)} \tag{10.35}$$

$$h = \frac{M\bar{c}}{\tau A} \tag{10.36}$$

The nominal values measured are

Diameter	D = 0.03 m
Length	L = 0.3 m

For the precision of your instruments, the following local Uncertainties were apparent:

$\delta t =$	0.0025	s
$\delta T =$	0.2	K
$\delta M =$	0.001	kg
$\delta \bar{c} =$	0.004 kJ/kg K	
$\delta D =$	0.025	mm
$\delta L =$	0.10	mm

10.6 Your job is to provide to the Bureau of Standards a value for thermal diffusivity. Thermal diffusivity is a measure of how quickly a temperature spreads through an object. As an analogy, thermal diffusivity is the reciprocal of thermal inertia; if an object has a high thermal diffusivity, the temperature permeates quickly; if an object has a low thermal diffusivity, a hot spot persists, spreading slowly. The definition for thermal diffusivity α is

$$\alpha \equiv \frac{k}{\rho \bar{c}} = \frac{k A D}{M \bar{c}}$$

Taking a peek at Eq. 10.8 above, this definition is nearly a reciprocal.

The Bureau of Standards will provide you with the accepted value of thermal conductivity k for the material, along with its Uncertainty. Using the same values as Exercise 10.5, provide an equation for the value α in terms of k. Also provide an equation for its 95% Uncertainty.

10.7 The value for the gravitational constant is often given as 9.8 m/s^2 in engineering courses. Some elementary physics classes use the convenient value 10 m/s^2, which begins with an Uncertainty of only 2%. The definition for the gravitation constant at the surface of the Earth, where M_E is the mass of the Earth and where R_E is the local radius of the Earth, is given by the equation

$$g = \frac{G M_E}{R_E^2}$$

The gravitational constant remains a current and relevant research area. Newton's gravitational constant $G = 6.67408 \times 10^{-11}$ $[m^3/kg\ s^2]$ has a Relative Uncertainty of about 50×10^{-6} or 50 ppm. Of all universal constants in physics, Newton's gravitational constant G is about the least precise. The reasons for G lacking precision are twofold: (1) every object with mass in the universe affects the mass we are studying (including every distant black hole, evidenced by the tide induced by the moon), and (2) gravity is much weaker than other physics forces (by almost 40 orders of magnitude).

For the Earth's values:

the nominal mass is $M_E = 5.9722 \times 10^{24}$ kg, with a Relative Uncertainty of 10^{-4}.

the nominal radius is R_E = 3963 miles at its equator sea level, add for tides and ocean waves or mountains, with a Relative Uncertainty of 10^{-2}.

Using these values, write your most accurate estimate for the surface gravitational constant g. Also provide its overall Uncertainty.

10.8 The basic equation for thermal radiation heat transfer is

$$\dot{Q} = \sigma \in A\left(T_1^4 - T_2^4\right)$$

where the Stefan–Boltzmann constant σ = 5.670367 $\times 10^{-8}$ [W/m^2 K^4] with a Relative Uncertainty of about 2.3 × 10^{-6} or 2.3 ppm.

the emissivity ε, in the range from 0 to 1, is a function of light wavelength.

The nominal values are:

T_{Sun}	5780	K
T_{Earth}	280	K
T_{Space}	2.725 K	

10.8.1 For incoming radiation from the sun to the Earth, the area is the sunlit frontal area πR_E^2. Let the emissivity ε =1. What percentage of the thermal radiation heat transfer between the sun and Earth is due to the sun? How much due to the Earth? What is the relative percentage?

10.8.2 For outgoing radiation from the Earth to space, the area is the full surface of the Earth $4\pi R_E^2$. Let the emissivity ε = 0.6. What percentage of the thermal radiation heat transfer between the Earth and space is due to the Earth? How much due to space? What is the relative percentage?

10.8.3 Write the overall Uncertainty equation for the basic equation.

11

Using Uncertainty Analysis in Planning and Execution

Uncertainty Analysis of a proposed experiment can pay big dividends in the planning stage, providing guidance for both the overall plan and for the execution of the details. When two or more approaches are available for the same task, a question frequently raised is, "Which method will produce the best data?" If the word "best" is interpreted to mean "least uncertain," the question can be answered before the experiment is ever run, using the techniques of Uncertainty Analysis. In this chapter, we consider two different approaches for performing a hypothetical experiment, either of which could be used. Uncertainty Analysis guides to the better method by estimating the Uncertainty which would result from each method. The steps forward in this chapter (by section) are:

11.1: Use Uncertainty Analysis to guide the choice of experimental plan as we deduce the Data-Reduction Equations (DREs).
11.2: Perform Uncertainty Analysis on the DREs; compare and choose.
11.3: Use Uncertainty Analysis in selecting instruments as we deduce the data interpretation program (DIP).
11.4: Use Uncertainty Analysis in debugging the experiment.
11.5: Organize the reporting of Uncertainties in the experiment.
11.6: Apply Multiple-Sample Uncertainty Analysis.
11.7: Coordinate with Uncertainty Analysis standards.
11.8: Describe the overall Uncertainty in a single measurement.
11.9: Apply additional statistical tools to reduce Uncertainty.

11.1 Using Uncertainty Analysis in Planning

The application of Uncertainty Analysis to a proposed test program is called "a priori Uncertainty Analysis" or "pre-test analysis" in Abernethy (1981). This chapter describes the use of *a priori* Uncertainty Analysis in choosing between two proposed methods for measuring the heat-transfer coefficient from a cylindrical rod to an air stream. Akin to the wind chill of an exposed arm in winter, this issue often arises in technology.

Our analysis starts with the DRE for each method (the equations by which the result is computed from raw data), along with estimated values of the Uncertainty in each piece of input data. The Uncertainties in the results are then calculated for representative

Planning and Executing Credible Experiments: A Guidebook for Engineering, Science, Industrial Processes, Agriculture, and Business, First Edition. Robert J. Moffat and Roy W. Henk.

Companion website: www.wiley.com/go/moffat/planning

points across the domain of the proposed experiment by each of the two methods. By comparing the Uncertainties of the two methods, the better choice becomes obvious.

This chapter focuses on the use of Uncertainty Analysis as an aid in decision-making. The techniques of Uncertainty Analysis itself were discussed in Chapter 10.

11.1.1 The Physical Situation and Energy Analysis

In weather forecasts, "wind chill" is a familiar term for how wind makes air feel colder. Imagine your arm exposed to winter wind at varying speeds. Your blood adds heat to the arm. The rate of heat loss from your arm increases as the apparent temperature of the wind decreases, hence "wind chill" temperature. Many assumptions go into calculating the wind chill temperature since the purpose is a simple way to warn the public, rather than for engineering design.

In engineering, rather than wind chill temperature, we measure the convective heat-transfer coefficient from an existing test specimen (a cylindrical rod) in an airstream over a range of velocities. The objective is to produce a reference-quality dataset describing the "constant properties" heat-transfer coefficient over a range of Reynolds numbers. The test is to be designed to achieve the minimum Uncertainty that can be realized using available instruments and equipment. All testing is to be conducted at approximately ambient conditions of temperature and pressure, using an existing test tunnel, but the velocity can be varied over a wide range.

Figure 11.1 shows the test specimen and illustrates the energy-flow terms recognized as "possibly important" in establishing its thermal response.

We begin from first principles. The basic law of physics which applies is conservation of energy, i.e. the first law of thermodynamics. The conservation law for this case in words is: the power input equals the rate of heat loss by (convection + radiation + conduction) + the rate that energy is stored in the cylinder (by temperature rise).

Considering a control volume that includes all of the test specimen but is infinitesimally outside its surface, an energy balance applying standard engineering models becomes Eq. 11.1:

$$hA(T-T_\infty)+A\varepsilon\sigma\left(T^4-T_W^4\right)+2K_e(T-T_\infty)+M\bar{c}\frac{dT}{d\theta}-P=0 \tag{11.1}$$

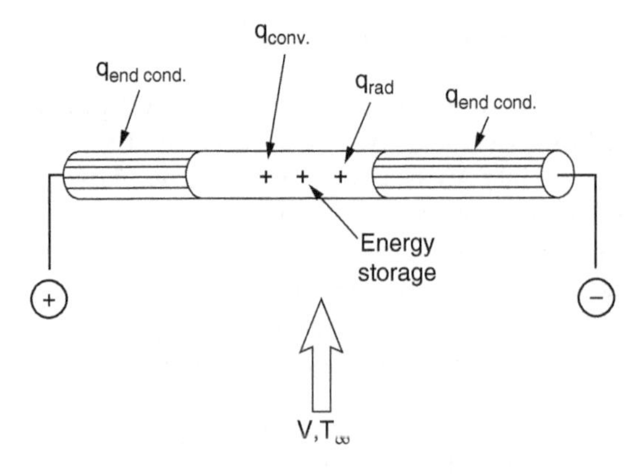

Figure 11.1 The test specimen.

where the terms are defined as:

h	Convective heat-transfer coefficient, W/m^2K
A	Frontal area = length · diameter, m^2
T	Surface temperature, °C
T_∞	Fluid stream temperature, °C
T_W	Duct wall temperature, °C
ε	Emissivity
σ	Stefan–Boltzmann constant, W/m^2K^4
Ke	Conductive heat-transfer coefficient, system, W/K
M	Mass of cylindrical rod specimen, kg
$\bar{c}$	Specific heat of cylindrical rod material, J/kg°C
θ	Time, s
P	Power added to heat the system, W

In this equation, it is assumed that the temperature inside the rod is uniform, so that the single symbol, T, represents both the surface temperature and the average internal temperature. This assumption limits the test range to "relatively low" h values, so that the internal temperature gradient is negligible compared to the temperature difference used in defining h. The usual method of assessing a negligible gradient (how uniform is the rod temperature) is to calculate the nondimensional Biot number hr/k and to require that it be 0.01 or less.

This energy-balance equation is the basis for both methods considered for the proposed experiment – the steady-state method and the transient method. For each method we will create a DRE, on which we perform Uncertainty Analysis.

11.1.2 The Steady-State Method

Equation 11.1 can be solved for h by simply rearranging it. If, at the same time, a linearized approximation is substituted for the radiation term and the temperature of the tunnel walls is assumed to be the same as the tunnel air, a particularly handy form of the equation emerges.

In Eq. 11.2, the first term represents the simplest approximation to h, and the following terms are the corrections which should be applied to account for radiation $\left(4\varepsilon\sigma T_m^3\right)$, conduction ($2K_e$), and unsteady test conditions, $\left(M\bar{c}dT/d\theta\right)/\left(A\left(T-T_\infty\right)\right)$.

$$h=\frac{P}{A\left(T-T_\infty\right)}-4\varepsilon\sigma T_m^3-2K_e-\frac{M\bar{c}}{A}\frac{dT/d\theta}{\left(T-T_\infty\right)} \tag{11.2}$$

To simplify the Uncertainty Analysis of this method, our goal of the moment, assume the conduction term as well as the radiation term are negligible. Furthermore, for steady state the unsteady term is dropped. The resulting equation is

$$h=\frac{P}{A\left(T-T_\infty\right)} \tag{11.3}$$

Equation 11.3 is the handy DRE for steady-state tests – familiar as the first equation students learn for convection heat transfer. The Uncertainty Analysis will be based on this equation.

11.1.3 The Transient Method

In the transient method, the specimen is first heated above the air temperature and then quickly inserted into the flow. (The intent is to produce the equivalent of a "step change" in temperature of the surroundings. There are several ways to accomplish this goal, some of which don't involve actually moving the test piece.) The temperature of the specimen is recorded during the cooling interval, and the heat-transfer coefficient deduced from the time–temperature history. Typically, there is no power applied during the cooling transient; hence $P = 0$.

The DRE for the transient method is derived from Eq. 11.1. By linearizing the radiation term and collecting all heat-transfer terms, an "overall h" is defined as

$$h_o = h + 4\sigma\varepsilon T_m^3 + \frac{2K_e}{A} \tag{11.4}$$

With the use of h_o, Eq. 11.1 becomes

$$h_o A(T - T_\infty) + M\bar{c}\frac{dT}{d\theta} = 0 \tag{11.5}$$

If h_o remains constant throughout the transient, this equation has a simple solution. Ignoring variable fluid properties effects, h can be expected to be constant if the Reynolds number is constant (i.e. if the flow is steady).

Values of $(T - T_\infty)$ versus time are shown in Figure 11.2 on semi-log coordinates. The "characteristic time" of the system, τ, can be found from

$$\tau = \frac{t_2 - t_1}{\ln\left(\dfrac{T_1 - T_\infty}{T_2 - T_\infty}\right)} \tag{11.6}$$

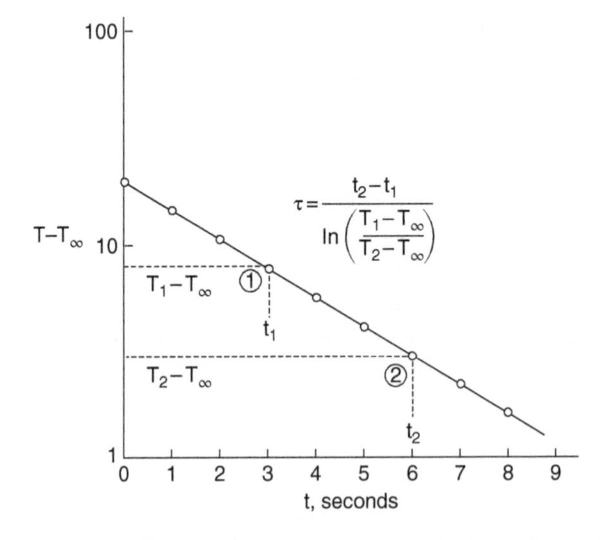

Figure 11.2 Semi-log plot of transient response of a first-order system.

Any two points within the linear portion of the curve can be used as points 1 and 2. This method is preferred over the use of a single measurement of the $1-e^{(-1)}=0.632$ completion time, because it uses more information from the test, is less susceptible to error, and is self-validating. (If the thermal response of the system cannot be described by a simple linear, first-order equation [such as Eq. 11.5], the line will not be straight.) If the plot exhibits curvature in these coordinates, this is evidence that the proposed experimental procedure is not applicable, and some other procedure must be sought.

The value of h_o is found using the definition of "characteristic time" and h found from h_o by subtracting the terms originally embedded in it (see Eq. 11.7 for those terms).

$$h = \frac{M\bar{c}}{\tau A} - 4\sigma\varepsilon T_m^3 - \frac{2K_e}{A} \tag{11.7}$$

For simplicity in illustrating the use of Uncertainty Analysis, only the first term of Eq. 11.7 will be used.

Thus, the DRE for the transient method becomes Eq. 11.8.

$$h = \frac{M\bar{c}}{\tau A} \tag{11.8}$$

11.1.4 Reflecting on Assumptions Made During DRE Derivations

Have you noticed, or been perturbed by, the many assumptions made during our derivations of the DREs? We ignored some terms and decided other contributing factors were negligible. We even assumed the convective heat-transfer coefficient h was constant and then proceeded to derive equations for how it changed. Indeed, h is known to range from less than one to several million, dependent upon flow conditions. Even for one simple flow, Figure 10.1 showed that the value of h can vary over three orders of magnitude; scatter around the fitted curve confirms the Uncertainty of the h value is quantifiable. Might you question the validity of the DREs we derived? Good for you!

Let's consider together. Perhaps some observations will assuage you:

- If your field is industrial design, agriculture, economics, biology, animal behavior, human behavior, chemistry, medicine, politics, etc., then as an experimentalist you are used to modeling equations of complex systems with a reduced set of pertinent factors. You may be amused that this area of physics does too.
- If your field is physics or a related engineering field, you derive equations from laws which cannot be violated by humans, creatures, or inanimate objects. Such laws include the second law of thermodynamics, conservation of energy, Einstein's laws, etc. Recall Feynman's challenge about turbulence being the unsolved area of classical physics, Section 2.4.6 and Panel 2.2. Thermo-fluid physics is complex. As workers in the field, we welcome your expert contribution.
- If your laboratory uses conditioned air, controlled temperatures, or cooled electronics, the engineering design depended on simplifications such as these.
- The simplifications actually do result in the useful ability to answer questions.

- As it is challenging classical physics, there will always be a need for research closer to first principles that neglect fewer effects. The equations we actually use in the laboratory are more complex than our simplified examples here. Here our goal was to show steps of Uncertainty Analysis in experiment planning
- Thermo-fluid physics is essential from nano-processes in cell life up to the evolution and life cycle of celestial stars. From breath (not inhaling the exact same air you exhaled), to oceans, to ocean life, to weather, to neutron stars, to cream in coffee cups, it applies. It provides beauty and surprise as well: flowing hair, flags, flames, clouds.

11.2 Perform Uncertainty Analysis on the DREs

11.2.1 Uncertainty Analysis: General Form

Execution of an Uncertainty Analysis requires knowledge of the Uncertainty in each measurement, as well as the value itself. The Uncertainty in each recorded piece of data is considered to be

i) Independent.
ii) From a Gaussian population of possible values.
iii) The Uncertainty interval quoted for each variable must be quoted at the same "odds."

Note that the data values need not be independent, only their Uncertainties. As introduced in Eq. 10.12, a complete entry is

$$x_i = x_i \pm \delta x_i \quad (20:1)$$

Recall that Eq. 10.12 means: "The observed value was x_i and 19 of 20 readings of x_i are expected to lie within $\pm\,\delta x_i$ of that value, if repeated readings were taken following the prescribed replication pattern." It is important that δx_i accurately reflects the Uncertainty in x_i for the considered level of replication.

For a result, Z, calculated from a set of input data ($x_1, x_2, x_3, \ldots, x_N$), Eq. 10.22 gave the Uncertainty as

$$\delta Z = \left\{\left(\frac{\partial Z}{\partial x_1}\delta x_1\right)^2 + \left(\frac{\partial Z}{\partial x_2}\delta x_2\right)^2 + \left(\frac{\partial Z}{\partial x_3}\delta x_3\right)^2 + \ldots + \left(\frac{\partial Z}{\partial x_N}\delta x_N\right)^2\right\}^{1/2}$$

For the special case where the result Z can be written as a product of terms, where each term is raised to some power, as in Eq. 10.23, we have

$$Z = C x_1^a x_2^b x_3^c x_4^d \ldots$$

then the Relative Uncertainty, $\delta Z/Z$, can be found as in Eq. 10.24:

$$\frac{\delta Z}{Z} = \left\{\left(a\frac{\delta x_1}{x_1}\right)^2 + \left(b\frac{\delta x_2}{x_2}\right)^2 + \left(c\frac{\delta x_3}{x_3}\right)^2 + \left(d\frac{\delta x_4}{x_4}\right)^2 + \ldots\right\}^{1/2}$$

This form is particularly useful if the Uncertainties in the input values are known as "percent of reading" or can easily be put in that form before the Uncertainty Analysis is done.

11.2.2 Uncertainty Analysis of the Steady-State Method

In the steady-state method, h is formally found using Eq. 11.2, but for demonstration of Uncertainty Analysis, Eq. 11.3 will be used. First, we see that the frontal area A equals the length L times the diameter D. We simply denote the temperature difference as $\Delta T \equiv (T - T_\infty)$. Then we may cast Eq. 11.3 in the power form (Eq. 10.23), obtaining

$$h = P^1 D^{-1} L^{-1} \Delta T^{-1} \tag{11.9}$$

Let's see how this allows for simple calculation of the Relative Uncertainty.

Applying Eq. 10.24 to the DRE for the steady-state case yields the following equation for the Relative Uncertainty in h:

$$\frac{\delta h}{h} = \left\{ \left(\frac{\delta P}{P}\right)^2 + \left(-1 \cdot \frac{\delta D}{D}\right)^2 + \left(-1 \cdot \frac{\delta L}{L}\right)^2 + \left(-1 \cdot \frac{\delta \Delta T}{\Delta T}\right)^2 \right\}^{1/2} \quad \text{or more simply}$$

$$\frac{\delta h}{h} = \left\{ \left(\frac{\delta P}{P}\right)^2 + \left(\frac{\delta D}{D}\right)^2 + \left(\frac{\delta L}{L}\right)^2 + \left(\frac{\delta \Delta T}{\Delta T}\right)^2 \right\}^{1/2} \tag{11.10}$$

In the present case, the existing specimen has the following fixed dimensions:

Diameter	$D = 0.0254\,\text{m}$
Length	$L = 0.254\ \text{m}$

For the instruments available, the Uncertainty estimates were assumed to be:

$$\begin{aligned} \delta P &= 0.5 \quad \text{W} \\ \delta D &= 0.025\,\text{mm} \\ \delta L &= 0.125\,\text{mm} \\ \delta \Delta T &= 0.2 \quad {}^\circ\text{C} \end{aligned}$$

The effect of these Uncertainties will depend on the values of P, D, L, and ΔT, and some of those values will change with test conditions. To estimate the Uncertainty before the test is run, some decisions must be made about the conduct of the test.

In heat-transfer measurements, it is good practice to keep the surface temperature constant throughout the test, in order to reduce the effect of changes in fluid properties. If ΔT is chosen to be 20 °C, the Relative Uncertainty in ΔT will be 1%, an acceptable level. The overall Uncertainty of the measurement of h will then depend on the power level, P, which will change as the air velocity changes. To estimate this effect, only an estimate of the range of expected h is needed. The Uncertainty in h can then be calculated as a function of h.

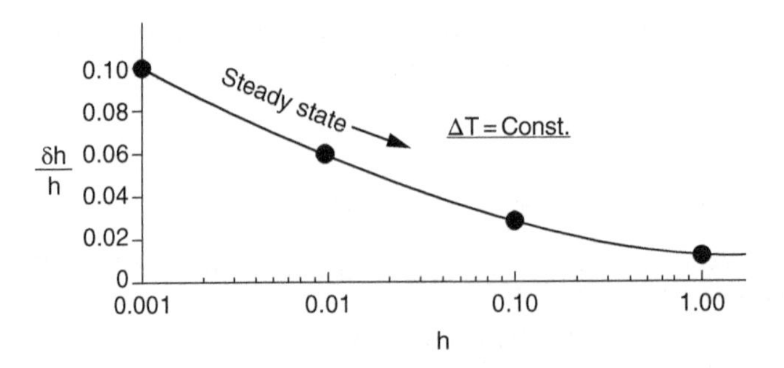

Figure 11.3 The Uncertainty is largest at low values of *h* when the test is run this way.

Based on results of related or similar tests, it was judged that *h* would probably lie between 10 and 1000 W/m^2 K. For the given geometry, for several possible values of *h*, the power, *P*, was calculated using Eq. 11.3. Then, for each of those values of *h*, the Relative Uncertainty was estimated using Eq. 11.10. The results are shown in Figure 11.3.

The Relative Uncertainty in *h* is largest at low values of *h*, where the Uncertainty in *P* dominates the calculation. As *h* increases, the Uncertainty decreases. It is important to keep in mind that this behavior arises from the decision to maintain a constant surface temperature on the specimen – it is that decision that controls the variation of Uncertainty. If a constant power test were selected, the result would be different.

It is important to "stay loose" when planning an experiment and not get locked into a particular approach without considering other options: the first idea is rarely the best one!

11.2.3 Uncertainty Analysis – Transient Method

In the transient method, *h* is found using Eq. 11.7, which includes several terms. To simplify the illustration of the Uncertainty Analysis, Eq. 11.8 will be used. As previously, we see that the frontal area *A* equals the length *L* times the diameter *D*. Then when we cast Eq. 11.8 in the power form (Eq. 10.23), we obtain Eq. 11.11. (The parentheses are used so the bar c does not look like it has "−1" exponent.)

$$h = M^1 (\bar{c})^1 \tau^{-1} D^{-1} L^{-1} \tag{11.11}$$

Once again, the form allows a simple determination of the Relative Uncertainty, $\delta h/h$, following Eq. 9.24:

$$\frac{\delta h}{h} = \left\{ \left(\frac{\delta M}{M}\right)^2 + \left(\frac{\delta \bar{c}}{\bar{c}}\right)^2 + \left(-1 \cdot \frac{\delta \tau}{\tau}\right)^2 + \left(-1 \cdot \frac{\delta D}{D}\right)^2 + \left(-1 \cdot \frac{\delta L}{L}\right)^2 \right\}^{1/2} \quad \text{or more simply}$$

$$\frac{\delta h}{h} = \left\{ \left(\frac{\delta M}{M}\right)^2 + \left(\frac{\delta \bar{c}}{\bar{c}}\right)^2 + \left(\frac{\delta \tau}{\tau}\right)^2 + \left(\frac{\delta D}{D}\right)^2 + \left(\frac{\delta L}{L}\right)^2 \right\}^{1/2} \tag{11.12}$$

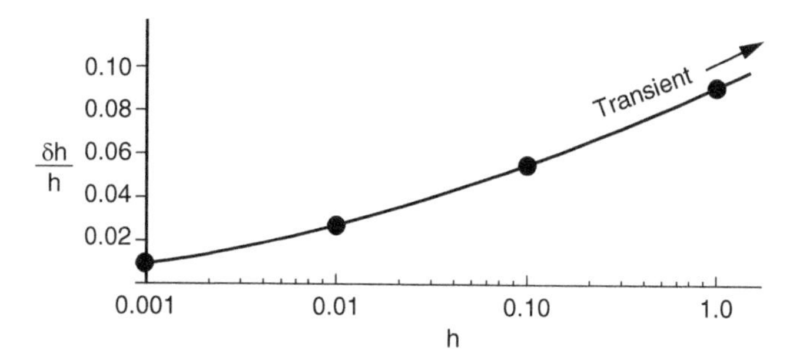

Figure 11.4 The Relative Uncertainty using the transient method is low at low values of h and becomes larger at high values.

The data list is quite different from that of the steady-state method, and some new Uncertainty estimates are needed. The following values were used:

$$\begin{aligned} \delta t &= 0.0025 \quad \text{s} \\ \delta T &= 0.2 \quad \text{K} \\ \delta M &= 0.001 \quad \text{kg} \\ \delta \bar{c} &= 0.004\,\text{kJ/kg K} \\ \delta D &= 0.025 \quad \text{mm} \\ \delta L &= 0.125 \quad \text{mm} \end{aligned}$$

For the range of h already defined, values of $\delta h/h$ were calculated using Eq. 11.8 and used to predict the Relative Uncertainty in h at each point in the operating domain of the experiment.

The results are shown in Figure 11.4. The Uncertainty increases as h increases because the ambiguity in determining the time constant becomes a larger fraction of the value of the time constant. There are various ways in which the magnitude of the Uncertainty might be changed and, to some extent, the shape of the curve, but the general shape is set by the physics of the situation.

11.2.4 Compare the Results of Uncertainty Analysis of the Methods

The objective of the exercise is to select one of the two methods as the plan for executing the experiment. Our criterion is to obtain the minimum Uncertainty, using the suite of instruments and the equipment available. The simplest way to make this comparison is simply to plot both Uncertainty curves on the same sheet and look for the cross-over point.

Planning Decisions

i) The Nth-Order Uncertainty was chosen for this comparison, since that is the value that will be reported in the publication of experimental results.
ii) Figure 11.5 shows the cross-over point to be at $h = 0.04\,W/cm^2C$.
iii) For tests where h is expected to be above the cross-over value, the steady-state method should be specified.
iv) For tests where h is expected to be below the cross-over value, the transient method would yield lower uncertainties.

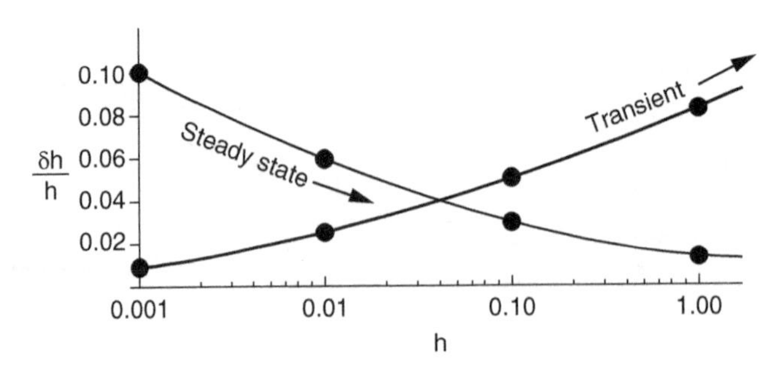

Figure 11.5 Direct comparison of the Uncertainties allows selection of the better method based on the expected range of the result.

In the present examples, the scope of the experimental variations was limited in order to channel the development quickly toward a single solution. In a more general case, other Degrees of Freedom might exist, and one might find that the dimensions of the test specimen, the test velocity, or the operating temperature difference could be changed to lower the Uncertainty. It may begin to appear that the possibilities are endless, but the important point to remember is that desk time is cheaper than test time. Uncertainty Analysis will frequently reveal that an apparently plausible test program simply cannot deliver the required precision.

11.2.5 What Does the Calculated Uncertainty Interval Mean?

The Uncertainty intervals used as input for the steady-state method were δP, δD, δL, and δT, while for the transient method, the inputs were $\delta\tau$, δD, δL, δM, δT, and δc_p. The individual estimates included allowances for both the reading Uncertainties and the calibration Uncertainties in the instruments involved; they are N^{th}-Order Uncertainty interval estimates. The calculated Uncertainty interval for the result is thus an estimate of the Standard Deviation of the population of all possible experiments whose data were reduced using these same equations, with these same corrections. Any person, in any lab, using any set of instruments, should produce results which agree with these predictions. If they don't, then there are differences in the experiments that need to be discovered.

The Nth-Order Uncertainty is the Uncertainty that should be quoted in reporting the work to the outside world.

The results of repeated trials with the same specimen and the same instruments will display much less scatter than the prediction just made. Repeated trials on the same system should show only the First-Order Uncertainty since only the variable Uncertainties are sampled. The calibration Uncertainties of the instruments don't introduce scatter at this level of replication.

Experimenters frequently calculate the Nth-Order Uncertainty interval for their tests and then are "pleased" that their results on repeated trials lie well within the calculated Uncertainty! This does not have much meaning. The scatter on repeated trials should not be judged against the Nth-Order Uncertainty but against the First-Order interval, which is always smaller and sometimes much smaller. Recall that we saw in Chapter 7,

Figure 7.7 how two different analyses resulted in vastly different Uncertainty intervals, with the more precise analysis being more credible.

The First-Order Uncertainty interval is used mainly in the "debugging" phase of an experiment, in which unexpectedly large scatter means that the experiment is unstable.

The Nth-Order Uncertainty interval is the only value that should be reported with the data for external users. It is this value which relates to the agreement which others may expect to find with the reported data.

11.2.6 Cross-Checking the Experiment

In a critical situation, an experiment such as this should be equipped to run by both methods, thus providing an opportunity for validating the experiment by cross-checking the two methods. In the vicinity of the cross-over point, both methods are equally certain; therefore, neither is penalized by a direct comparison. If the two results agree within the expected interval, this is strong evidence that the results are correct. This follows from the fact that the two methods have no data in common except the surface area of the specimen; there is no other single parameter which can affect both methods in the same way. If the two methods agree, they are probably both correct.

The test for agreement between two results is to compare the difference between the two with the calculated Uncertainty interval for the difference ($Z_1 - Z_2$). This is the interval within which the observed difference would lie, 19 times out of 20, by chance alone. If the observed difference lies outside this interval, there is only one chance in 20 that the results are the same.

Since the Relative Uncertainty in each result is already known, the calculation is simply:

$$\delta(Z_1 - Z_2) = \left\{(\delta Z_1)^2 + (\delta Z_2)^2\right\}^{1/2} \tag{11.13}$$

Cross-checking the wind chill example discussion: since blood is unlikely to maintain full body temperature of the arm, the full Eq. 11.2 is needed to include both energy input and temporal cooling of the cylinder. Wind chill is not presented at 20:1 odds, so don't bet your life or your arm on the forecast. Be warned and keep warm. For the wind chill Uncertainty interval, we advise the following two-step calculation:

i) Calculate the difference dT between the static temperature and the wind chill temperature.
ii) Estimate the Uncertainty interval of the wind chill is uneven $+0.5\ dT$; $-2\ dT$.

11.2.7 Conclusions

Uncertainty Analyses should be done as a routine part of the planning for experiments, not simply as go/no-go tests on the plan, but to contribute to the decision process concerning the test methods and objectives.

The Uncertainty interval for most experiments is dominated by the behavior of a few terms – frequently one or two. For planning purposes, it is generally sufficient to work with these dominant terms only.

Uncertainty estimates should be made at Zeroth-, First-, and Nth-Order, to assess the proposed instruments, to provide proper guidance in debugging, and to ensure proper evaluation for reporting.

11.3 Using Uncertainty Analysis in Selecting Instruments

In most experiments, some measurements are more important than others in determining the accuracy of the final result. It makes good sense to ensure that the most critical instruments are of the highest quality and to apportion the instrumentation budget according to the role of the instruments.

Uncertainty Analysis can be used to formalize the process of identifying the critical instruments. The key is to identify the sensitivity coefficient for each measurement and put the highest-quality instruments on the measurements with the high sensitivity coefficients.

This means that the DIP must be laid out before the instruments are selected. It is only through the equations used to calculate the result that one can determine the sensitivity coefficients.

Consider the evaluation of effectiveness of a heat exchanger. Assuming that the cold fluid is the fluid of minimum capacity rate, the effectiveness can be calculated using a simple temperature rise ratio. The DIP is

$$\xi = \frac{T_{cold,out} - T_{cold,in}}{T_{hot,in} - T_{cold,in}} \tag{11.14}$$

The cold fluid inlet temperature appears twice in this equation, but, more than likely, it is measured only once, and the same value inserted into both locations. If that is so, then this equation has only three variables (not four), and the Relative Uncertainty in effectiveness can be found from

$$\frac{\delta\xi}{\xi} = \frac{1}{\left(T_{hot,in} - T_{cold,in}\right)}\left\{\left(\frac{\xi-1}{\xi}\delta T_{cold,in}\right)^2 + \left(\frac{1}{\xi}\delta T_{cold,out}\right)^2 + \left(\delta T_{hot,in}\right)^2\right\}^{1/2} \tag{11.15}$$

For a given set of inlet temperatures, the coefficients of the terms inside the parentheses can be used as sensitivity coefficient. Note that they each involve the value of the effectiveness, ξ. We should then expect that the relative importance of the different measurements would depend on whether we are testing a high-effectiveness exchanger or a low-effectiveness one. This is demonstrated in Table 11.1.

If the tests will be run mainly on low-effectiveness heat exchangers, it is far more important to have accurate measurements on the cold fluid temperatures, $T_{cold,in}$ and $T_{cold,out}$, than on the hot fluid. At high effectiveness, it is more important to get $T_{hot,in}$ and the $T_{cold,out}$ temperatures as accurately as possible.

As these results show, it is not always obvious where the most accurate instruments are needed. The only way to tell is to examine their effects on the result, and the best way to do that is by executing an Uncertainty Analysis on the proposed data interpretation equations before the instruments are specified.

Table 11.1 Sensitivity coefficients.

Effectiveness	$T_{cold,in}$	$T_{cold,out}$	$T_{hot,in}$
ξ	$\frac{\xi-1}{\xi}$	$\frac{1}{\xi}$	-1
0.1	−9.0	10.0	−1.0
0.2	−4.0	5.0	−1.0
0.4	−1.5	2.5	−1.0
0.6	−0.66	1.66	−1.0
0.8	−0.25	1.25	−1.0
1.0	–	1.00	−1.0

11.4 Using Uncertainty Analysis in Debugging an Experiment

The word "debugging" is used to describe the second phase of development of a new experiment. The first phase is the initial "shakedown" period in which one stops leaks, improves the speed control, and generally "makes it run." In the next phase, one makes it "run right." This requires cleaning up the hardware and software so the output from the system (hardware plus software) can be trusted. This process of removing errors from an experiment is called "debugging."

Uncertainty Analysis is the most powerful aid an experimenter can wield in the debugging process. The debugging process hinges on distinguishing significant events from insignificant ones. The question is, "How can one tell a significant change in the result from an insignificant one?" The answer, as you might suspect, lies in using Uncertainty Analysis to determine whether or not the observed change can be accounted for by chance alone, based on the already-recognized Uncertainties in the data.

11.4.1 Handling Overall Scatter

Repeated trials of an experiment will always display scatter. The more versatile the replications, and the larger the time interval between trials, the greater will be the scatter. Whether that scatter is meaningful or not depends upon how large it is compared with the First-Order Uncertainty interval. If 19 of 20 trials lie within the First-Order interval, the system is behaving as expected. If not, if the scatter is significantly larger than expected, then some significant variable is not being properly controlled or accounted for in the Data Reduction Program.

It is helpful, in discussing the debugging process, to look at the entire experimental system as a single instrument, designed to measure the intended result. From that view, it is intuitively reasonable to ask, "How will I calibrate this instrument?" This viewpoint is illustrated in Figure 11.6 (which was first seen in Chapter 1, as Figure 1.1).

As with any instrument, the answer is, "You calibrate the experiment by comparing its output with a standard result."

Before an experiment can be calibrated, it must at least provide repeatable results. So, in debugging an experiment, the first task is to make the system behave repeatably. If

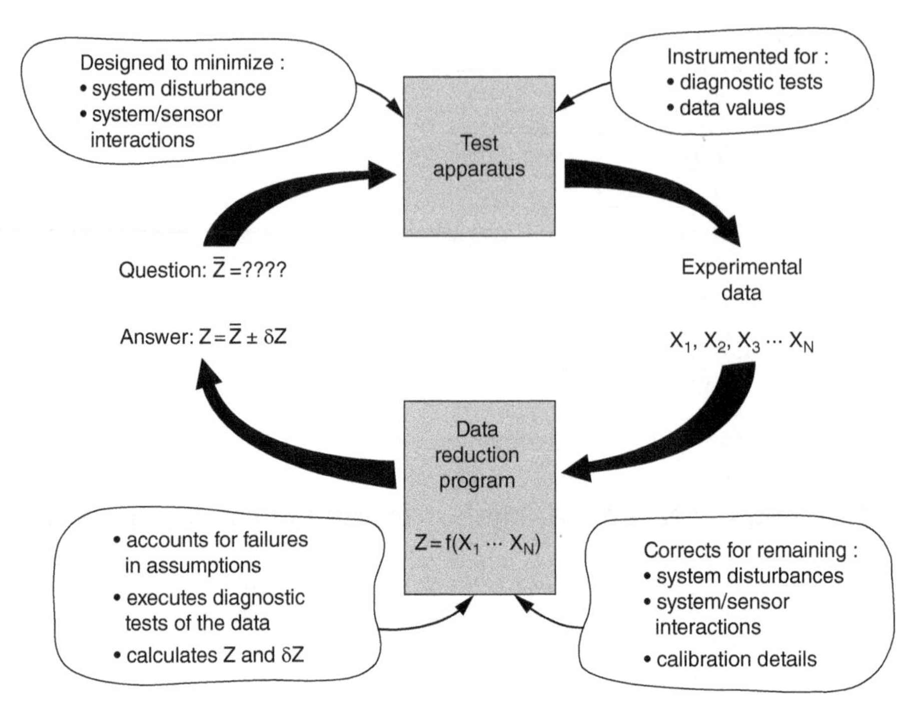

Figure 11.6 The experiment as an instrument. Employ Uncertainty Analysis at each step.

the system doesn't produce the same result, within the required accuracy, on repeated trials, there is no point in proceeding.

Some processes are inherently not repeatable (e.g. the details of turbulence, individual golf and bowling scores, etc.), and those have to be handled statistically, generally with large datasets. Most engineering experiments, however, are expected to produce the same results every time, within some Uncertainty interval.

The first question to ask is, "How much scatter should I expect on repeated trials with this same apparatus and instrumentation?"

11.4.2 Sources of Scatter

Scatter comes from three sources:

i) Random error (noise) in the instruments.
ii) Short-term unsteadiness in the process.
iii) System responses to small changes in uncontrolled variables.

The random error introduced by the instruments can be assessed by a series of end-to-end calibration tests using accurately known inputs. Suppose, for example, that 30 sets of "pseudodata" are taken, over a reasonable period of time, by applying appropriately large, accurately known input signals to each channel of the instrumentation system. These 30 "datasets" can be processed through the Data Reduction Program and the Standard Deviation, σ_o, of the result found by simple statistics. The square of the Standard Deviation, σ_o^2, is the variance introduced by the instrumentation, and $2\sigma_o$

is the Uncertainty interval expected to cover 19 of 20 trials, due to instrument noise alone. Approximately the same results can be obtained by numerically propagating the expected random error characteristics of the instruments through the Uncertainty Analysis program, but the more direct way is preferred. The random error characteristics of the instruments may already be available, since they may have been used before and their characteristics documented.

The short-term unsteadiness of the system can be assessed by taking 30 time-series datasets over a representative time period with the system running at a stable set point (preferably, one which is important to the program). Inspect visually and perform a frequency spectrum analysis as in Section 10.1.2.5. The Standard Deviation of the results, σ_1, can be easily found by simple statistics. The square of the Standard Deviation, σ_1^2, is the variance of the data due to the combined effects of short-term unsteadiness and instrument noise. Subtracting the variance introduced by the instrumentation noise, σ_o^2, from the total variance, σ_1^2, yields the variance contribution of the process.

Nineteen of 20 trials are expected to lie within $\pm 2\sigma_1$ on repeated trials. If the process is too unstable, you must reduce that instability before proceeding or inflate your tolerance for scatter.

Sometimes the trouble will be in the control system: it is not holding the parameters well enough. Sometimes the problem could arise from inherent instabilities like turbulence or combustion instability. If the scatter cannot be reduced, this fact should be documented carefully because variability in the output of the experiment will limit the resolution with which you can identify other problems. It may be necessary, for example, to switch to a multiple-sample experiment design and report only the mean values of sets of data rather than single-sample values. Many replications may be needed at each set point to get acceptable data from a rig that has high random variations in its output.

Once the system is stable and repeatable under normal conditions, it is very useful to "push the envelope" of the system by running at a few nonstandard set points, changing the temperature level, pressure level, or flow rate over some reasonable range. If the results are still within the acceptable band for repeatability, the system is ready to move ahead.

11.4.3 Advancing Toward Calibration

Once the equipment is repeatable, and stable, it can be calibrated. One name for this process is "baselining."

In this part of the process, a standard test procedure is run using a "documented specimen, at documented conditions" and the results compared with accepted values – the values everyone agrees should be returned for those conditions. There will always be a difference – experiments rarely run without error. The issue is whether or not the observed difference between the present value and the reference value is within the expected interval for the difference, considering the Uncertainties involved in both experiments.

To assess that agreement, it is best to formally express the Uncertainty in the difference in the results in terms of the individual Uncertainties of the two results, using the Root-Sum-Square (RSS) combination:

$$\delta\left(Z_{now} - Z_{ref}\right) = \left\{\left(\delta Z_{now}\right)^2 + \left(\delta Z_{ref}\right)^2\right\}^{1/2} \tag{11.16}$$

With a perfect result the value of (Z_{now}–Z_{ref}) would be zero but, in the face of the acknowledged Uncertainties in the two values, squared then summed, we cannot expect to see the zero to better than $\pm\delta(Z_{now}-Z_{ref})$ as predicted from Eq. 11.16.

11.4.4 Selecting Thresholds

There is another issue to be addressed in making this test: "What level of Uncertainty estimate should be used in making this comparison?" The answer depends on where the reference dataset came from. If the reference dataset is from the same apparatus, using the same instruments, then both Uncertainties should be estimated at First-Order. If the reference dataset came from a different apparatus, or was measured using a different set of instruments, then both Uncertainties should be evaluated at N^{th}-Order. The difference arises from the possibility of fixed errors affecting the apparent difference in results. If both datasets came from the same instruments, then the fixed errors will not affect the measured difference. If, however, the two datasets were measured using different sets of instruments, then their fixed errors could be different, and that could affect the measured difference and thus must be acknowledged.

11.4.5 Iterating Analysis

If one finds a larger difference than can be explained by the recognized sources of Uncertainty, then the "detective work" begins.

The Data Reduction Program equations should be reviewed to determine whether or not the parameters measured are really the ones called for. As an example, suppose the task is to measure the heat-transfer coefficient on an electrically heated specimen. Presumably, the electrical power input would be measured with a wattmeter. The first approach might be to claim that all of the power measured by the wattmeter went off as heat transfer. If that approach resulted in a failure of the baseline check, it might be necessary to insert a subroutine which estimated the conduction and radiation losses from the specimen. Or perhaps steady state was assumed (but no check made to see that the system was acceptably steady). Elaboration of the DIP is often necessary to bring the realities of the lab into the data interpretation process. Sometimes, the solution is to be found in fixing the hardware: reducing heat losses, improving uniformity, improving stability, etc. Sometimes the solution is found by elaborating the software. In general, both hardware and software fixes are necessary.

11.4.6 Rechecking Situational Uncertainty

A final issue: Situational Uncertainty. This refers to the question of whether or not the situation created by the test hardware is really the same as that used in the reference dataset. Sometimes, different rigs produce different results for "nominally" the same conditions. In heat-transfer tests, for example, differences in turbulence level can cause significant differences in measured heat-transfer coefficients, even though the test specimens are identical, and the mean flow conditions are carefully matched.

These debugging tests will all be evaluated in the light of the expected Uncertainty intervals. When the results agree within the expected intervals, then development work will stop. If the Uncertainty intervals are too high, the debugging process will be stopped too soon – while there are still significant differences to be dealt with. For example, if N^{th}-Order Uncertainty intervals are used to assess the scatter on repeated trials from the same system, results that should be rejected will be accepted. It is important to understand the significance of the different levels at which Uncertainties can be estimated and to keep clear as to which should be used for what purpose.

It is also important to recognize that it is not "being conservative" to quote your Uncertainty intervals higher than necessary – that doesn't make you appear "modest" or "conservative," it prevents you from cleaning up your experiment as much as you might have. We saw a similar situation in Chapter 7 when we practiced modeling experimental data; an imprecise model resulted in unnecessarily large Uncertainty intervals, Eq. 7.10, while a precise model resulted in truly credible Uncertainty intervals, Eq. 7.11. The difference is clearly visible in Figure 7.7.

The most conservative experimenter is the one who expects the smallest Uncertainty from his measurements – and gets it!

11.5 Reporting the Uncertainties in an Experiment

Authors are expected to report the measurement Uncertainty in at least their principal result. The Uncertainty can be estimated using either the single-sample or multiple-sample technique, depending on the nature of the experiment.

The Nth-Order Uncertainty estimate is the only value that should be reported, unless the author is discussing both the level of the results and the scatter observed on repeated trials. In that case, both the Nth-Order and the First-Order estimates should be presented.

Whichever technique is used, the Uncertainty statement should be accompanied by enough discussion, and enough supporting data, that the reader can properly interpret its significance. This includes, at least, presenting the equations by which the result was calculated and the Uncertainties assigned to the data. If a program or a spreadsheet was used to estimate the Uncertainties, the critical statements from that work should be shown.

If "error bars" are shown, they should be identified as First-Order or N^{th}-Order.

The most useful Uncertainty estimate, from a reader's viewpoint, is one which has the following meaning:

> *If the present experiment were repeated using the same apparatus and techniques, and similar (but not the same) instruments, 19 of 20 repeated trials would produce results within ± XX (or PP%) of the present value.*

or, equivalently:

> *The reported value is the best estimate for the result, and, with 95% confidence, the True Value is believed to lie within ± XX of that value.*

Such statements allow the reader to assess the significance of the difference between the newly reported work and other sources in the literature.

If the final results happen to have a log-normal distribution, then the results would report an uneven Uncertainty interval similar to +2 XX; −0.5 XX.

Note: Authors not familiar with the benefits of formal Uncertainty Analysis sometimes arbitrarily inflate the Uncertainty intervals assigned to their data according to what others seem to claim for the same type of work, even though they may, privately, feel their work is more accurate. This may be done to avoid arguments, or perhaps from fear of being considered boastful or naive. In either case, this inflation is counterproductive. When comparing results between two laboratories, the two datasets will be judged to "agree" if their difference is less than the expected Uncertainty in that difference (the RSS of the Uncertainties of the two datasets). Thus, overestimating the Uncertainty intervals makes it difficult to spot real differences and allows bad data to stay unchallenged.

11.5.1 Progress in Uncertainty Reporting

Airy (1879) recognized early the necessity of evaluating and reporting Uncertainty. One may wonder whether the quote "If your experiment needs statistics, you ought to have done a better experiment" attributed to fellow countryman E. Rutherford impeded progress on its application. Nonetheless, another fellow countryman, Ronald A. Fisher (1921, 1925), set the application of statistics to agricultural experiments on firm ground. Fisher (1935) furthermore established the statistical basis for design of experiments.

Eventually other fields, including ours of classical thermo-fluid physics, began to incorporate experimental Uncertainty - Kline and McClintock (1953). We here note the expanding influence of Uncertainty Analysis across fields of engineering: Abernethy (1981), Abernethy and Thompson (1973), Abernethy et al. (1985), Moffat (1973, 1978, 1982, 1985), Kline (1985), J.L. Taylor (1986), Coleman and Steele (C&S; see preface), Range Commanders Council (2007), and Kurz (2013).

Subsequently, many engineering societies, journals, and publications have specified requirements for Uncertainty levels to be reported: ANSI/ASME (1985), ISO (1987), Dieck (1992), Wells (1992), Taylor and Kuyatt (1994), and Possolo (2015, 2016).

The above articles and books constitute just a small sample of publications highlighting Uncertainty in experiments. The spread of Uncertainty Analysis has reached across fields of engineering and science. Sadly, a few journals remain which don't require that Uncertainties be reported. Wherever Uncertainty is reported, those publications have strengthened their credibility.

11.6 Multiple-Sample Uncertainty Analysis

11.6.1 Revisiting Single-Sample and Multiple-Sample Uncertainty Analysis

In this era of high-speed data acquisition, it is important to be clear about the meaning of "single-sample" and "multiple-sample." Consider a 100 kHz data acquisition system set up to measure the temperatures, pressures, and flows in an engine running at a steady engine condition. A burst of data taken in 0.01 second contains 1000 observations. Is that a single-sample or a multiple-sample dataset?

It could be regarded as either! Those 1000 observations comprise both a multiple-sample of the high-frequency components of the random error in the instrumentation system and, at the same time, they comprise a single-sample of the behavior of the engine.

Throughout these Single-Sample Uncertainty notes there have been repeated calls for "30 observations over a representative time interval." The purpose of the words "over a representative time interval" is to avoid a situation in which a high-speed burst of data provides 30 observations that sample only the instrumentation variability and not the process variability. It is the process timescale that is important.

Multiple-sample tests or experiments are those in which enough data are taken at each test point, over a sufficient length of time, to support a sound statistical interpretation of the characteristics of the process being studied. Multiple-sample experiments are usually characterized by a sparse set of test points, each being held for a long enough time to acquire a large number of observations.

11.6.2 Examples of Multiple-Sample Uncertainty Analysis

Examples of situations in which a multiple-sample experiment design would be used are

- Measuring oil or gas flow rate in a pipeline for custody transfer purposes.
- Testing an aircraft engine to demonstrate compliance with fuel consumption warranties.

Such testing is common in industry and usually follows thoroughly standardized test procedures. Standardization of the tests is important to avoid arguments between buyer and seller as to the validity of the test method. In addition, it is important to have a thoroughly standardized procedure for calculating measurement Uncertainty in such cases, to avoid arguments between buyer and seller as to the meaning of "borderline" test results. Multiple-Sample Uncertainty Analysis has responded to these pressures by leaning heavily on well-accepted statistical methods and, in some instances, forcing the structure of the experiment to conform to the requirements dictated by statistical theory.

Multiple-Sample Analysis concentrates on processing the means of sets of observations of each measurand. This reduces the visible effects of random errors, both due to the instrumentation system and to system unsteadiness, since those are "averaged out." As a consequence, Multiple-Sample Uncertainty estimates are dominated by "fixed errors," in contrast to single-sample experiments which are (usually) dominated by variable errors. As a result of this approach, Multiple-Sample Uncertainty Analysis focuses on the calibration and traceability hierarchy in detail and uses statistical methods to allow the combination of calibration results with test results in a meaningful way. These features of Multiple-Sample Analysis could be "borrowed" into Single-Sample Analysis for situations where the calibration errors contribute significantly to the overall Uncertainty.

As a consequence of its use in industrial acceptance testing, Multiple-Sample Uncertainty Analysis relies heavily on statistical theory for methods of dealing with datasets of different numbers of observations, for example. The state of the art in specifications for Multiple-Sample Uncertainty Analysis has progressed from ANSI/ASME PTC 19.1-1985, *Measurement Uncertainty*. See the NIST Reference. The PTC 19.1

document is a direct descendent of the early work of Abernethy and Thompson (1973) and clearly reflects their statistical approach in its statement of purpose:

> ...(to identify)...corrective action required to achieve test objective.
> ...(to provide)...test validation.
> ...(to reduce)...risk of making erroneous decisions.
> ...(to demonstrate)...compliance with agreements.

There are some differences of viewpoint and nomenclature which distinguish Multiple-Sample Uncertainty Analysis from Single-Sample Analysis, and these should be pointed out at the outset.

11.6.3 Fixed Error and Random Error

In multiple-sample experiments, the reported value of the result is calculated from the mean values of the set of N observations at the test point, not from individual observations. The Uncertainty interval quoted for the result is the Uncertainty in that "averaged" result over the N observations and has the following meaning: "If you run another set of data, with N observations, the average result should agree with the present result within..."

The "fixed error" and the "random error" in the result, Z, are estimated separately, and then the Uncertainty in the result, $U_{0.95}$, calculated from those two estimates as an RSS combination, with the "random error" estimated at 95% confidence for the number of Degrees of Freedom in its estimate. The "Uncertainty" in a result is thus found as a weighted sum of the "fixed error" and the "random error" of the result.

The term "Uncertainty" is used only in connection with the final result. Individual measurements used in calculating the result are described as having "fixed" and "random" errors but are not said to have "Uncertainties" per se, unless the measurement itself constitutes a result. For example, a measurement of "average gas temperature," as a result in its own right, would be said to have an Uncertainty but, if this value was being used as one input to the calculation of a more complex parameter, only its "fixed" and "random" errors would be described. All errors which are not "fixed" are presumed to be "random."

Perhaps because Multiple-Sample Uncertainty Analysis is so frequently concerned with standardized procedures, little emphasis has been given to the evolution of diagnostic techniques to help the experimenter in developing new experiments. The only outputs generated under Multiple-Sample Uncertainty Analysis, except for the overall Uncertainty in the result, are the bias limit and precision index of the result. This is not to say that a skillful practitioner cannot extract useful diagnostics from a Multiple-Sample Uncertainty Analysis – the information is certainly there – but simply to point out that Single-Sample and Multiple-Sample Uncertainty Analyses have somewhat different objectives and, hence, somewhat different structures.

11.7 Coordinate with Uncertainty Analysis Standards

The material in this section is derived from ASME PTC 19.1-1985, except where otherwise noted. That document describes the statistical basis for Multiple-Sample Uncertainty Analysis and covers situations ranging from simple to complex.

The present discussion is limited to "simple" experiments, defined as follows. A "simple" experiment is one in which each sensor has been previously calibrated, a measuring system established by connecting appropriate sensors, amplifiers, and recorders, and N datasets taken by sweeping through all the measurement channels N times. A single dataset consists of one observation of each variable, and the set of N observations constitutes one experiment. The mean value of each variable is calculated from the set of N observations and used to calculate some result, Z. The objective of the Uncertainty Analysis is to describe the Uncertainty in the result.

The "simplicity" of this experiment arises from two sources: first, the calibration of the sensors is considered "done" before the present experiment starts, and second, each variable is measured the same number of times. Some experiments, particularly calibrations, involve simultaneous consideration of calibration, data acquisition, and data interpretation and involve different numbers of observations of different parts of the experiment. This complicates the statistical interpretation considerably. The average user should not try to interpret those situations, either from this document or from PTC 19.1-1985, but should consult an expert in the statistical analysis of experimental error.

11.7.1 Describing Fixed and Random Errors in a Measurement

The fixed and random errors of each measurement are described by its bias limit and its precision index, respectively.

The information needed for a Multiple-Sample Uncertainty Analysis consists of

- The mean value of a set of N observations of the measurement.
- The bias limit of each measurement, B_X.
- The precision index of the mean, Sx_N, for each set of measurements – an estimate of the Standard Deviation which would be observed if the means of a large number of sets containing N observations each were compared.
- The number of Degrees of Freedom of the precision index of the mean.
- The set of equations necessary to calculate the result, Z, from the data.

11.7.2 The Bias Limit

Any error which will not change during the conduct of the experiment is treated as a "fixed error." Such errors arise from sources such as manufacturer's tolerances on instruments (e.g. "accurate within ± 0.001 volts," or tolerances on the setup of instrument channels (e.g. "...set the output to 1.00 V ± 0.001 for a 1.000 mV input"). These fixed errors which remain embedded in a measurement, while "knowable," are not known – if they were known, a correction would have been applied and the "error" would no longer exist, just its Residual Uncertainty. These Residuals must be estimated from manufacturer's specifications or by recording the tolerances used in the end-to-end calibrations run in the test cell (the setup tolerances on the instruments). The estimated value of the fixed error is called the "bias limit." This estimate should be chosen as a best estimate of the value which would include 95% of the realizations. It will be used as though it represented a 2σ estimate of the fixed error: zero centered, and with a Normal Distribution. The phrase "zero centered" means that the possible fixed error is equally likely to be positive as negative and that the distribution of possible error is symmetric about zero.

The bias limit of a measurement is an estimate of the probable value of its fixed error, usually estimated at 95% confidence (i.e. a value large enough that a prudent person would bet 20:1 against it being exceeded).

Consider a single variable, x_i, which is being measured in a multiple-sample manner. There may be several sources of fixed error in that measurement. Consider, first, the instrument, or measurement system channel dedicated to that measurement – that instrument has been calibrated, and its calibration has an estimated fixed error, B_{ical}. The calibration has no variable error, since an error in calibration cannot cause scatter on repeated readings. Other fixed errors may be introduced during the data acquisition and data reduction process, represented by B_{iacq} and B_{icorr}. The overall bias limit of the measurement, B_i, is the RSS combination of all its fixed error components:

$$B_i = \left\{ \left(B_{i,cal}\right)^2 + \left(B_{i,acq}\right)^2 + \left(B_{i,ses}\right)^2 + \ldots\ldots + \left(B_{i,dist}\right) \right\}^{1/2} \tag{11.17}$$

11.7.2.1 Fossilization

Each of the terms on the right-hand side of Eq. 11.17, such as $B_{i,cal}$, for example, can be interpreted in two ways, depending on the structure of the experiment, although this point is not discussed in PTC 19.1-1985. If all of the steps (calibration, acquisition, and reduction) are being considered as part of one measurement act, then the terms on the right side of Eq. 11.17 represent only the fixed errors introduced by each source, and the random errors from each source will be picked up in the calculation of the precision index. If, as is more usually done, the calibration information has been "handed down" from a previous step, then the $B_{i,cal}$ term on the right side of Eq. 11.17 represents the "fossilized" overall Uncertainty of that calibration. In this latter case, the only random errors in the present experiment would be those apparent in the present dataset – all earlier "random errors" would have been gathered into the fossilized bias limit estimate $B_{i,cal}$. Shifting terms from precision to bias, through fossilization, does not alter $B_{i,cal}$ since the RSS combination is unaffected but does alter the balance between fixed and random errors. If the precision index and the bias limit are being used separately as diagnostics, the effects of fossilization must be kept in mind.

11.7.2.2 Bias Limits

Bias limits can be assessed either experimentally or analytically, following the procedures discussed earlier concerning fixed errors. Experimentally, the acceptance tolerance for less-than-perfect response in a pretest system calibration check constitutes the bias limit for that channel. The experimental approach is by far to be preferred. If it is not practical to do an end-to-end calibration off each channel of the system, then the fixed errors will have to be estimated from manufacturer's specifications. Analytically, Taylor's worksheet for cataloging the errors in a measurement, as shown in Figure 10.5, can be used. The symbols B and S in that worksheet stand for bias errors (B) and precision errors (S). If, in a measurement channel, the sensor has been previously calibrated, its overall calibration Uncertainty (the RSS combination of the fixed and random errors of that calibration) should be regarded as having been "fossilized" and the overall Uncertainty entered as the fixed error of calibration. No random error component would be entered for the sensor calibration. The other components of the channel, however, being involved in the "present act" of measurement, would contribute both fixed and random errors.

The bias limit is estimated at 95% confidence, hence serves in an equivalent manner to a 2σ estimate of the fixed error. It is combined with the 2σ (or the 95% confidence interval based on the Student's t-statistic) estimate of the random error in calculating the overall Uncertainty. Thus, although the term has no statistical basis, it is used, as a practical matter, as though it did.

11.7.3 The Precision Index

The precision index of a measurement is a measure of its random error, and in Multiple-Sample Uncertainty Analysis, it can be estimated from the test data alone without an external reference, if the dataset is sufficiently large. The precision index of a variable is equal to its Standard Deviation, S_N, deduced from a set of observations. To be useful in estimating Uncertainty, S_N must be accompanied by knowledge of the number of Degrees of Freedom it represents, $Df = N-1$, so that an appropriate value of the Student's t-multiplier can be used to describe the 95% confidence interval.

The precision index of a set of measurements is calculated from the data of the present act of measurement; it is not necessary to estimate it. Sometimes, however, one wishes to estimate what the precision index of a set of measurements would be, to evaluate the performance of a proposed measuring system. It is also helpful to understand how the contributions to random error arise in a system and interact. For these two reasons, it is worthwhile to consider the overall precision index of a measurement as the RSS combination of the precision indices of all of the subordinate measurements considered to be part of the present measurement. A worst-case situation would involve calibration, acquisition, and data interpretation, all being considered part of one operation. For that complex experiment:

$$S_X = \left\{ \left(S_{\text{cal}}\right)^2 + \left(S_{\text{acql}}\right)^2 + \left(S_{\text{intl}}\right)^2 \right\}^{1/2} \tag{11.18}$$

Each of the three terms on the right-hand side of Eq. 11.18 such as S_{cal}, for example, may also be the RSS combination of several random error components contributed by the measurements used in the process at that level, providing they are all considered to be part of the same act of measurement.

In a more usual situation, there is only one term to consider, S_{acq}, since random errors can arise only within the "present act" of measurement – random errors arising in previous steps are "fossilized." The issue of "fossilization" is not addressed in PTC 19.1, except indirectly, but is important to understand. The term can be illustrated by the following example. When an instrument is calibrated against a laboratory standard, the resulting calibration has an Uncertainty composed of two parts: the fixed error (bias limit) of the calibration, and the random error (precision index) of the calibration. These can be combined to form $U_{0.95}$, the overall Uncertainty of the calibration.

When that instrument is subsequently applied to a measurement task, the overall Uncertainty in its calibration must be regarded entirely as fixed error for the new task: there is no way in which the random error component of the calibration Uncertainty can cause scatter in repeated measurements using the calibrated instrument in the new task – the precision error of the calibration is said to have been "fossilized," that is "converted to a fixed error."

The usual multiple-sample measurement situation involves an instrument whose calibration Uncertainty has been "fossilized" being applied to a new task. The only sources of random error in this new act of measurement are those which arise from the new task. The precision index of the measurements is determined directly from the new measurements themselves:

$$S_X^2 = \frac{1}{N-1}\left\{\sum_{i=1}^{N}(x_i - \bar{x})^2\right\}^{1/2} \tag{11.19}$$

The precision index of the mean is what is needed, however, in the analysis of Uncertainties, since the resulting Uncertainty will be associated with the mean value as in Section 7.1.9, not an individual value:

$$S_{\bar{X}} = \frac{S_X}{\sqrt{N}} \tag{11.20}$$

11.7.4 The Number of Degrees of Freedom

The number of Degrees of Freedom, *Df*, associated with the precision index of a variable (based on one set of *N* observations) is *N*–1. The same value applies to the precision index of the mean. This is the value to be used in what is described above as "the usual situation," where the random errors of previous steps have been fossilized, and the only components of randomness are visible in the data itself.

When complex measurements are considered (i.e. when the all of the data from calibration, acquisition, and reduction are being processed as one measurement act), calculating the number of Degrees of Freedom is more complicated, involving use of the Welch–Satterthwaite equation, and the full text of PTC 19.1-1985 should be consulted.

11.8 Describing the Overall Uncertainty in a Single Measurement

In Multiple-Sample Uncertainty Analysis, the Uncertainty of a single measurement is not usually calculated if it is used only as a component of a future calculation; instead, three pieces of information are provided about each measurement:

Its bias limit, B_X.
The precision index of its mean, $S_{\bar{X}}$.
The number of Degrees of Freedom, *Df*.

11.8.1 Adjusting for a Single Measurement

The Uncertainty for a single measurement which is considered to be an end in itself is calculated by

$$U_{0.95}(x) = \left\{(B_x)^2 + (tS_x)^2\right\}^{1/2} \tag{11.21}$$

Here, "t" is the Student's t-multiplier for 95% confidence and ν Degrees of Freedom. Note that $S_{\bar{X}}$, the precision index of the mean, is used in calculating $U_{0.95}$.

The final statement describing a variable and its Uncertainty would be:

The best estimate of x_i is the mean value, $\bar{x}_i$, plus or minus its Uncertainty interval $U_{0.95}$.

11.8.2 Describing the Overall Uncertainty in a Result

The result, Z, of a multiple-sample experiment is assumed to be calculated from a set of measurements using a DIP (by hand or by computer) of the form given earlier:

$$Z = Z(x_1, x_2, x_3, \ldots, x_N) \tag{11.22}$$

At this point, each measurement has a known bias limit and precision index associated with it, and the number of Degrees of Freedom for each precision index is known. The first step in estimating the overall Uncertainty in the result is to calculate the bias limit and the precision index of the result.

The bias limit of each measurement affects the bias limit of the calculated result in proportion to its sensitivity coefficient (the partial derivative of Z with respect to x_i is the "sensitivity coefficient" of the result Z, with respect to the measurement x_i).

If only one variable used in calculating Z had a nonzero bias limit, the bias limit of Z would be

$$B_Z = \frac{\partial Z}{\partial x_i} B_{xi} \tag{11.23}$$

When several independent variables are used in the function Z, and each has a nonzero bias limit, the individual terms are combined by an RSS method:

$$B_Z = \left\{ \sum_{i=1}^{N} \left(\frac{\partial Z}{\partial x_i} B_{xi} \right)^2 \right\}^{1/2} \tag{11.24}$$

The precision index of the result, S_Z, is affected by the precision index of each measurement in the same way, and the same form is used in calculating it:

$$S_Z = \left\{ \sum_{i=1}^{N} \left(\frac{\partial Z}{\partial x_i} S_{xi} \right)^2 \right\}^{1/2} \tag{11.25}$$

The number of Degrees of Freedom, Df, associated with the precision index of the result is N–1, where N is the number of observations in the dataset – assuming each variable was measured the same number of times. For more complex situations, where different numbers of observations were taken on some of the variables, the Welch–Satterthwaite relation must be used to assess the statistics. For this situation, the full text of PTC 19.1-1985 should be consulted.

The Student's t-multiplier (using the two-tailed distribution) is next found, at the 95% confidence level for the appropriate number of Degrees of Freedom, and the Uncertainty in the result evaluated:

$$U_{0.95}(Z)=\left\{(B_Z)^2+(tS_Z)^2\right\}^{1/2} \tag{11.26}$$

This is the interval, around the mean value, Z, within which the True Value is believed to lie.

Note that $U_{0.95}$ is the Uncertainty in the mean value of a set of N observations – it is not the Uncertainty in a single observation.

Further discussions of Uncertainty have resulted in distinguishing those estimates that are based on statistical evidence taken from the data itself from estimates that are based on experience or judgment but that have no statistical evidence to support them.

11.8.3 Adding the Overall Uncertainty to Predictive Models

If the Motivating Question and the objectives of your experiment include generating a predictive model, the overall Uncertainty of the measurement system must be added to the modeling equation.

In Chapter 7, when we practiced modeling experimental data, we did not include any Uncertainty due to the measurement system in the data. The modeling process investigated candidate factors to explain data variation and eliminated candidate factors which did not survive the Null Hypothesis. Candidate factors not included in the final model, along with unrecorded factors, are bundled together in noise according to the model. The final model, evaluated by the R language through the linear model function lm(mod7) and anova(mod7), resulted in Residuals and Standard Errors for each surviving factor. We adjusted these Residuals and Standard Errors to generate Uncertainties for each factor as shown in Eq. 7.11 and in Figure 7.7. In some cases, Uncertainties were combined using the RSS method.

How does the Uncertainty Analysis presented in Chapters 10 and 11 impact predictive models? First, we check how the overall Uncertainty corresponds to each surviving factor in the final model. We then combine the model Uncertainty for each factor with the overall Uncertainty for the same factor using the RSS method. This gives the predictive model (including all Uncertainties) to be reported, as agreed between you and your client.

11.9 Additional Statistical Tools and Elements

11.9.1 Pooled Variance

The variance of a sample is difficult to assess with precision unless the sample contains many elements. Few experiments are replicated often enough to provide a good estimate of the variance in the output for each different set point of the apparatus.

With only moderate optimism, one can assert that the variance in the data is determined by the conditions at the "set point" of the apparatus and that the variance would be the same every time that particular set point was re-established. With that assumption, we can store information about the variance in observations made at a particular set point and use the same value each time that set point is repeated. With only a bit more optimism, we can "borrow" estimates of the variance from "nearby" set points to help estimate the variance of the entire dataset – we don't have to document the variance at every set point. This procedure should be used with caution when the experiment is wide ranging. The variance at one end of the experimental domain may be much different than at the other. As an example, consider testing an axial flow compressor across its flow range: at some flow the compressor may be near its surge line, and the variance in the pressure and flow readings would go up dramatically. One could not "pool" those data with other data from more quiescent conditions.

The pooled variance is estimated by the following:

$$S^2 = \frac{(N_1 - 1)S_1^2 + (N_2 - 1)S_2^2 + (N_3 - 1)S_3^2 + \ldots + (N_k - 1)S_k^2}{(N_1 - 1) + (N_2 - 1) + (N_3 - 1) + \ldots + (N_k - 1)} \tag{11.27}$$

where S_i^2 = variance in y at $x = x_i$, and N_i = number of y measurements at $x = x_i$.

The denominator represents the "Degrees of Freedom" in the pooled variance estimate. Since the pooled variance has a larger number of Degrees of Freedom than any of its constituents, it has a higher precision.

The pooled Standard Deviation is simply the square root of S^2.

11.9.1.1 Student's t-Distribution – Pooled Examples

To illustrate the use of the t-statistic with a pooled variance estimate, consider the two data points at $x = 20$ from the dataset listed in the beginning of Section 11.9.1:

$$y_{20,1} = 1.4$$

$$y_{20,2} = 1.5$$

$$\bar{y} = 1.45$$

from the earlier example on pooled variance, $S_{\text{pooled}} = 0.11$, with 4 Degrees of Freedom.

From Table 11.2, using a confidence level of 0.950, we find $t = 2.78$, from which, using Eq. 11.27, we find:

$$\mu = 1.45 \pm 2.78\left(\frac{0.11}{\sqrt{2}}\right) \tag{11.28}$$

This is the best estimate of the mean of the population of y values for $x = 20$ which can be deduced from the present data. Note that using pooled variance is equivalent to assuming the variance in the data is independent of the x value at which it is measured.

11.9.2 Estimating the Standard Deviation of a Population from the Standard Deviation of a Small Sample: The Chi-Squared χ^2 Distribution

The principal difficulty in much experimental work is estimating σ, the Standard Deviation of the population, from fewer than 30 observations. The Standard Deviation of the population is different from the Standard Deviation of the set of observations but can be estimated from it. The relationship between the Standard Deviation of a sample, S_N, and that of its parent population is described by the Chi-Squared statistic. This statistic must be used in estimating σ from the auxiliary data of a Single-Sample Uncertainty Analysis, as was pointed out by Taylor (1986).

Table 11.2 Probability points of the double-sided student's t-distribution.

Df	P = 0.995	P = 0.990	P = 0.950	P = 0.900
1	127.00	63.70	12.70	6.31
2	14.10	9.92	4.30	2.92
3	7.45	5.84	3.18	2.35
4	5.60	4.60	2.78	2.13
5	4.77	4.03	2.57	2.01
6	4.32	3.71	2.45	1.94
7	4.03	3.50	2.36	1.89
8	3.83	3.36	2.31	1.86
9	3.69	3.25	2.26	1.83
10	3.58	3.17	2.23	1.81
11	3.50	3.11	2.20	1.80
12	3.43	3.05	2.18	1.78
13	3.37	3.01	2.16	1.77
14	3.33	2.98	2.14	1.76
15	3.29	2.95	2.13	1.75
16	3.25	2.92	2.12	1.75
17	3.22	2.90	2.11	1.74
18	3.20	2.88	2.10	1.73
19	3.17	2.86	2.09	1.73
20	3.15	2.85	2.09	1.72
25	3.08	2.79	2.06	1.71
30	3.03	2.75	2.04	1.70
40	2.97	2.70	2.02	1.68
60	2.91	2.66	2.00	1.67
120	2.86	2.62	1.98	1.66
∞	2.81	2.58	1.96	1.64

Table 11.3, adapted from Lindgren (1976) citing Airy (1879), lists the minimum and maximum values of σ/S_N, where S_N is the Standard Deviation of the sample of N measurements, and σ is the Standard Deviation of the population from which the set of N samples was taken. The parameter *Df* is the number of Degrees of Freedom in the estimate of S_N, equal to $N-1$ where N is the number of values in the set.

Given a set of 30 readings from which a sample Standard Deviation, S_{30}, had been calculated, the most likely value for σ, the Standard Deviation of the parent population, is the measured value, S_{30}, but σ might lie anywhere between $1.34S_{30}$ and $0.80S_{30}$, the maximum and minimum values from Table 11.3 for $Df = 30$. Either value could be used for σ, depending on whether one wished to use the most likely, the largest, or the smallest estimate of the random Uncertainty component in the measurement. The choice should be governed by the relative risks involved in underestimating or overestimating σ.

Table 11.3 Estimating sigma from S_N.

$Df = N-1$	σ/S_N (maximum)	σ/S_N (minimum)
1	31.62	0.45
2	6.26	0.52
5	2.45	0.63
10	1.75	0.70
20	1.44	0.76
30	1.34	0.80
40	1.28	0.82
50	1.24	0.84
60	1.22	0.85

In a single-sample experiment, the auxiliary experiment by which σ is estimated is usually conducted at the beginning of the main experiment, as part of the shakedown and debugging process. Using the value of σ from the auxiliary test to interpret the single-sample observations of the main experiment is akin to pooling the variance over the experiment. If experimental conditions change a great deal over the range of tests planned, then σ should be estimated at several points across the range.

References

Abernethy, R.B., Benedict, R.P., and Dowdell, R.B. (Jun 01, 1985). ASME measurement uncertainty. *J. Fluids Eng.* 107 (2): 161–164. ASME paper 83-WA/FM-3. doi:https://doi.org/10.1115/1.3242450153-160.

Abernethy, R.B. and Thompson, J. W. Jr, "Handbook Uncertainty in Gas Turbine Measurements," AEDC-TR-73-5, AD 755356, Feb. 1973.

Abernethy, R.B. (1981). SAE in-flight propulsion measurement committee E-33. Its life and work. *SAE Aerospace Journal.*

Airy, S.G.B. (1879). *Theory of Errors of Observation.* London, UK: Macmillan and Company.

ANSI/ASME. Measurement Uncertainty. Supplement to ASME Performance Test Codes, PTC 19.1-1985. American Society of Mechanical Engineers, New York. 1985.

Coleman, H.W. and Steele, W.G. (2018), *Experimentation, Validation, and Uncertainty Analysis for Engineers,* 4e. Wiley.

Dieck, R. H., Measurement Uncertainty, Instrument Society of America, Research Triangle Park, North Carolina, 1992.

Fisher, R.A. (1921). On the 'probable error' of a coefficient of correlation deduced from a small sample. *Metron* 1: 3ff.

Fisher, R.A. (1925). *Statistical Methods for Research Workers.* Edinburgh: Oliver and Boyd.

Fisher, R.A. (1935). *The Design of Experiments.* Edinburgh: Oliver and Boyd.

ISO. Fluid Flow Measurement Uncertainty. International Organization for Standardization, Technical Committee TC30 SC9. Draft Revision of ISO/DIS 5168. May 1987.

Kline, S.J. (Jun 01, 1985). The purposes of uncertainty analysis. *J. Fluids Eng.* 107 (2): 153–160. https://doi.org/10.1115/1.3242449.
Kline, S.J. and McClintock, F.A. (February, 1953). Describing the Uncertainties in single sample experiments. *Mech. Eng.* 75: 3–8.
Rainer Kurz, "Gas Turbine Performance and Maintenance," 42nd Turbomachinery Symposium, October 2013, Houston.
Lindgren, B.W. *Statistical Theory*, 3rd ed. p. 575. MacMillan Publishing Company, 1976.
Moffat, R.J. (1973). The Measurement Chain and Validation of Experimental Measurements. *Acta IMEKO*: 45.
Moffat, R. J., "Planning Experimental Programs," Stanford University, Thermosciences Division Lecture Notes, 1978.
Moffat, R.J. (June, 1982). Contributions to the theory of single-sample uncertainty analysis. *J. Fluids Eng.* 104: 250–260.
Moffat, R.J. (Jun 01, 1985). Using uncertainty analysis in the planning of an experiment. *J. Fluids Eng.* 107 (2): 173–178. https://doi.org/10.1115/1.3242452161-164.
The NIST Reference on Constants, Units, and Uncertainty. www.physics.nist.gov/cgi-bin/cuu/Info/index.html.
Possolo, A., "Simple Guide for Evaluating and Expressing the Uncertainty of NIST Measurement Results" NIST Technical Note 1900. United States Department of Commerce Technology Administration, National Institute of Standards and Technology, October 2015.
Possolo, A., "Evaluating and Quantifying Uncertainty," NIST Presentation. United States Department of Commerce Technology Administration, National Institute of Standards and Technology, January 14, 2016.
Range Commanders Council, Telemetry Group. "Uncertainty Analysis Principles and Methods," RCC Document 122-07, September 2007. http://www.dtic.mil/dtic/tr/fulltext/u2/a619551.pdf.
Taylor, B.N. and Kuyatt, C.E., "Guidelines for Evaluating and Expressing the Uncertainty of NIST Measurement Results" NIST Technical Note 1297, 1994. United States Department of Commerce Technology Administration, National Institute of Standards and Technology.
Taylor, J.L. (1986). *Computer-Based Data Acquisition Systems: Design Techniques.* Instrument Society of America.
Wells, C.V., "Principles and Applications of Measurement Uncertainty Analysis in Research and Calibration" National Renewable Energy Laboratory. 1992, NREL/TP-411-5165.

Homework

We hope that you will take a look at Exercise 11.6, which highlights a current research area.

Exercises 11.1 through 11.3 require the ideal gas law, written as $\frac{P}{\rho} = RT$, where

P is pressure [Pa = N/m^2] or [psi]
ρ is density [kg/m^3]
R is the specific gas constant, specific to each gas
T is temperature [K]

Gas	Specific gas constant, R [J/kg K]	Molecular weight [g/mol]
Air	287.05	28.965
Carbon dioxide	188.92	44.01
Helium	2077.1	4.003
Hydrogen	4124.7	2.016
Universal R	8.31446 [J/mol K]	

11.1 Your laboratory has a tank equipped with a thermometer which can read from −30 °C to 150 °C, marked in intervals of 2 °C. The tank absolute pressure gauge can read from 50 to 500 kPa in intervals of 10 kPa. The tank contains 370 kPa air at room temperature of 22 °C. What is the density inside the tank? Quantify the Uncertainty of your density estimate. Justify your values. What are your answers if the tank contains Carbon dioxide? (cf. Exercise 10.1.)

Your laboratory has sufficient funds to replace one of the instruments mentioned. Which instrument will you propose to replace to have greatest impact at reducing the Uncertainty of density? Rank the instruments and each impact on Uncertainty to justify your answer. Find and provide specs for an off-the-shelf scientific instrument that will improve density Uncertainty. Estimate the improved Uncertainty of the density estimate.

11.2 Your laboratory has a tank equipped with a thermometer which can read from −30 °C to 150 °C, marked in intervals of 2 °C. The tank pressure gauge can read from 0 to 3000 kPa in intervals of 50 kPa. A barometer in the room is reading 29.8 in Hg and a thermometer is reading 77 °F. The tank contains 550 kPa air at room temperature. What is the density inside the tank? Quantify the Uncertainty of your density estimate. Justify your values. What are your answers if the tank contains Carbon dioxide? (cf. Exercise 10.2.)

Your laboratory has sufficient funds to replace one of the instruments mentioned. Which instrument will you propose to replace to have greatest impact at reducing the Uncertainty of density? Rank the instruments and each impact on Uncertainty to justify your answer.

Exercises 11.3 through 11.5 require the following definitions

h	Convective heat-transfer coefficient [W/m^{2}K]
k	Thermal conductivity [W/m K]
A	Frontal area = length · diameter = $L \cdot D$ [m^2]
T	Surface temperature [°C]
T_∞	Fluid stream temperature [°C]
M	Mass of cylindrical rod specimen [kg]
$\bar{c}$	Specific heat capacity of cylindrical rod material [J/kg °C]
t	Time [s]
τ	Characteristic time [s]
P	Power added to heat the system [W]

11.3 The convective heat-transfer coefficient h is a convenient model for engineering designs. When the temperature of a solid surface and flowing fluid are both known, as well as the exposed surface area, then h allows the designer to estimate the amount of heat transferred from the surface to the fluid.

Your job, using an experiment designed to mimic the situation, is to provide the value h to the engineer (cf. Exercise 10.4). Along with the model value, you must also provide its 95% Uncertainty. The equation to estimate h is given as (Eq. 11.3)

$$h = \frac{P}{A(T - T_\infty)}$$

The nominal values measured are

Power	$P = 200.0$ W
Diameter	$D = 0.03$ m
Length	$L = 0.3$ m

For the instruments available, the Uncertainties are

$\delta P =$	0.5 W
$\delta D =$	0.02 mm
$\delta L =$	0.1 mm
$\delta \Delta T =$	0.2 °C

Your laboratory has sufficient funds to replace one of the instruments mentioned. Which instrument will you propose to replace to have greatest impact at reducing the Uncertainty of h? Rank the instruments and each impact on Uncertainty to justify your answer.

11.4 When a hot object is suddenly immersed in a fluid (for example, quenching or diving), a transient technique is used to experimentally estimate the convective heat-transfer coefficient h. Your job, using an experiment designed to mimic the actual situation, is to provide the value h to the design engineer (cf. Exercise 10.5).

Along with the model value, you must also provide its 95% Uncertainty. The equation to estimate h is given as (Eqs. 11.6 and 11.8)

$$\tau = \frac{t_2 - t_1}{\ln\left(\frac{T_1 - T_\infty}{T_2 - T_\infty}\right)}$$

$$h = \frac{M\,\bar{c}}{\tau A}$$

The nominal values measured are

Diameter	$D = 0.03$ m
Length	$L = 0.3$ m

For the precision of your instruments, the following local Uncertainties were apparent:

$\delta t =$	0.0025	s
$\delta T =$	0.2	K
$\delta M =$	0.001	kg
$\delta \bar{c} =$	0.004 kJ/kg K	
$\delta D =$	0.025	mm
$\delta L =$	0.10	mm

Your laboratory has sufficient funds to replace one of the instruments mentioned. Which instrument will you propose to replace to have greatest impact at reducing the Uncertainty of h? Rank the instruments and each impact on Uncertainty to justify your answer.

11.5 Your job is to provide to the Bureau of Standards a value for thermal diffusivity. Thermal diffusivity is a measure of how quickly a temperature spreads through an object. As an analogy, thermal diffusivity is the reciprocal of thermal inertia – if an object has a high thermal diffusivity, the temperature permeates quickly; if an object has a low thermal diffusivity, a hot spot persists, spreading slowly (cf. Exercise 10.6).

The definition for thermal diffusivity α is

$$\alpha \equiv \frac{k}{\rho \bar{c}} = \frac{k A D}{M \bar{c}}$$

Taking a peek at Eq. 11.8 above, this definition is nearly a reciprocal.

The Bureau of Standards will provide you with the accepted value of thermal conductivity k for the material, along with its Uncertainty. Using the same values as Exercise 10.5, provide an equation for the value α in terms of k. Also provide an equation for its 95% Uncertainty.

Your laboratory has sufficient funds to replace one of the instruments mentioned. Which instrument will you propose to replace to have greatest impact at reducing the Uncertainty of thermal conductivity α? Rank the instruments and each impact on Uncertainty to justify your answer.

11.6 The value for the gravitational constant is often given as 9.8 m/s^2 in engineering courses. Some elementary physics classes use the convenient value 10 m/s^2, which begins with an Uncertainty of only 2%. The definition for the gravitation constant at the surface of the Earth, where ME is the mass of the Earth and where RE is the local radius of the Earth, is given by the equation

$$g = \frac{G M_E}{R_E^2}$$

The gravitational constant remains a current and relevant research area. Newton's gravitational constant G = 6.67408×10^{-11} [m^3/kg s^2] has a Relative Uncertainty of about 50×10^{-6} or 50 ppm. Of all universal constants in physics,

Newton's gravitational constant G is about the least precise. The reason for G lacking precision is twofold: (i) every object with mass in the universe affects the mass we are studying (including every distant black hole, evidenced by the tide induced by the moon), and (ii) gravity is much weaker than other physics forces (by almost 40 orders of magnitude) (cf. Exercise 10.7).

For the Earth's values,

a) The nominal mass is $M_E = 5.9722 \times 10^{24}$ kg, with a Relative Uncertainty of 10^{-4}.
b) The nominal radius is $R_E = 3963$ miles at its equator sea level, add for tides and ocean waves or mountains, with a Relative Uncertainty of 10^{-2}.

Using these values, write your most accurate estimate for the surface gravitational constant g. Also provide its overall Uncertainty.

Which of the key values above would have the greatest impact at reducing the Uncertainty of g?

12

Debugging an Experiment, Shakedown, and Validation

12.1 Introduction

Debugging is the process of eliminating the errors in an experiment. Errors are distinct from Uncertainty, which can be reduced but not eliminated. Shakedown and validation are companion tasks to debugging, with the purposes to eliminate errors and to reduce Uncertainty.

Eliminating errors begins with identifying the sources of error. One of the most useful tools to look for variable errors is time-series analysis of the data. Plotting a sequence of measurements as a function of time will often show a correlation that will identify the source of the error. The power of this process is greatly enhanced by intentionally perturbing the operating conditions –pushing the system outside its normal operating parameters.

12.2 Classes of Error

There are three classes of error we have to deal with:

- Fixed.
- Random.
- Variable but deterministic.

Truly random errors are rare. Some instrumentation errors may appear to be random, because the processes that generate the error are of such high frequency that observations at normal data rates yield only a very sparse sample. Generally, what appear to be "random errors" are the result of sampling a variable error too slowly to see the changes as continuous. No physical processes in the macroscopic, continuum world are random if looked at with sufficiently time-resolved data. Macroscopic physical systems cannot change state in zero time: conservation laws for mass, momentum, and energy must be obeyed and all rate equations have finite coefficients.

> *For data to be truly random, each observation would have to be entirely independent of the preceding and following observations, as happens when rolling unbiased dice.*

Planning and Executing Credible Experiments: A Guidebook for Engineering, Science, Industrial Processes, Agriculture, and Business, First Edition. Robert J. Moffat and Roy W. Henk.

Companion website: www.wiley.com/go/moffat/planning

When measurements in a macroscopic, continuum situation appear random, this almost always means that the time interval between the observations was too long, longer than the autocorrelation time of the process.

The two-choice classification (error and Uncertainty) is still sufficient for describing the final results of an experiment to its end user, either from an instrument calibration or a system test. It is not adequate, however, for the process of "debugging" the experiment itself.

Debugging is the process of reducing the error and variance of an experiment. Either more accuracy (less error in the mean) and/or higher precision (less scatter in the results) allow for better discrimination. When the purpose of the tests is to evaluate a design change, for example, better accuracy and higher precision allow more confident identification of the best of several competing designs.

12.3 Using Time-Series Analysis in Debugging

The debugging process involves comparing the means and variances achieved in diagnostic tests with expectations based on an Uncertainty Analysis (or with previous results), looking for differences between the observed error and the expected error. Most of the variations in test data come from variations in the test conditions, and that variation is never random. With modern, high-data-rate instrumentation, variations in the data can be seen as smooth, continuous functions of time. The time sequence of those variations, and the phase relationship between different data channels, can reveal a great deal about the source of the variations. Inspect and do a spectrum analysis as in Section 10.1.2.5. If the data are randomized by pooling and averaging and then described only statistically, all of that useful diagnostic information is lost. Thus, it is important to preserve the time sequence of the data and to recognize errors that vary, but not randomly. When errors appear to be random, the sampling rate should be increased.

12.4 Examples

Some examples may be helpful in illustrating the concept.

The source of a variable (but deterministic) error can often be determined by perturbing the operating point of the test system or by examining a time series of data at nominally steady conditions. Data showing a trend or a cyclic variation with time should mean something entirely different to the experiment director than a sequence of measurements with a steady mean and random scatter.

Once the source has been identified, variable (but deterministic) errors can always be dealt with either by changing the hardware or the software of the experiment.

12.4.1 Gas Temperature Measurement

Consider a thermocouple inserted into a gas stream with the intent of measuring the gas temperature, as was illustrated in Figure 10.6. The temperature of a thermocouple (in general) will differ from the temperature of the gas stream due to several factors.

One of these factors is radiant heat transfer from the probe to the walls surrounding it. The measurement error due to radiation depends on the wall temperature and, if the wall temperature changes during a test, the error will change. Then the indicated temperature will change, even if the actual gas temperature does not.

For example, consider a sequence of gas temperature measurements showing unexplainable changes while the process appeared to be steady. One clue to the source of this variation would be the correlation between wall temperature and apparent gas temperature. If the apparent gas temperature goes down when the wall temperature goes down, this might indicate that radiation error was important. The experimenter could then either install a correction algorithm to acknowledge radiation error, or put shielding around the probe, or insulate the walls of the duct to reduce the error source. Note that it helps if the experiment has some diagnostic data channels built in, in addition to the output channels. Let hard-earned experience be your guide.

Tight control of an experiment is not always a good thing. Stabilizing the test conditions does not remove the variable (but deterministic) errors, it makes them more difficult to find! During the shakedown and debugging phase of an experiment, the "peripheral" test conditions (e.g., room temperature) should be intentionally pushed to values outside the normal envelope, to learn whether or not the system is robust. If the same results can be obtained over a wide range of "peripheral" conditions, then the variable (but deterministic) errors have probably all been identified and fixed.

12.4.2 Calibration of a Strain Gauge

Consider calibration of a strain gauge using a simple dead-weight beam deflection test: a weight suspended from the end of a beam on which a strain gauge has been mounted. The strain at the gauge location is calculated from the known properties of the material and the applied load. The excitation of the gauge is monitored and the gauge factor calculated from its output. The intent of the experiment is to document the long-term stability of the gauge calibration under load.

The following events might be noted during the shakedown and debugging phase of this experiment.

During a one-hour test with a high sampling rate, the output appears steady except for the round-off error in the display: the system seems to have only a small random error. The gauge factor reported from this test might differ slightly from its expected value, but by the same amount at each observation. A "fixed error" in gauge factor would be suspected.

Between 6:00 a.m. and 9:00 a.m. and again between 4:00 p.m. and 6:00 p.m., larger variances might be observed. After continued experience reveals that the scatter always occurs within the same time window, the problem might be attributed to vibrations induced by traffic on a nearby freeway (the morning and evening "rush hour"). Increasing the mass of the table and installing seismic mounts could eliminate this problem.

Intentionally changing the room temperature might show that variation in the room temperature caused a change in the calculated calibration factor, due to the temperature coefficient of gain on the amplifiers. One solution would be to set more strict tolerances on the room temperature controller. One consequence of this finding would be to recognize that at least some of the "fixed error" reported from the short-term test was, in fact, a consequence of the particular room temperature which existed at the time of those observations. Another solution (if it were not possible to stabilize the room temperature) would have been to apply a correction factor to the amplifier gain based on the measured room temperature.

Once the short-term and diurnal variations were under control, the variance in the overall data pool would be significantly reduced, and attention could focus on longer-term stability. For example, with a sufficiently high resolution, there might have been an observable variation on a 29-day period (one lunar period), variations caused by the lunar tidal attraction – akin to a periodic variation in local gravity.

12.4.3 Lessons Learned from Examples

There are several points to be retained from this section:

- What might be regarded as a "fixed" error on a short-term test is often just one particular realization of a variable (but deterministic) error.
- The source of an error which is "fixed" in one time frame can often be identified by running relatively high-rate sequences of observations over a longer time period.
- Knowing the timewise behavior of an error is the first step toward finding its source.
- Stability is not always a good thing. In the debugging phase, an experiment should be pushed outside its normal envelope. This is a powerful approach to finding variable errors.

In the following discussions, when the term "variable error" is used, it will refer to any error that changes with time – either randomly or deterministically. When the term "random" error is used, it will refer to an error that is truly random for the experiment being considered.

12.5 Process Unsteadiness

When data taken at the same set point display scatter that exceeds plus or minus one in the least significant digit displayed, this is evidence of instability in the system. This instability could originate within the instrumentation or within the process being studied.

Many processes are unsteady and have outputs that fluctuate at a high frequency, even though the process may be regarded as "steady in the mean." Combustion is like that: the output temperature of a combustion chamber may fluctuate at frequencies from 0.1 Hz to 10 kHz, even in "steady" operation. If such a system is sampled continuously at a high rate, it will be obvious that the system is unsteady. If it is sampled by an instrument of the same frequency capability, but with a repetitive sample rate of only one observation per second, the continuous nature of the time variation will be lost, and the unsteadiness will be regarded as "scatter" in the data. If the same system were viewed through an instrument with only very low frequency response, it might be regarded as "steady" and certainly would appear to display less scatter. How much of the system unsteadiness comes through the instrument system is determined by the characteristics of the instrumentation system and the subsequent data processing (i.e., averaging). This affects the classification of errors between "fixed" and "variable," data which vary with time, due solely to the process variations during the "total period of observation."

12.6 The Effect of Time-Constant Mismatching

The phrase "time-constant mismatching" refers to the use of instruments having different time constants in the same system, or also to the use of a high-frequency instrument to observe the behavior of a system with a large time constant.

One example, discussed below, would be the use of a fast-reading wattmeter to measure the power into a heat-transfer specimen that had a time constant of several minutes. Values of the heat-transfer coefficient h calculated from the data from such a system will show a great deal of scatter unless the power delivered to the specimen came from a well-stabilized power source.

Time-constant mismatching can introduce scatter in many experiments. There are two issues to consider:

i) When the same aspect of a system is being "observed" by instruments that have different frequency response characteristics.
ii) When different aspects of a system are being monitored.

It is well recognized that, when the same aspect of a system is being observed by instruments with different frequency characteristics, the data from the instruments will be out of phase. This phase error will cause problems when the data from the different instruments are used to calculate sums, differences, or products.

It is less often recognized that an equally serious problem arises when different aspects of the system are being monitored. For example, consider a heat-transfer test using a fairly heavy, slowly responding specimen, electrically heated from the building power source. Instantaneous values of temperature and power are used to calculate the heat-transfer coefficient from the specimen to the surrounding air flow. The measured temperature of the specimen might reflect the average energy balance over the past 20 minutes (assuming a system time constant of 4 minutes or so), while the measured power is truly "instantaneous." If there are significant fluctuations in building line voltage, an instantaneous power measurement will not be representative of the average power over the preceding 20 minutes. The result of calculating h using instantaneous power and time-averaged temperature will be the appearance of scatter in the measured heat-transfer coefficient. This type of situation can be handled either by time averaging the instantaneous power measurements over an interval equal to five time constants of the specimen, or by using a temperature sensor that has the same time constant as the specimen.

Wind tunnels that draw their flow from the room and return it to the room (open-return tunnels) have a particularly difficult time with temperature instability. The room air temperature is far from uniform. When a package (a volume) of warm air is drawn into the tunnel, it will be stretched into the flow channel and may take several seconds to pass through the tunnel. During this period, the tunnel temperature may be one or two degrees warmer than average. Techniques are available for designing "thermal attenuators" (bundles of copper tubing placed in the tunnel inlet) that can reduce these fluctuations to an acceptable level. Another option is to embed the air temperature sensor in a block of material having the same time constant as the specimen.

12.7 Using Uncertainty Analysis in Debugging an Experiment

The word "debugging" is used to describe the second phase of development of a new experiment. The first phase is the initial "shakedown" period in which one stops leaks, improves the speed control, and generally "makes it run." In the next phase, one makes it run right. This requires cleaning up the hardware and software so the output from the system (hardware plus software) can be trusted. This process of removing errors from an experiment is called "debugging."

Uncertainty Analysis is the most powerful aid an experimenter can have in the debugging process. The debugging process hinges on distinguishing significant events from insignificant ones. The question is:

How can one tell a significant change in the result from an insignificant one?

The answer, as you might suspect, lies in using Uncertainty Analysis to determine whether or not the observed scatter can be accounted for from the already-recognized Uncertainties in the data.

Repeated trials of an experiment will always display scatter. The more versatile the replications, and the larger the time interval between trials, the greater will be the scatter. Whether that scatter is meaningful or not depends upon how large it is compared with the First-Order Uncertainty interval. If 19 of 20 trials lie within the First-Order interval, the system is behaving as expected. If the scatter is larger than expected, then some significant variable is not being properly controlled or accounted for in the data interpretation program (DIP).

It is helpful, in discussing the debugging process, to look at the entire experimental system as a single instrument, designed to measure the intended result. From that view, it is intuitively reasonable to ask: "How will I calibrate this instrument?" This viewpoint is illustrated below.

12.7.1 Calibration and Repeatability

As with any instrument, you calibrate the experiment by comparing its output with a standard result, as shown in Figure 12.1 (first seen in Chapter 1, as Figure 1.1).

Before an experiment can be calibrated, it must be repeatable. If the system does not repeat, within the required accuracy, there is no point in proceeding. Some processes are inherently not repeatable (e.g. turbulence, golf and bowling scores, etc.) and those have to be handled statistically, generally with large datasets. Most engineering experiments, however, are expected to produce the same results every time, within some Uncertainty interval. The first question is, then:

How much scatter should I expect on repeated trials with this same apparatus and instrumentation set?

Scatter comes from three sources:

i) Random error (noise) in the instruments.
ii) Short-term unsteadiness in the process.
iii) Sensitivity to uncontrolled variables.

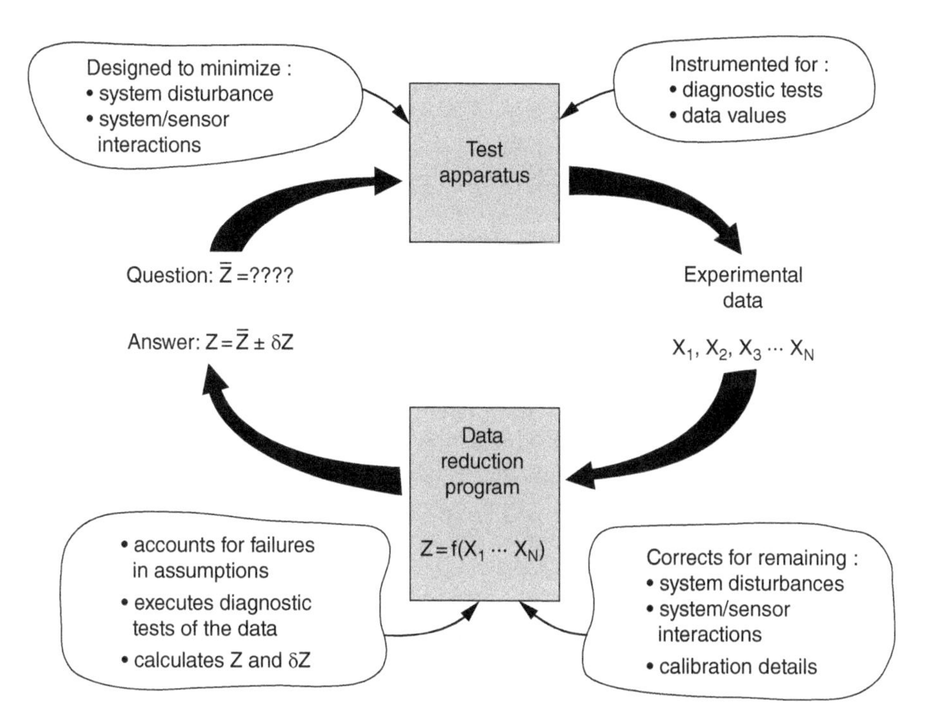

Figure 12.1 The general calibration process applies to whole experiments as it does to individual instruments.

The random error introduced by the instruments can be assessed by a series of end-to-end calibration tests using accurately known inputs. Suppose, for example, that 30 sets of "data" are taken, over a reasonable period of time, by applying appropriately large, accurately known input signals to each channel of the instrumentation system. These 30 "datasets" can be processed through the DIP and the Standard Deviation, σ_o, of the result found by simple statistics. The square of the Standard Deviation, σ_o^2, is the variance introduced by the instrumentation, and $2\sigma_o$ is the Uncertainty interval needed to cover 19 of 20 trials, due to instrument noise alone. The same results can be obtained by propagating the random error characteristics of the instruments through the Uncertainty Analysis program, but the more direct way is preferred. The random error characteristics of the instruments may already be available, since they have probably been used before, and their characteristics are probably known.

The short-term unsteadiness of the system can be assessed by taking 30 datasets over a reasonable time period with the system running at a stable set point (preferably, one which is important to the program). The Standard Deviation of the results, σ_1, can be easily found by simple statistics. The square of the Standard Deviation, σ_1^2, is the variance of the data due to the combined effects of short-term unsteadiness and instrument noise. Subtracting the variance introduced by the instrumentation noise, σ_o^2, from the total variance, σ_1^2, yields the variance contribution of the process, and $2\sigma_1$ is the Uncertainty interval within which 19 of 20 trials would be expected to lie on repeated trials.

If the system has acceptable short-term repeatability, it is time to move ahead to the next level of checking. If not, then some changes must be made. Sometimes the trouble will be in the control system, not holding the parameters well enough. Sometimes the problem could arise from inherent instabilities like turbulence or combustion instability. If the scatter cannot be reduced, it should be documented quite carefully because variability in the output of the experiment will limit the resolution with which you can identify other problems. It may be necessary, for example, to require some minimum number of repeated trials at each set point in order to get average data of the quality needed from a rig which has high random variations in its output.

12.7.2 Stability and Baselining

Once the system is stable and repeatable under normal conditions, it is useful to perturb the system by running at a few nonstandard set points, changing the temperature level, pressure level, or flow rate over some reasonable range. If the results are still within the acceptable band for repeatability, the system is ready to move ahead.

Next comes baselining. In this part of the process, a standard test procedure is run, using "documented hardware," and the results compared with accepted values – the values everyone agrees should be returned for those conditions. There will always be a difference – experiments rarely run without error. The issue is whether or not the observed difference between the present value and the reference value is within the expected interval for the difference, based on the Uncertainties involved in both experiments.

To assess that, it is best to formally express the Uncertainty in the difference in the results in terms of the individual Uncertainties of the two results, using the Root-Sum-Square combination:

$$\delta\left(Z_{\text{now}} - Z_{\text{ref}}\right) = \left\{\left(\delta Z_{\text{now}}\right)^2 + \left(\delta Z_{\text{ref}}\right)^2\right\}^{1/2} \tag{12.1}$$

With an imagined perfect result, the value of $(Z_{\text{now}} - Z_{\text{ref}})$ would be zero but, in the face of the acknowledged Uncertainties in the two values which are squared then summed, we cannot expect to see the zero to better than $\pm\delta$ as predicted from Eq. 12.1.

There is another issue to be addressed in making this test:

What level of Uncertainty estimate should be used in making this comparison?

The answer depends on where the reference dataset came from. If the reference dataset is from the same apparatus, using the same instruments, then both Uncertainties should be estimated at First-Order. If the reference dataset came from a different apparatus, or was measured using a different set of instruments, then both Uncertainties should be evaluated at N^{th}-Order. The difference arises from the possibility of fixed errors affecting the apparent difference in results. If both datasets came from the same instruments, then the fixed errors cannot cause scatter. If, however, the two datasets are measured using different sets of instruments, then their fixed errors could be different, thus could cause scatter, and thus must be acknowledged.

If one finds a larger difference than can be explained by the recognized sources of Uncertainty, then the "detective work" begins.

12.8 Debugging the Experiment via the Data Interpretation Program

The DIP takes the acquired raw data, with all the associated factors, and converts the raw data to the parameters needed to answer the Motivating Question for the experiment. By adopting a three-tier strategy using the DIP to debug the experiment, the experimentalist prepares him- or herself to successfully execute the experiment. We present the three tiers from highest level to base level:

i) Perform a drill of the complete experiment through the DIP to final simulated report. Debug the experiment.
ii) Check the data and human interfaces of the DIP. Confirm that the DIP supplies the output parameters required to answer the motivating question. Validate every piece of equipment. Debug the main block of the DIP.
iii) (base level) Debug the DIP code. Test every subroutine, function, and internal loop that is called by the DIP. The DIP is no stronger than its weakest link; the DIP cannot be more accurate than the largest Uncertainty of its components. In recent physics news, a multiyear feud of contrasting results between research teams was reconciled by sharing routines at this level. One team found where the other had implemented a shortcut in the model routine to shorten run time. The feud was replaced by a need for more research to decide the proper model routine.

12.8.1 Debug the Experiment via the DIP

Plan for these steps as you debug:

- Perform a drill of the complete experiment.
- Generate simulated raw data for inputs to every piece of equipment.
- Process raw data through the DIP and display results.
- Inspect whether the output of the DIP is reasonable.
- Write a simulated final report.
- Debug the experiment.
- Perform another drill at an extreme limit of operating map.
- Refine the experiment.
- Iterate.
- Do this at limits of all measurable factors.
- Refine and debug your experiment plan before collecting archival data.

Your window of opportunity on the rig for the actual experiment to collect archival data may be tightly constrained. Be ready to enter when allowed. Heed the advice of the Glenn Miller song "In the Mood."

12.8.2 Debug the Interface of the DIP

The DIP equations must be reviewed to determine whether or not the parameters measured are really the ones called for. As an example, suppose the task is to measure the heat-transfer coefficient on an electrically heated specimen. Presumably, the electrical power input would be measured with a wattmeter. The first approach might be to claim that all of the power measured by the wattmeter went off as heat transfer. If that approach resulted in a failure of the baseline check, it might be necessary to insert a subroutine which estimated the conduction and radiation losses from the specimen. Or perhaps, steady state was assumed (but no check made to see that the system was acceptably steady). Elaboration of the DIP is often necessary to bring the realities of the lab into the data interpretation process. Sometimes, the solution is to be found in fixing the hardware: to reduce heat losses, improve uniformity, etc. Sometimes the solution is found by elaborating the software. In general, both hardware and software fixes are necessary.

12.8.3 Debug Routines in the DIP

Debug the DIP code. Test every

- Subroutine
- Function
- Internal loop

that is called by the DIP.

Just as each tool operates best when designed properly and used according to instructions, each routine in the DIP must be debugged and documented for correct use. Each routine must be debugged so that it correctly processes the input data to provide required output.

> *We recommend refusing to use any computer program as a black box, uncritically accepting its output.*

The DIP should be thoroughly documented by the original programmers. It is much more challenging but not impossible for someone who did not participate in the original coding to debug. Someone with detailed understanding of the experiment must be involved with the shakedown and validation of the DIP.

> *Do due diligence.*

Debugging is a high-skill task which requires someone with knowledge of the grammar of the programming language. We know of cases where it took months to debug a DIP.

The very production of this text required similar debugging and shakedown to that described in this chapter. With the publishing team on three continents and the authors in yet separate nations from the publishing team and each other, multiple paths existed to introduce errors. Even the format conversion team created errors which forced the authors to again shakedown and debug. Along the way, we authors attempted to revalidate each equation, figure, paragraph, and reference. We take responsibility for the errors we did not catch. We covet your corrections and feedback for improvements.

12.9 Situational Uncertainty

A final issue: Situational Uncertainty refers to the question of whether or not the situation created by the test hardware is really the same as that used in the reference dataset. Sometimes, different rigs produce different results for "nominally" the same conditions. In heat-transfer tests, for example, differences in turbulence level can cause significant differences in measured heat-transfer coefficients, even though the test specimens are identical, and the mean flow conditions are carefully matched.

These tests will all be compared in the light of the expected Uncertainty intervals. When the results agree within the expected intervals, then development work will stop. If the Uncertainty intervals are too high, the debugging process will be stopped too early – while there are still significant differences to be dealt with. For example, if N^{th}-Order Uncertainty intervals are used to assess the scatter on repeated trials from the same system, results that should be rejected will be accepted. It is important to understand the significance of the different levels at which Uncertainties can be estimated and to keep clear as to which should be used for what purpose.

13

Trimming Uncertainty

13.1 Focusing on the Goal

Chapters 1 through 5 posed a number of questions to discern and define the type of experiment to be planned. Your answers to these questions culminated in the Motivating Question, emphasized in Chapter 4. After several iterations, you and your client have refined and decided upon the Motivating Question your experiment must answer. Your client has agreed that answering this question is worth the resources and time required to perform the experiment. We presume that you have posted the Motivating Question prominently in your lab, keeping the target in view, and helping course correct when a major earthquake (like the Loma Prieta[1]) or other disaster disrupts your progress.

Following Chapter 9, you prioritized and collected the output data necessary to answer the question, as well as peripheral data and backup data. You may still have opportunity to collect more data. You may persuade your client of the benefits of continuing data collection (tighter confidence intervals, fortifying credibility). You may train the client's team how to improve their techniques and predictions.

Strengthen your case by answering the Motivating Question as early as possible.

13.2 A Motivating Question for Industrial Production

In Section 8.5, we encountered a factory tasked with producing material at strength. Our client, the factory manager, has identified six candidate factors on the factory production line which might affect material strength. This company has its own client, the super-factory, which requires that material strength exceed 60 MPa in its finished product. Our client wishes to be the exclusive supplier. In order to be an exclusive supplier, the super-factory requires Six Sigma 6σ manufacturing compliance and, of course, a guarantee of on-time production rate.

1 The Loma Prieta earthquake upset more than our experiments at Stanford. Sixty-seven people across the Bay Area perished. Thousands were injured, more were traumatized. Gas fires erupted. Major commuting bridges collapsed horrifically. As the earthquake struck just before a World Series baseball game in San Francisco, the quake was widely televised. Many buildings on the Stanford campus suffered major structural damage, including our buildings and laboratories.

Planning and Executing Credible Experiments: A Guidebook for Engineering, Science, Industrial Processes, Agriculture, and Business, First Edition. Robert J. Moffat and Roy W. Henk.

Companion website: www.wiley.com/go/moffat/planning

Table 13.1 Candidate factors for factory production.

Label	Factor	−	+	Units	Factor type
X1	Tension control	Manual	Automatic	Discrete	
X2	Temperature	200	250	°C	Continuous
X3	Throughput	10	20	kg/min	Continuous
X4	Mixing	Single	Double		Discrete
X5	Machine used	#1	#2		Discrete
X6	Moisture	20	80	%	Continuous
Y	**Response**	Strength (etc.)		MPa	

Table 13.1 relists the candidate factors available for our experiment (cf. Table 8.5). We and our client agreed to initially report 95% confidence intervals for experimental results. Our client requests that we investigate the possibility of Six Sigma 6σ production which requires tighter confidence intervals.

13.2.1 Agreed Motivating Questions for Industrial Example

We and our client agreed to the following Motivating Questions for our experiment:

1) Which machines, and at what operating conditions, can achieve the required material strength? (MQ1)
2) What is the predicted Uncertainty (95% confidence) of material strength across operating map conditions? (MQ2)
3) Which factors significantly impact material strength and need attention? (MQ3) In statistics lingo, find factors for which we "reject the Null Hypothesis." Corollary: Which candidate factors are lost in operational noise (essentially have Null effect; in statistics lingo, find factors for which we "cannot reject the Null Hypothesis")?
4) For what operating conditions does this factory have the capability of guaranteeing production with Six Sigma 6σ reliability (less than four defects per one million tests)? (MQ4)

13.2.2 Quick Answers to Motivating Questions

The following are quick answers to the Motivating Questions:

1) Both machines can produce the material, as shown in Figure 9.5, at the required strength, 60 MPa. Machine 1 can achieve the required strength only near maximum operating conditions. Machine 2 can achieve the required strength across the whole operating map. We recommend using only Machine 2.
2) The predicted Uncertainty (95% confidence) of material strength, at key points on the operating map for each machine, is given in Table 13.2. This table shows overall max and min Uncertainties. The Uncertainty for the strongest material produced on Machine 1, at the corner near maximum operating conditions, is so large that we

Table 13.2 Uncertainty of predicted material strength, for each machine, at key points of the operating map.

Strength 95% Uncertainty [MPa], total		
Machine	**At center**	**At corner**
1	15.94	18.32
2	10.97	12.61

predict nearly 50% would fail required strength. Unless a 50% failure rate is acceptable, we advise never using Machine 1 for this product.

3) From the candidate factors listed in Table 13.1, three factors notably impact material strength. These three factors (in order of dominance) are: machine (X5), throughput (X3), and temperature (X2); for these we rejected the Null Hypothesis. The effects of the other candidate factors were each at the level of measurement noise: tension (X1), mixing (X4), and moisture (X6).
4) This factory may be able to guarantee production with Six Sigma 6σ reliability on Machine 2, when operated at maximum throughput. With each future production run on Machine 2, the additional data are likely to improve Uncertainty intervals and assure a factory guarantee.

13.2.3 Challenge: Precheck Analysis and Answers

Please review this experiment in Sections 8.5 and 8.6.7. In those sections, there are enough results for you to provide our client with useful tentative answers to Motivating Questions 1, 2, and 3. Before proceeding, would you challenge yourself to formulate answers?[2]

The remainder of this chapter elaborates in detail how we deduced answers to all Motivating Questions, as given in Section 13.2.2, plus more.

First targeting Motivating Question #3, we began the experiment with a Plackett–Burman (PB) 12-run Screening Design. The factory made the 12 production runs according to the conditions we specified and for each run measured the material strength. In contrast with the PB design, a full-factorial two-level experiment design for six factors would have required $2^6 = 64$ runs. We find that answers to all of our motivating questions are accessible from our more economical 12-run PB test data.

13.3 Plackett–Burman 12-Run Results and Motivating Question #3

The process of the Plackett–Burman experiment design and the initial PB results were detailed in Chapter 8. Here we analyze the PB results specifically to answer the Motivating Questions.

2 We would be delighted to hear your insights. We are always learning.

A PB experiment is designed to estimate first-order linear effects. So our initial model included all factors, without interaction or higher-order terms. We executed the R-language commands below to generate the model and find the results of Analysis of Variance (anova).

```
apb=read.csv('mh13PB.csv')     # Load the csv file with experimental results into R.
head(apb)                      # Inspect the top six lines of the data file.
mod0=lm(Y~X1+X2+X3+X4+X5+X6,data=apb)     # Linear model the main factors.
summary(mod0)                  # Display analysis for coefficients of main factors.
anova(mod0)                    # Display analysis of variance for main factors.
```

The anova Table 13.3 summarizes how each factor explains variation in the results. Here the anova table lists Variance (defined as sum of squares) in the SumSq column (cf. Table 8.7). Each row shows how much of the total Variance in Y, the experimental material strength, is explained by each factor (X1, X2, ..., X6). Our primary interest in Table 13.3 is the Fisher statistic, listed in the F-value column. The F-value is a ratio of the benefit of a factor divided by the cost of including the factor in analysis, as shown conceptually in Eq. 13.1. Generally, we are interested in factors with an F-value larger than 1.

$$\text{F-value} \propto \frac{\text{Variation explained by factor}}{\text{Cost of including factor in analysis}} \tag{13.1}$$

We have found that it is easier to explain to clients how to make decisions based on the F-value (Fisher Statistic). This anova Table 13.3 reveals three dominant terms by the F-value. In order of dominance, the factors are X5, X3, and X2. Some researchers are more familiar with the p-value. Abuse of the p-value is frequently criticized in the literature. We can think of the p-value as the probability that the factor has Null effect (the Null Hypothesis). If the p-value is small, the factor has low probability of Null effect, that is, we can "reject the Null Hypothesis." Via the p-values, we "reject the Null

Table 13.3 Example 8.2. Analysis of Variance (anova) as parsed for each model factor.

Label	Df	SumSq	MeanSq	F-value	p-value
X1	1	40	40	0.216	0.66181
X2	1	1728	1728	9.244	0.02874
X3	1	2408	2408	12.883	0.01572
X4	1	280	280	1.500	0.27527
X5	1	10208	10208	54.610	0.00071
X6	1	12	12	0.064	0.81007
Residuals	5	935	187		

Hypothesis" for factors X5, X3, and X2, the same three factors we accepted by the F-value. Furthermore, by using the F-value we avoided abuse of the p-value.

How does the anova Table 13.3 give us a hint about other causes for product strength to vary? In other words, what unreported factors might affect product strength? One clue: three factors explain a variance of 14344 $MPa^2 = 10208 + 2408 + 1728$; the leftover experimental variance had SumSq = 1267 $MPa^2 = 280 + 40 + 12 + 935$. The fourth largest variance (SumSq = 280) belongs to the discrete factor mixing (X4). Clearly, mixing types did not alter strength enough to obtain a high F-value. What other factors might account for leftover variance? May we suggest factors such as raw material freshness (age), material supplier, product recipe, contamination (or purity), others? If the company could provide records with such information, we could continue to investigate more broadly without additional experimental production runs.

From the anova Table 13.3, we provide the following answers to Motivating Question #3:

3A. The dominant candidate factors from the PB screening, in order of dominance, are:

X5: Machine used (discrete: 1 or 2), variance explained = 10208.
X3: Throughput (continuous), variance explained = 2408.
X2: Temperature (continuous), variance explained = 1728.

3B. The remaining candidate factors, which we will continue to monitor and record, yet which we will neglect in future analysis, allowing them to become noise, are:

X1: Tension control (discrete: automatic or manual).
X4: Mixing (discrete: single or double).
X6: Moisture (continuous, ranging from 20% to 80%).

By looking at the coefficients in `summary(mod0)` for these dominant factors (cf. Sections 8.6.6 and 8.6.7), the results reveal that Machine 2 produced better strength than Machine 1 (X5), fast throughput was better than slow (X3), and higher temperature improved strength (X2).

13.4 PB 12-Run Results and Motivating Question #1

With the reduced set of dominant factors, we proceed to seek an answer to Motivating Question #1. We need a predictive equation for the material strength (the response *Y*) in terms of the dominant factors. We will continue R-language analysis to generate and evaluate predictive models. Recall that Motivating Question #1 asks:

MQ #1. Which machines, and at what operating conditions, can achieve the required material strength?

We have already learned that some preprocessing improves predictive models. As we saw in Section 8.6.2 for continuous factors, the Standard Error of model terms tends to be minimized when the Response Surface is centered around the mean of each and every continuous term (whether or not included in the model). Therefore, for our first predictive model of this factory, which we label mod9, we will preprocess the results within R, thereby keeping the integrity of the original data. Note that we ordered the factors (in the model) according to dominance, which further improves F-values. *The order of factors in the model affects F-values.*

We executed the R-language commands below for preprocessing and to generate an overall predictive model from the PB Screening Design.

```
> apb=read.csv('mh13PB.csv') # Load the csv file with experimental results into R.
> x2m=mean(apb$X2); x2m        # Find the mean temperature (continuous factor X2).
                                 Display it.
> x3m=mean(apb$X3); x3m        # Find the mean throughput (continuous factor X3).
                                 Display it.
> x6m=mean(apb$X6); x6m
> apb$X2c=apb$X2-x2m           # Add a shifted factor X2c, centered about its mean.
> apb$X3c=apb$X3-x3m           # Add a shifted factor X3c, centered about its mean.
> apb$X6c=apb$X6-x6m
> head(apb)                    # Inspect the top six lines of the augmented datatable.
> mod9=lm(Y~X5+X3c+X2c, data=apb)        # Linear model only dominant factors.
> summary(mod9)                # Display analysis with coefficients for main factors.
```

13.4.1 Building a Predictive Model Equation from R-Language Linear Model

Our client, the factory manager, happens to be mathematically savvy. So you prepare to explain how to build a predictive model equation from the R-language linear model results, which are stashed in mod9. We displayed the results by the command `summary(mod9)`.

Just as a line has a slope and an Intercept (the value at zero), the linear model finds a slope for each continuous variable x_i. Here, one of our continuous variables is temperature (T), which was recorded in the file as X2. The slope of material strength Y in terms of temperature is given by a partial derivative

$$\frac{\partial Y}{\partial T} = 0.4800\left[\frac{\text{MPa}}{{}^\circ\text{C}}\right]$$ which in Table 13.4 is the `Estimated Coeff` for factor $X2$c.

Likewise, the slope of material strength Y in terms of throughput is given by

$$\frac{\partial Y}{\partial \text{ThruPut}} = 2.8333\left[\frac{\text{MPa}}{\text{kg/min}}\right]$$ which is the `Estimated Coeff` for factor $X3$c.

The R language also sets a model zero for all factors, the zero location at which the linear model finds an Intercept. The model zero naturally incorporates the zero of each continuous variable. What about discrete variables which don't have a natural zero? In our PB design, the factor machine (X5) is a discrete variable with levels M1 and M2. In this example, the R language chose the zero of M1; therefore, the `Estimated Coeff` for Intercept in Table 13.4 gives the mean strength when Machine 1 was used. The `Estimated Coeff` of factor X5, level M2, gives the additional mean strength (beyond the M1 Intercept) when Machine 2 was used.

Table 13.4 R results of dual predictive model, mod9.

Call: lm(formula = Y ~ X5 + X3c + X2c, data = apb)				
Coeffs:	Estimate	Std.Error	t-value	Pr(>|t|)
(Intercept)	40.8333	5.1384	7.947	4.58e-05 ***
X5 M2	58.3333	7.2667	8.027	4.26e-05 ***
X3c	2.8333	0.7267	3.899	0.00455 **
X2c	0.4800	0.1453	3.303	0.01081 *
Residual Standard Error: 12.59 on 8 Degrees of Freedom				

In Table 13.4, the Intercept value shows the strength (40.83 MPa) of material produced on Machine 1 at central operating conditions. Since the X5 M2 `Estimated coeff` is positive, Machine 2 yields stronger product (an additional 58.33 MPa) than Machine 1.

Using the procedure described above, from Table 13.4 we write the dual predictive model equation for the PB screening test of this factory (compare values to `summary(mod9)`) as

$$Y\,[strength] = 40.83[\mathrm{MPa}] + \begin{bmatrix} 0.00, \text{if M1} \\ 58.33, \text{if M2} \end{bmatrix}[\mathrm{MPa}] + 2.833\left[\frac{\mathrm{MPa}}{\mathrm{kg/min}}\right](flow - 15.)[\mathrm{kg/min}]$$
$$+\, 0.480\left[\frac{\mathrm{MPa}}{^\circ\mathrm{C}}\right](Temp - 225.)[^\circ\mathrm{C}]. \tag{13.2}$$

Next, we continue R analysis using the "Analysis of Variance" command `(anova)` which revealed how variance in the response was captured by the selected dominant factors.

```
> anova(mod9)                  # Display analysis of variance for dominant factors.
Analysis of Variance Table
Response: Y
          Df  Sum Sq Mean Sq F-value    Pr(>F)
X5         1 10208.3 10208.3  64.440 4.26e-05 ***
X3c        1  2408.3  2408.3  15.203 0.004551 **
X2c        1  1728.0  1728.0  10.908 0.010815 *
Residuals  8 1267.3   158.4
```

Although the variance `(in SumSq)` is identical to the values in Table 13.3, each F-value has increased.

13.4.2 Parsing the Dual Predictive Model Equation

Might your client prefer to view the same dual predictive equation as a pair of model equations, one for each machine? For Machine 1, the dual predictive equation is

$$Y_{M1}\left[strength\right]=40.83\left[\text{MPa}\right]+2.833\left[\frac{\text{MPa}}{\text{kg/min}}\right]\left(flow-15.\right)\left[\text{kg/min}\right]$$
$$+0.480\left[\frac{\text{MPa}}{°\text{C}}\right]\left(Temp-225.\right)\left[°\text{C}\right]. \tag{13.3a}$$

For Machine 2, the dual predictive equation is

$$Y_{M2}\left[strength\right]=99.17\left[\text{MPa}\right]+2.833\left[\frac{\text{MPa}}{\text{kg/min}}\right]\left(flow-15.\right)\left[\text{kg/min}\right]$$
$$+0.480\left[\frac{\text{MPa}}{°\text{C}}\right]\left(Temp-225.\right)\left[°\text{C}\right]. \tag{13.3b}$$

From the R analysis, we can also derive Uncertainties for each of the values. Remember that good scientific practice (and respectable journals) requires Uncertainties for reported values.

13.4.3 Uncertainty of the Intercept in the Dual Predictive Model Equation

First, we will determine the Uncertainty of the Intercept for Machine 1, which has the value 40.833 MPa. Since our preprocessing placed zero at the mean of all continuous variables, essentially the center of the operating map, the Intercept is a central value. We refer to Section 7.1, specifically the Uncertainty of the mean. We use that equation, swapping the "Standard Error" of the Intercept divided by the square root of Degrees of Freedom (DoF). Mathematically,

$$\delta\text{Intercept}=Multiplier\frac{\text{Standard Error}}{\sqrt{\text{DoF}}} \tag{13.4}$$

Here the *Multiplier* comes from the "Student's t" for the desired confidence level and the DoF. For a 95% confidence level and DoF = 8, the *Multiplier* = 2.31. Therefore, for the 0.95 Uncertainty of the Intercept for Machine 1, we obtain

$$\delta\text{Intercept}_{M1,0.95}=2.31\frac{5.1384}{\sqrt{8}}\left[\text{MPa}\right]=4.197\left[\text{MPa}\right] \tag{13.5a}$$

This means that for material produced on Machine 1, we claim 95% confidence that the mean strength at the center of the operating map has a value in the range 40.833 ± 4.197 MPa, or equivalently, $36.64\text{ MPa} < \text{mean strength}_{M1} < 45.03\text{ MPa}$.

The Uncertainty of the Intercept for Machine 2 is more complicated. First, we must note that the Intercept for Machine 2 = 40.8333 + 58.3333 = 99.17 MPa. The respective Standard Errors add according to Root-Sum-Squares (RSS), since the terms are mutually independent. Therefore, for the 0.95 Uncertainty of the mean for Machine 2, we obtain

$$\delta \text{Intercept}_{M2,0.95} = 2.31 \frac{\sqrt{5.1384^2 + 7.2667^2}}{\sqrt{8}} [\text{MPa}] = 7.269 [\text{MPa}] \quad (13.5b)$$

This means that for material produced on Machine 2, we claim 95% confidence that the mean strength has a value in the range 99.167 ± 7.269 MPa, or equivalently, 91.90 MPa < mean strength$_{M2}$ < 106.44 MPa.

13.4.4 Mapping an Answer to Motivating Question #1

Equations 13.3a and 13.3b along with Uncertainties in Eqs. 13.5a and 13.5b are key steps to providing an answer to Motivating Question #1. Now we use the equations to predict strength over the operating map conditions. Recall that process temperature, factor X2, falls in the range 200–250 °C. The process throughput, factor X3, falls in the range 10–20 kg/min.

Table 13.5 shows the strength [MPa] predicted by the dual predictive model equation, Eqs. 13.2, 13.3, and 13.5, at key operating conditions. Recall that the super-factory client requires a guaranteed strength of 60 [MPa]. On Machine 1, it is clear that only a very narrow region of the operating map achieves the required strength. If the factory decides to produce the material on Machine 1, it must operate near maximum throughput and maximum temperature.

Table 13.5 Predicted strengths over the operating map using the dual predictive model.

Machine 1	Intercept	ThruPut	Temp [°C]	Predicted [MPa]
Center	40.83	15.00	225.00	40.83
Max	40.83	20.00	250.00	67.00
BorderLine	40.83	20.00	235.42	60.00
BorderLine	40.83	17.53	250.00	60.00
Min	40.83	10.00	200.00	14.67
MinMax	36.64	20.00	250.00	62.80

Machine 2	Intercept	ThruPut	Temp [°C]	Predicted [MPa]
Center	99.17	15.00	225.00	99.17
Max	99.17	20.00	250.00	125.33
Corner	99.17	20.00	200.00	101.33
Corner	99.17	10.00	250.00	97.00
Min	99.17	10.00	200.00	73.00
MinAll	91.90	10.00	200.00	65.73

On Machine 2, the whole operating map achieves the required strength. Even if the true Intercept was at the minimum of the 95% confidence interval, and other factors were at operational minimum, Machine 2 is predicted to produce acceptable material.

13.4.5 Tentative Answers to Motivating Question #1

The good news for your client is that according to the dual predictive model, the factory Machine 2 appears able to achieve the required strength at all operating conditions. We note that our calculated Uncertainties were for only 2.31 σ (Standard Errors).

The less good news is that factory Machine 1 appears to be able to achieve the required strength only near maximum operating conditions.

Before giving more definitive answers to Motivating Question #1, the Uncertainties of the dual modeling equation must be included.

13.5 Uncertainty Analysis of Dual Predictive Model and Motivating Question #2

Recall MQ #2. What is the predicted Uncertainty (95% confidence) of material strength across operating conditions?

Please review Section 7.6 to write a model equation including Uncertainties with 95% confidence. For convenience, we copy the key R results of the dual predictive model in Table 13.4:

```
> mod9=lm(Y~X5+X3c+X2c, data=apb) # Linear model only dominant factors.
> summary(mod9)                   # Display analysis with coefficients for main factors.
```

Call: lm(formula = Y ~ X5 + X3c + X2c, data = apb)				
Coeffs:	Estimate	Std.Error	t-Value	Pr(>|t|)
(Intercept)	40.8333	5.1384	7.947	4.58e-05 ***
X5 M2	58.3333	7.2667	8.027	4.26e-05 ***
X3c	2.8333	0.7267	3.899	0.00455 **
X2c	0.4800	0.1453	3.303	0.01081 *

Residual Standard Error: 12.59 on 8 Degrees of Freedom

(copy of) Table 13.4 R results of dual predictive model, mod9.

13.5.1 Uncertainty of the Constant in the Dual Predictive Model Equation

The Uncertainty of the constant in the dual predictive model equation differs from the Uncertainty of the Intercept (mean). There are similarities, however, and overlap. We have already demonstrated how to determine the Uncertainty of the Intercept in Section 13.4.3.

The overall Uncertainty of the predictive model equation has its minimum at zero, the location corresponding to the Intercept. Aside: When there is a discrete variable with multiple levels, the R language will choose one of the levels as a zero. If there are multiple discrete variables, each contributes one level to the zero. When there is a continuous variable, the R language places the Intercept at its zero. This aside explains why we recommend choosing a reasonable central value for every continuous variable to become a zero for the predictive model; the central value could be the mean, median, or another value preferred by client.

The difference for the Uncertainty of the constant in the model equation is due to addition of the Residual Standard Error. The Residual Standard Error is a constant Uncertainty that applies to all predictions using the model equation.

The Uncertainty of the constant for Machine 1 includes two orthogonal terms, the Standard Error for the Intercept of M1 and the Residual Standard Error. The two Standard Errors add according to RSS, Pythagoras, since the terms are mutually independent. Mathematically,

$$\delta\text{Constant} = \textit{Multiplier}\sqrt{\Sigma(\text{Standard Error})^2} \tag{13.6}$$

As previously, the *Multiplier* comes from the "Student's t" for the desired confidence level and the DoF. For a 95% confidence level and DoF = 8, the *Multiplier* = 2.31.

Therefore, for the 0.95 Uncertainty of the constant for Machine 1, we obtain

$$\delta\text{Constant}_{M1,0.95} = 2.31\sqrt{5.1384^2 + 12.59^2}\,[\text{MPa}] = 31.41\,[\text{MPa}] \tag{13.7a}$$

This means that for material produced on Machine 1, the minimum predicted 95% Uncertainty is 31.41 MPa. The minimum overall Uncertainty occurs at the zero corresponding to the Intercept. At any other location in the operating map, the Uncertainty is higher.

Notice that the Uncertainty of the constant for Machine 2 includes three orthogonal terms, the Standard Errors for the Intercepts of M1 and M2 and the Residual Standard Error. The three Standard Errors add according to RSS, Pythagoras.

Therefore, for the 0.95 Uncertainty of the Constant for Machine 2, we obtain

$$\delta\text{Constant}_{M2,0.95} = 2.31\sqrt{5.1384^2 + 7.2667^2 + 12.59^2}\,[\text{MPa}] = 35.62\,[\text{MPa}] \tag{13.7b}$$

Thus, for material produced on Machine 2 the minimum predicted 95% Uncertainty is 35.62 MPa. Note: the Relative Uncertainty for Machine 2 is less than for Machine 1. The minimum overall Uncertainty occurs at the Zero corresponding to the Intercept. At any other location in the operating map, the Uncertainty is higher.

13.5.2 Uncertainty of Other Factors in the Dual Predictive Model Equation

The Uncertainty due to a factor coefficient is simply the Standard Error multiplied by the confidence level *Multiplier*. As previously, the *Multiplier* comes from the "Student's t" for the desired confidence level and the DoF. For a 95% confidence level and DoF = 8, the *Multiplier* = 2.31. Mathematically, where the experimental strength result is *Y*, we seek

$$\frac{\partial Y}{\partial \text{Factor}} = Multiplier\left(\text{Standard Error}_{\text{Factor}}\right) \tag{13.8}$$

Therefore, for the 0.95 Uncertainty of the throughput coefficient, we obtain

$$\frac{\partial Y}{\partial \text{ThruPut}},_{0.95} = 2.31 \cdot 0.7267 = 1.679\left[\frac{\text{MPa}}{\text{kg / min}}\right] \tag{13.9a}$$

And for the 0.95 Uncertainty of the temperature coefficient, we obtain

$$\frac{\partial Y}{\partial T},_{0.95} = 2.31 \cdot 0.1453 = 0.336\left[\frac{\text{MPa}}{°\text{C}}\right] \tag{13.9b}$$

With the values above, we now have all values needed to answer Motivating Question #2.

13.5.3 Include All Coefficient Uncertainties in the Dual Predictive Model Equation

When the dual predictive model equations for each individual machine include the 95% confidence interval for each coefficient, we obtain the following pair of model equations.

For Machine 1,

$$Y_{\text{M1}}\left[strength\right] = (40.83 \pm 31.41)[\text{MPa}] + (2.833 \pm 1.679)\left[\frac{\text{MPa}}{\text{kg / min}}\right](flow - 15.)[\text{kg / min}] + (0.480 \pm 0.336)\left[\frac{\text{MPa}}{°\text{C}}\right](Temp - 225.)[°\text{C}]. \tag{13.10a}$$

And for Machine 2,

$$Y_{\text{M2}}\left[strength\right] = (99.17 \pm 35.62)[\text{MPa}] + (2.833 \pm 1.679)\left[\frac{\text{MPa}}{\text{kg / min}}\right](flow - 15.)[\text{kg / min}] + (0.480 \pm 0.336)\left[\frac{\text{MPa}}{°\text{C}}\right](Temp - 225.)[°\text{C}]. \tag{13.10b}$$

Should we be suspicious that this model predicts that each machine behaves exactly the same for the factors throughput and temperature? We shall keep our suspicions in check until after we have answered all the Motivating Questions possible for this dual predictive model.

13.5.4 Overall Uncertainty from All Factors in the Predictive Model Equation

The overall Uncertainty has already been explained in Section 11.8. Measured Uncertainties occur independently, so Uncertainties add by the RSS. We seek an equation for the overall Uncertainty at any set of conditions in the operating map domain. We recall Eq. 11.26

$$U_{0.95}(Y)=\left\{(B_Y)^2+\Sigma\left(t\,S_Y\right)^2\right\}^{1/2}$$

where t is the 0.95 value from the Student's t-distribution. Adjusting Eq. 11.25 for the current PB experiment with the dual predictive model equation, we obtain

$$U_{0.95}(Y)=\left[\left(\delta\text{Constant}_{M,0.95}\right)^2+\left(\left[\frac{\partial Y}{\partial\text{ThruPut}},_{0.95}\right]\Delta\text{ThruPut}\right)^2+\left(\left[\frac{\partial Y}{\partial T},_{0.95}\right]\Delta T\right)^2\right]^{1/2} \tag{13.11}$$

For all of our predictive model equations to this point, we have centered the operating map at the exact center of the continuous variables temperature [200, 250 °C] and throughput [10, 20 kg/min]. Comparing Eq. 13.13 to Eq. 13.11 (or 13.12), we note the following relations, using absolute values:

$$\Delta\text{ThruPut}=\left|flow-15.\right|\left[\text{kg/min}\right],\quad \Delta\text{ThruPut}_{\text{max}}=5.0\left[\text{kg/min}\right] \tag{13.12a}$$

$$\Delta T=\left|Temp-225.\right|\left[°\text{C}\right],\quad \Delta T_{\text{max}}=25.0\left[°\text{C}\right] \tag{13.12b}$$

Finally, we obtain an equation for overall Uncertainty across the operating domain. For Machine 1

$$U_{M1,0.95}(Y)=\left[(31.41)^2+\left([1.679]\Delta\text{ThruPut}\right)^2+\left([[0.336]\Delta T\right)^2\right]^{1/2}[\text{MPa}] \tag{13.13a}$$

And for Machine 2

$$U_{M2,0.95}(Y)=\left[(35.62)^2+\left([1.679]\Delta\text{ThruPut}\right)^2+\left([0.336]\Delta T\right)^2\right]^{1/2}[\text{MPa}] \tag{13.13b}$$

Both of these equations take the form of a hyperbolic surface.

When reporting the predictive modeling equation and the predicted Uncertainty equation, we recommend using Eq. 13.3a alongside Eq. 13.13a due to ease of calculation, plotting, and statistical accuracy. Likewise, use Eq. 13.3b alongside Eq. 13.13b for reporting Machine 2. We hesitate to recommend Eq. 13.10a and 13.10b format because it leads novice experimentalists (I once was one) to neglect RSS and to over-estimate with the worst-case error.

For comparison to Table 13.2, we can report the predicted overall Uncertainty at key points of the operating map. The predictions we found in this section are shown in Table 13.6 as dual Machine 1 and dual Machine 2. We note how these Uncertainties are much larger than in Table 13.2. In Section 13.6, we show how to obtain the improved (smaller) Uncertainties which were preemptively listed in Table 13.2.

Table 13.6 Uncertainty of dual prediction model material strength, for each machine, at key points.

Strength 95% Uncertainty [MPa], total		
Machine	**At center**	**At corner**
Dual: 1	31.41	33.58
Dual: 2	35.62	37.54

13.5.5 Improved Tentative Answers to Motivating Questions, Including Uncertainties

The good news for your client is that according to the dual predictive model with Uncertainties, the factory Machine 2 appears able to achieve the required strength at operating conditions above a throughput of 15 kg/min and temperature above 225 °C. We note that our calculated Uncertainties were for only 2.31σ (Standard Errors). Generally, Machine 2 produces a much higher material strength than required.

The bad news, when we include Uncertainties, is that factory Machine 1 cannot guarantee the required strength at any operating condition.

Furthermore, with these large Uncertainties the factory cannot guarantee 6σ level of production. The failure rate would appear to far exceed four events per million, even for factory Machine 2.

13.5.6 Search for Improved Predictive Models

Do the large Uncertainties in the dual predictive model equations disturb you? Can we do better? Yes, we can. We saw in Section 7.7 that we can "Decrease Uncertainty (Improve Credibility) by Isolating Distinct Groups." If we can reduce the Uncertainties, our answers to Motivating Questions will change. Knowing this, we have kept our answers to MQ#1 and MQ#2 tentative.

In Section 13.6, we will create individual models for Machine 1 and Machine 2 and see how it impacts our answers to the Motivating Questions. In any case, by looking at the coefficients for these dominant factors (Table 13.4), the results reveal that Machine 2 produced better strength than Machine 1 (X5), fast throughput was better than slow (X3), and higher temperature improved strength (X2). All the equations so far have helped, in particular Eqs. 13.2, 13.3a and 13.3b, and 13.13a and 13.13b. Next we find a better way.

13.6 The PB 12-Run Results and Individual Machine Models

In Section 7.7, we recommended two techniques to credibly decrease Uncertainty:

i) Avoid extrapolation by translating each quantitative factor to a reasonable zero. For the factory PB data, we did this in Section 13.4 by centering each continuous factor about a central value, the mean.
ii) Reduce the number of groups in a model. Isolate and model each distinct group based on combinations of the categorical factors.

After the initial screening for dominant factors in Section 13.3, our factory PB data included only one discrete factors X5, for Machine 1 and Machine 2. We have already centered all continuous factors around their respective means. We use the `subset` command to isolate each group. Then we analyze each group separately.

```
> a1 = subset(apb,X5==1)      ! Form a datatable that includes only Machine 1 data.
> a2 = subset(apb,X5!=1)      ! Form a datatable that excludes only Machine 1 data.
> head(a1)          # Inspect the top six lines of the datatable with only Machine 1 data.
> head(a2)
```

Machine 1

```
> mod1=lm(Y~X3c+X2c, data=a1)       # Linear model only dominant factors.
> summary(mod1)                     # Display analysis for Machine 1 data.
Call: lm(formula = Y ~ X3c + X2c, data = a1)
Coefficients:
             Estimate   Std.Error   t-value      Pr(>|t|)
(Intercept) 40.83333    1.89419     21.56      0.000218 ***
X3c          0.67500    0.40182      1.68      0.191576
X2c          0.81500    0.08036     10.14      0.002043 **
Residual Standard Error: 4.64 on 3 Degrees of Freedom

> anova(mod1)                       # Display analysis of variance for Machine 1.
Analysis of Variance Table
Response: Y
           Df      Sum Sq      MeanSq    F value     Pr(>F)
X3c         1      620.17      620.17     28.808   0.012660 *
X2c         1     2214.08     2214.08    102.848   0.002043 **
Residuals   3       64.58       21.53
```

Machine 2

```
> mod2=lm(Y~X3c+X2c, data=a2)       # Display analysis for Machine 2 data.
> summary(mod2)
Call: lm(formula = Y ~ X3c + X2c, data = a2)
Coefficients:
             Estimate     Std.Error  t-value      Pr(>|t|)
(Intercept) 99.1667      1.3035     76.078      5.01e-06 ***
X3c          4.2750      0.2765     15.461      0.000588 ***
X2c          0.3850      0.0553      6.962      0.006081 **
Residual Standard Error: 3.193 on 3 Degrees of Freedom

> anova(mod2)
Analysis of Variance Table
Response: Y
Df          Sum Sq     Mean Sq    F value      Pr(>F)
X3c            1     1980.17    1980.17     194.240   0.0007998 ***
X2c            1      494.08     494.08      48.466   0.0060808 **
Residuals      3       30.58      10.19
```

Note that, in the analysis, we ordered the factors according to dominance in the earlier Section 13.3, rather than according to the F-values in this section. Machine 1 has surprises: Temperature X2 dominates ThroughPut X3 via both F-value and p-value. For Machine 1, anova guides us to retain both factors, whereas `summary(mod1)` guides us to only retain temperature. We follow anova's guidance. For Machine 2, there are no surprises.

Echoing the lesson in Section 7.7, we see that the Standard Errors of the coefficients dramatically decreased by modeling individually for Machine 1 and then for Machine 2. This is good.

13.6.1 Individual Machine Predictive Model Equations

Using the R output above we obtained the following pair of individual predictive model equations.

On Machine 1,

$$Y_{M1}[strength] = 40.83[\text{MPa}] + 0.675\left[\frac{\text{MPa}}{\text{kg/min}}\right](flow - 15.)[\text{kg/min}] + 0.815\left[\frac{\text{MPa}}{°\text{C}}\right](Temp - 225.)[°\text{C}]. \quad (13.14a)$$

And on Machine 2,

$$Y_{M2}[strength] = 99.17[\text{MPa}] + 4.275\left[\frac{\text{MPa}}{\text{kg/min}}\right](flow - 15.)[\text{kg/min}] + 0.385\left[\frac{\text{MPa}}{°\text{C}}\right](Temp - 225.)[°\text{C}]. \quad (13.14b)$$

13.6.2 Uncertainty of the Intercept in the Individual Predictive Model Equations

For the individual predictive model equations, the situation is simpler than in Section 13.4.3. Recall that the Uncertainty of the Intercept (mean) is given by Eq. 13.4:

$$\delta\text{Intercept} = Multiplier\frac{\text{Standard Error}}{\sqrt{\text{DoF}}}$$

As before the *Multiplier* comes from the "Student's t" for the desired confidence level and the DoF. In this case DoF = 3; then for a 95% confidence level the *Multiplier* = 3.18. Therefore, for the 0.95 Uncertainty of the Intercept for Machine 1, we obtain

$$\delta\,\text{Intercept}_{M1,0.95} = 3.18\frac{1.8942}{\sqrt{3}}[\text{MPa}] = 3.478[\text{MPa}] \quad (13.15a)$$

This means that for material produced on Machine 1, the Mean at the center of the Response Surface has a value in the range 37.36 MPa < Strength <44.31 MPa with 95% certainty.

Likewise, for the 0.95 Uncertainty of the Intercept for Machine 2, we obtain

$$\delta\text{Intercept}_{M2,0.95} = 3.18\frac{1.3035}{\sqrt{3}}[\text{MPa}] = 2.393[\text{MPa}] \quad (13.15b)$$

Thus, for material produced on Machine 2, the mean at the center of the Response Surface has a value in the range 96.77 MPa < Strength <101.56 MPa with 95% confidence.

13.6.3 Uncertainty of the Constant in the Individual Predictive Model Equations

The Uncertainty of the constant includes two orthogonal terms, the Standard Error for the Intercept and the Residual Standard Error (cf. Section 13.5.1). The two Standard Errors add according to Pythagoras, i.e. RSS. Mathematically, we recall (Eq. 13.6)

$$\delta\text{Constant} = \textit{Multiplier}\sqrt{\Sigma\left(\text{Standard Error}\right)^2}$$

As in Section 13.6.2, the *Multiplier* comes from the "Student's t" for the desired confidence level and the DoF. For a 95% confidence level and DoF = 3, the *Multiplier* = 3.18.

Therefore, for the 0.95 Uncertainty of the constant for Machine 1, we obtain

$$\delta\text{Constant}_{\text{M1,0.95}} = 3.18\sqrt{1.894^2 + 4.64^2}\,[\text{MPa}] = 15.94[\text{MPa}] \tag{13.16a}$$

This means that for material produced on Machine 1, the constant term across the Response Surface has a value in the range 24.90 MPa < Strength <56.77 MPa with 95% certainty.

Likewise, for the 0.95 Uncertainty of the constant for Machine 2, we obtain

$$\delta\text{Constant}_{\text{M2,0.95}} = 3.18\sqrt{1.3035^2 + 3.193^2}\,[\text{MPa}] = 10.97[\text{MPa}] \tag{13.16b}$$

Thus, for material produced on Machine 2, we are 95% certain that the constant term across the Response Surface has a value in the range 88.20 MPa < Strength <110.13 MPa.

13.6.4 Uncertainty of Other Factors in the Individual Predictive Model Equation

The Uncertainty due to a factor coefficient is just the Standard Error multiplied by the confidence level *Multiplier*. Contrast with Section 13.5.2. As previously, for a 95% confidence level and DoF = 3, the *Multiplier* = 3.18. Mathematically, where the experimental strength result is *Y*, we recall (Eq. 13.8)

$$\frac{\partial Y}{\partial\text{Factor}} = \textit{Multiplier}\left(\text{Standard Error}_{\text{Factor}}\right)$$

13.6.4.1 Uncertainties of Machine 1

For Uncertainties of the other factors, we must now consider each machine separately. For Machine 1, we obtain the 0.95 Uncertainty of strength due to the throughput to be

$$\frac{\partial Y}{\partial\text{ThruPut}}{}_{,\text{M1,0.95}} = 3.18\cdot 0.40182 = 1.278\left[\frac{\text{MPa}}{\text{kg / min}}\right] \tag{13.17a}$$

Furthermore, the 0.95 Uncertainty of strength due to temperature on Machine 1 is

$$\frac{\partial Y}{\partial \mathrm{T}},_{\mathrm{M1,0.95}} = 3.18 \cdot 0.08036 = 0.256\left[\frac{\mathrm{MPa}}{°\mathrm{C}}\right] \tag{13.17b}$$

13.6.4.2 Uncertainties of Machine 2

Likewise for Machine 2, we obtain the 0.95 Uncertainty of strength due to throughput to be

$$\frac{\partial Y}{\partial \mathrm{ThruPut}},_{\mathrm{M2,0.95}} = 3.18 \cdot 0.2765 = 0.879\left[\frac{\mathrm{MPa}}{\mathrm{kg/min}}\right] \tag{13.18a}$$

Furthermore, the 0.95 Uncertainty of strength due to temperature on Machine 2 is

$$\frac{\partial Y}{\partial \mathrm{T}},_{\mathrm{M2,0.95}} = 3.18 \cdot 0.0553 = 0.176\left[\frac{\mathrm{MPa}}{°\mathrm{C}}\right] \tag{13.18b}$$

With the values above, we now have all values needed to answer Motivating Question #2.

13.6.4.3 Including Instrument and Measurement Uncertainties

As discussed in Chapters 10 and 11, Measurement Uncertainty includes instrument Uncertainty, precision Uncertainty, and calibration, fixed, and random errors. All of these various Uncertainties are presumably independent of Residual Standard Errors and model Standard Errors. Therefore, the Uncertainties and Standard Errors add according to RSS (Pythagoras).

13.6.5 Include All Coefficient Uncertainties in the Individual Predictive Model Equations

When the individual predictive model equations for each individual machine include the 95% confidence interval for each coefficient, we obtain the following pair of model equations. Contrast with Section 13.5.3.

On Machine 1,

$$Y_{\mathrm{M1}}[strength] = (40.83 \pm 15.94)[\mathrm{MPa}] + (0.675 \pm 1.278)\left[\frac{\mathrm{MPa}}{\mathrm{kg/min}}\right](flow - 15.)[\mathrm{kg/min}] + (0.815 \pm 0.256)\left[\frac{\mathrm{MPa}}{°\mathrm{C}}\right](Temp - 225.)[°\mathrm{C}]. \tag{13.19a}$$

And on Machine 2,

$$Y_{\mathrm{M2}}[strength] = (99.17 \pm 10.97)[\mathrm{MPa}] + (4.275 \pm 0.879)\left[\frac{\mathrm{MPa}}{\mathrm{kg/min}}\right](flow - 15.)[\mathrm{kg/min}] + (0.385 \pm 0.176)\left[\frac{\mathrm{MPa}}{°\mathrm{C}}\right](Temp - 225.)[°\mathrm{C}]. \tag{13.19b}$$

It was correct to be suspicious in Section 13.5.3, when coefficients of so many modeling terms were identical. Here, for individual machine models, every term differs between Eqs. 13.19a and 13.19b.

13.6.6 Overall Uncertainty from All Factors in the Individual Predictive Model Equations

Since this section follows Section 13.5.4 exactly, we will skip steps here. We seek an equation for the overall Uncertainty at any set of conditions in the operating map domain. We recall (Eq. 13.11)

$$\mathcal{U}_{0.95}(Y)=\left[\left(\delta \text{Constant}_{\text{M},0.95}\right)^2+\left(\left[\frac{\partial Y}{\partial \text{ThruPut}},_{0.95}\right]\Delta \text{ThruPut}\right)^2+\left(\left[\frac{\partial Y}{\partial T},_{0.95}\right]\Delta T\right)^2\right]^{1/2}$$

Having centered the operating map at the exact center of the continuous variables, we note the same relations as shown in Eqs. 13.12a and 13.12b.

Finally, we obtain equations that predict the overall Uncertainty at any point across the operating domain for each machine. On Machine 1, we obtain

$$\mathcal{U}_{\text{M1},0.95}(Y)=\left[(15.94)^2+\left([1.278]\Delta \text{ThruPut}\right)^2+\left([0.256]\Delta T\right)^2\right]^{1/2}[\text{MPa}] \tag{13.20a}$$

And on Machine 2 we obtain

$$\mathcal{U}_{\text{M2},0.95}(Y)=\left[(10.97)^2+\left([0.879]\Delta \text{ThruPut}\right)^2+\left([0.176]\Delta T\right)^2\right]^{1/2}[\text{MPa}] \tag{13.20b}$$

Both equations take the form of a hyperbolic surface.

By comparing the predicted overall Uncertainty of individual modeling in Eqs. 13.20a and 13.20b with the overall Uncertainty of dual modeling in Eqs. 13.13a and 13.13b, we see that the predicted Uncertainties have improved (reduced) considerably for both machines. In Table 13.7, we can compare overall Uncertainties between the dual model and the individual models.

Table 13.7 Overall Uncertainty at key operating points. Compare dual and individual models.

Strength 95% Uncertainty [MPa], total		
Machine	**At center**	**At corner**
Dual: 1	31.41	33.58
Dual: 2	35.62	37.54
1	15.94	18.32
2	10.97	12.61

When reporting the predictive modeling equation and the predicted Uncertainty equation for Machine 1, we recommend using Eq. 13.14a alongside Eq. 13.20a due to ease of calculation, plotting, and accuracy. Likewise for reporting Machine 2, use Eq. 13.14b alongside Eq. 13.20b. Although Eqs. 13.19a and 13.19b display relationships between values more directly, one must beware of overestimating the error.

13.6.7 Quick Overview of Individual Machine Performance Over the Operating Map

Table 13.8 shows the strength [MPa] predicted by the individual machine predictive model equations, Eq. 13.14a and 13.14b, at key operating conditions. Recall that the super-factory requires a guaranteed strength of 60 [MPa].

For Machine 1, the narrow region of the operating map has shrunk. Including the Uncertainties, we still predict that about 50% of material produced on Machine 1 would fail to achieve the required material strength. We further diagnose that for Machine 1, process temperature has more impact than throughput.

For Machine 2, the whole operating map achieves the required strength. We highlight that for Machine 2, the throughput has much more impact than process temperature. When Machine 2 is operated at the highest throughput setting, the predicted material strength is sufficient to assure the client super-factory that this factory has already achieved Six Sigma production levels, with average strength twice the required strength. How satisfying is that!?!

Table 13.8 Predicted strengths over the operating mmap using individual machine predictive models.

Machine 1	Intercept	ThruPut	Temp [°C]	Predicted [MPa]
Center	40.83	15.00	225.00	40.83
Max	40.83	20.00	250.00	64.58
BorderLine	40.83	20.00	244.38	60.00
BorderLine	40.83	13.21	250.00	60.00
Min	40.83	10.00	200.00	17.08
Min @Max	37.36	20.00	250.00	61.11

Machine 2	Intercept	ThruPut	Temp [°C]	Predicted [MPa]
Center	99.17	15.00	225.00	99.17
Max	99.17	20.00	250.00	130.17
Corner	99.17	20.00	200.00	110.92
Corner	99.17	10.00	250.00	87.42
min	99.17	10.00	200.00	68.17
MinAll	96.77	10.00	200.00	65.77

13.7 Final Answers to All Motivating Questions for the PB Example Experiment

Note: All answers refer exclusively to the initial list of six factory variables shown in Table 13.1. Other possible factors were not considered, as they were not reported for analysis.

13.7.1 Answers to Motivating Question #1

MQ1. Which machines, and at what operating conditions, can achieve the required material strength (60 MPa)?

Predicted material strengths as produced on both machines are shown in Table 13.8.

We recommend the factory never use Machine 1 for this contract. Factory Machine 1 is predicted to produce, only 50% of the time, the required material strength when near maximum temperature and throughput operating conditions. The predictive model equation for Machine 1 is given in Eq. 13.14a.

We recommend the factory only use Machine 2. Factory Machine 2 can reliably exceed the required material strength at all operating conditions. The predictive model equation for Machine 2 is given in Eq. 13.14b.

We recommend that the factory produce this material only on Machine 2.

13.7.2 Answers to Motivating Question #2

MQ2. What is the predicted Uncertainty (95% confidence) of material strength across operating map conditions?

Tabulated overall Uncertainties on both machines are shown in Tables 13.2 and 13.7.

The overall predicted Uncertainty for Machine 1 across the operating map is given in Eq. 13.20a. The coefficient Uncertainties are given in Eqs. 13.16a, 13.17a, and 13.17b.

The overall predicted Uncertainty for Machine 2 across the operating map is given in Eq. 13.20b. The coefficient Uncertainties are given in Eqs. 13.16b, 13.18a, and 13.18b.

13.7.3 Answers to Motivating Question #3 (Expanded from Section 13.3)

MQ3. Which factors significantly impact material strength and need attention? That is, find factors for which we "reject the Null Hypothesis." Corollary: Which candidate factors allow operational flexibility (essentially have Null effect; that is, find factors for which we "cannot reject the Null Hypothesis")?

3A. The dominant candidate factors from the PB screening, in order of dominance, are:

X5: Machine used (discrete: 1 or 2).
X3: Throughput (continuous).
X2: Temperature (continuous).

3B. The remaining candidate factors, which we will continue to monitor and record, yet which we will allow to become noise in future analysis, are:

X1: Tension control (discrete: automatic or manual).
X4: Mixing (discrete: one or two times).
X6: Moisture (continuous: ranging from 20% to 80%).

3C. We recommend recording other factors beyond those recorded here. For example:

Freshness of raw materials: delivery date, supplier. Or age of raw materials.
Machine operator: name, title, experience.
Material contaminants. Or material purity.
Other factors, as agreed by factory management.
Data on typical factory charts: Pareto charts, run charts, control charts, etc.
Date, day, time production run begins and ends.
Record these factors, as well all factors initially listed, for every production run.

Warning: If your client, the factory management, wants any other possibly relevant factor to be considered in analysis, that factor must be recorded at the time of experiment (during production and during material property measurement: strength, density, etc.).

13.7.4 Answers to Motivating Question #4

MQ4. For what operating conditions does this factory have the capability of guaranteeing production with Six Sigma 6σ reliability (less than four defects per one million tests)?

Good news. Factory Machine 2 demonstrates the capability for Six Sigma 6σ production, even with only six tests during the initial PB experiment. Each future production run will provide additional results to refine the predictive equation and predicted Uncertainty. We estimate that Uncertainties will reduce with each additional production run.

Good news. Factory Machine 2 produces the strongest material at the maximum throughput rate. At the maximum throughput rate, it appears that factory Machine 2 will produce less than four defects per one million tests.

Dedicated testing according to Six Sigma 6σ protocols is required to confirm compliance. The factory management must decide whether recognition as Six Sigma 6σ compliant is worth the effort.

13.7.5 Other Recommendations (to Our Client)

i) Re: Future Production Runs

- Record more candidate factors which may impact product strength. Consider factors such as those in Section 13.7.3, item 3C.
- If new factors are chosen, then: Generate a new (second-stage) PB Screening Design. Incorporate new candidate factors along with current dominant factors. Glean a new set of dominant factors from a linear model using the R language on all collected data.
- Consider a predictive modeling equation with higher-order factors, interaction terms between factors, quadratic terms. Repeat until all dominant factors are identified.
- Use Gosset to select conditions for new production runs. The Gosset model will include interaction terms, quadratic terms, and higher-order terms. Choose at least two sets of operating conditions for runs selected by you, based on your expertise.
- Combine all results from future and finished runs to improve predictive equations and to obtain tighter confidence intervals.

- After optimal operating conditions are found, continue factory runs while recording all factors. Use the R language to continue improving equations and confidence intervals.

ii) Re: Quality Control

- Recheck materials science. Is there a known maximum strength for the product recipe? If so, the known maximum strength provides a standard against which to judge all products.
- In order to guarantee the required material strength, the factory is supplying product with significantly stronger material (with a mean about twice the required strength). The reason for producing overly strong material is that the overall Uncertainty is large.
- Test multiple-samples for each production run. Record strength and other properties of interest. Analyze using R with models of all dominant factors.

iii) Re: Machine 1

- Find another use for factory Machine 1.
- Inspect factory Machine 1. What aspect of Machine 1 degrades material strength of the product? Factory Machine 2 produces material with strength more than twice Machine 1.
- To assure reliable production, would factory management consider redundancy on the production line with the addition of another machine superior (or equal) to factory Machine 2?

iv) Re: Continual Improvement

- Regularly audit factory production runs. We are happy to continue working with you.
- Combine data and results from future production runs with current data to improve predictive equations and to obtain tighter confidence intervals.

13.8 Conclusions

We found definitive answers to the Motivating Questions (MQ1, MQ2, and MQ3), even with results from a 12-run PB Screening Design. For Motivating Question #4, although the factory cannot yet guarantee Six Sigma 6σ production, our results appear promising.

We recommended that the factory record other factors which might improve model predictions.

The general form of a predictive model equation (for a factory machine) is

$$Y[strength] = \text{Intercept}[\text{MPa}] + \left(\frac{\partial Y}{\partial \text{ThruPut}}\right)\left[\frac{\text{MPa}}{\text{kg / min}}\right](\text{ThroughPut} - 15.)[\text{kg / min}]$$
$$+\left(\frac{\partial Y}{\partial T}\right)\left[\frac{\text{MPa}}{°\text{C}}\right](\text{Temperature} - 225.)[°\text{C}]. \quad (13.21)$$

Table 13.9 Side-by-side comparison of all model values derived in this chapter.

	Label	Model values [MPa]			
	Machine	Machine 1		Machine 2	
Category	Model type	Dual M1	M1	Dual M2	M2
Value	Mean, intercept	40.833	40.833	99.167	99.167
	Throughput	2.833	0.675	2.833	4.275
	Temperature	0.480	0.815	0.480	0.385
Standard Error	Intercept	5.138	1.894	7.267	1.304
	Throughput	0.727	0.402	0.727	0.277
	Temperature	0.145	0.080	0.145	0.055
95% Uncertainty	Constant	31.412	15.937	35.616	10.967
	Throughput	1.679	1.278	1.679	0.879
	Temperature	0.336	0.256	0.336	0.176
	Overall min	31.41	15.94	35.62	10.97
	Overall max	33.58	18.32	37.54	12.61

The general form of the overall Uncertainty equation (for a factory machine) forms a hyperbolic surface according to (Eq. 13.11):

$$U_{0.95}(Y) = \left[\left(\delta \text{Constant}_{M,0.95}\right)^2 + \left(\left[\frac{\partial Y}{\partial \text{ThruPut}}{}_{,0.95} \right] \Delta \text{ThruPut} \right)^2 + \left(\left[\frac{\partial Y}{\partial T}{}_{,0.95} \right] \Delta T \right)^2 \right]^{1/2}$$

Table 13.9 collects and compares all model values found in this chapter, dual model versus individual model, Machine 1 versus Machine 2.

Individual models are much improved over the dual model, with substantially tighter overall Uncertainty intervals.

The R statistical language provided a quick, powerful way to process experimental results and to create predictive models.

By incorporating data from future production runs, prediction equations will likely improve.

Homework

13.1 Expand the individual predictive model for Machine 2. Include in the linear model three factors: throughput, temperature, and mixing. Find a new predictive model equation. Plot the original data using the contributed R-package scatterplot3d, as we demonstrated in Chapter 9. Discuss how your results compare with the individual model for Machine 2 in this chapter. How did the addition of the mixing factor improve or degrade the model predictions?

13.2 Expand the individual predictive model for Machine 2. Include in the linear model three factors: throughput, temperature, and mixing. Find an equation for the overall Uncertainty. Discuss how your results compare with the individual model for Machine 2 in this chapter. How did the addition of the mixing factor improve or degrade the overall Uncertainty?

13.3 Use Gosset to design an experimental plan with a model including interaction and higher-order terms for an additional 20 production runs on Machine 2. Select at least two points (sets of operating conditions) in the operating map that you would like to test.

13.4 Foreshadowing teaser to section 14.2.3 about typewriters.
Premise: A mind is necessary and sufficient to create information.
Given: A mind is sufficient to create information.
As for necessary, does our human experience throughout history know of any contradictory evidence? Lack of evidence, however, is not evidence of lack. DNA in a typical human cell contains more information than an encyclopedia. How would you debate for or against the premise? How shall we wonder?

14

Documenting the Experiment

Report Writing

Documenting the experiment begins on the first day of the project, when you start the project's logbook. It ends when your report, paper, or thesis has been published and copies have been sent to everyone on your list.

14.1 The Logbook

You should start a new logbook for each project. It should be bound, not loose-leaf. You should work in the logbook, or at least copy into it the good parts of what you do on scratch paper. You should initial and date every page. If something patentable arises, have that page witnessed by someone with a written statement that he or she understood the significance of what was witnessing at the time he or she witnessed it.

The logbook is not entirely a legalistic game, although patent rights may well be determined by the date on a signed page in such a book.

The larger purpose is to provide you with a historical record of all of the thinking, analysis, interpretation, and insight that go into the experiment. It may be a year or more before you write your final report. If you have no record of all of the effort you put into the work, and all of the good reasons you had for all of the small decisions you made along the way, you will not have anything to say at the end of the year except, "The rig is shown in Figure 1. The data are shown in Figure 2." And that is not a very convincing report.

The notion of a logbook fits very well into the general scheme of experiment planning. It provides a place where you can talk to yourself, raise questions, examine ideas, store references, and record observations that, at a cursory glance, seem to have no sensible explanation.

All of this feeds content into the final report.

14.2 Report Writing

Technical information is not "sold," like cars or fresh fruit, by copywriters: as the saying goes, "You can't make a silk purse out of a sow's ear." No amount of skill in writing can convert a poor job of engineering into a good job, but unfortunately many fine engineering achievements have died in the archives because of weak reporting. In your

Planning and Executing Credible Experiments: A Guidebook for Engineering, Science, Industrial Processes, Agriculture, and Business, First Edition. Robert J. Moffat and Roy W. Henk.

Companion website: www.wiley.com/go/moffat/planning

professional activity your reports will, in many cases, be your most important means of communication with your superiors and subordinates. Your reports, plus your personal logbook, will contain the only permanent record of your professional achievements. It is important that your reports be clear, complete and concise. This takes practice.

14.2.1 Organization of the Reports

There is no universally accepted structure for a technical report. The format described here will be acceptable for many purposes, however, and provides a framework on which more elaborate structures can be constructed, if required. The major divisions of the report are:

- Title page.
- Abstract.
- Foreword.
- Objectives.
- Results and conclusions.
- Discussion.
- References.
- Figures.
- Tabular data.
- Appendices.

The text sections (abstract, foreword, results and conclusions, discussion, and appendices) should each have the same internal organization, similar to that used in newspaper writing: *The most important topic comes first, and the most important topic gets the biggest "play."* This means it gets the most space. The reader will judge what you thought most important by what you talk about first and what you talk about most. Be sure the distribution of "bulk" matches the distribution of importance.

14.2.2 Who Reads What?

The sections differ in the kinds of information they contain and in their depth of coverage. They also differ in that they may be read by different classes of readers. Two typical readership hierarchies are:

i) For an "internal" company report:
   - Your immediate supervisor will read it all, as will some of your contemporaries.
   - The department head will read the abstract, the foreword, and the results and conclusions.
   - The vice-president will read the abstract.
   - At the end of the year, the title will be listed in the annual progress report.

ii) For a paper published in a technical journal:
   - Subscribers to the journal, who also happen to be working in your field, will read it all.
   - Subscribers to the journal, who wish to keep in touch with your field, will read the abstract.
   - The title and abstract may be reprinted by an abstracting service for wider circulation.

- Readers will judge, on the basis of your abstract, whether or not to read your entire paper.
- Most of the engineering world will never hear of your paper or at most will see its title only.

Keep these audience groups in mind in writing the report, so that you can have the biggest impact with your work. You have to attract the reader by showing the substance of your work.

14.2.3 Picking a Viewpoint

The first duty of the author is to decide what is important about his or her work – and that depends on the audience you think you are dealing with.

To illustrate, let's assume that you are about to write a report describing a series of engine tests of a new carburetor. What are the significant results of the test? Different readership groups will have different views.

TECHNICIAN: "I had to add three quarts of oil during that series; the rear shaft seal let go. I'll have to replace that before we run again."

TEST ENGINEER: "That new carburetor looks pretty good, fuel consumption was down 3% and exhaust aldehydes were down nearly 20%."

ENGINE CYCLE ANALYST: "Improving the homogeneity of the air–fuel mixture raised the engine's thermal efficiency to within 3% of its predicted value."

VICE-PRESIDENT, MARKETING: "We are pleased to announce a major breakthrough in our battle against smog."

Quality technical writing involves not only the technical skills but also the ability to recognize the significance of what has been accomplished. Do you remember the story about the difference between Shakespeare and an infinite number of monkeys writing since eternity past?[1]

> "If an infinite number of monkeys each played on a typewriter for eternity, sooner or later one would come up with a complete script for Hamlet." Sure he would, but, having done that, the monkey would have continued right on typing "a;kljhfdpoiguypoiwqe-o8;lksajn oiupq4oyupoisdf; ljnda;lkjn;aslkn..."
>
> Shakespeare, however, sat back and said to himself, "Hey, that's pretty good stuff!"

14.2.4 What Goes Where?

Everyone is familiar with the need for subheadings to organize a report. Many times, however, it looks to me as though the authors simply stick subheadings in at random, to break up the pages – the material following a subheading is not at all related to the subheading! The notion of sticking to the subheading is not a hard concept to grasp. It

1 Just imagine what experiment you could do with unlimited, infinite resources! Seconding Shakespeare, reality debunks this familiar thought experiment. We consider probabilistic resources: the age of the universe is 13.8 billion years = 4.5×10^{17} sec; the estimated number of atoms in the universe is about 10^{80}, so there would be fewer monkeys. If each monkey randomly typed 100 key punches per second, probability tells us not one monkey would even have written this short footnote. Let your mind create and clarify, like Shakespeare. See HW 13.4. Your task and privilege is to explain how your experiment credibly answers the Motivating Question.

is, however, hard to live by. The tendency is, when pursuing a subject, to follow its implications and collateral topics on a meandering path that ends up far away from the original point. The end result is a lot of material that is not related to the subheading it is under. This is not good writing, but it is human nature. When you have "finished" writing a section, go back over it and cut out all of the stuff that fits better under a different subheading and move it to where it belongs. Then "wrap it" in whatever text is necessary so it fits in the new location. That is a more productive approach than trying to "not write" what seems natural to write next.

14.2.4.1 What Goes in the Abstract?

Nowhere is the need for careful sorting out so evident as in the abstract: there is no place to hide. With only a few hundred words to work with, even one misdirected sentence will stand out. Authors must identify their own viewpoints, select what they believe to be the most important aspect of the report, and write with that in mind.

The abstract should provide specific, substantive answers to the following questions: What was done? Why? What happened? So what?

For example:

> A 283 cu. inch engine was tested using a prototype Model X-1 carburetor to investigate possible effects on fuel economy and hydrocarbon emission. Fuel consumption decreased an average of 3%, and hydrocarbon emission decreased an average of 20% over the speed and load range, compared to results obtained using a standard carburetor. Measured hydrocarbon concentration was 12% below the acceptable hydrocarbon emission specification. Further development of this carburetor principle is recommended.

These 71 words are, obviously, a brutal condensation of what might be a 20-page report. The abstract must rely on phrases like "an average of…" and "over the speed and load range" to convey the thrust of the work without being misleading.[2] It is the author's responsibility that the few, brief statements in the abstract properly reflect the true significance of the report.

14.2.4.2 What Goes in the Foreword?

The foreword should contain the background material necessary to introduce a technically competent reader to the subject matter and its terminology and to show the motivation for the present study. Reports that are part of a routine series frequently reduce the foreword to a mere skeleton. Reports that must stand alone, or that introduce new areas, will require a more extensive introduction. Keep your audience in mind. Note that there should be no "feedback" from the present program into the foreword. The foreword must end up looking like it could have been written the day before the work it introduces (that's why it's called the FOREword – the words that go before).

Here are a couple of brief examples to illustrate the content of a foreword.

---

2 Charles Goren, a famous bridge player and teacher, said that the main task in bridge bidding was "To always tell the least damaging lie." This is similar to the problem of writing a very compact abstract. You can't say everything, so what do you say?

Foreword of a routine test

Mr. A. Lucas, of Carburetor Development, requested testing of an X-1 prototype for fuel economy and hydrocarbon emission, under standard conditions.

Foreword of a test that is not routine

Investigation of the fuel–air ratio, on a cylinder-by-cylinder basis, suggested that nonuniformity might contribute significantly to excess fuel consumption and hydrocarbon emission. Tests conducted using laboratory equipment to provide an entirely uniform mixture have shown substantial improvements both in fuel economy and hydrocarbon emission. As a result of these tests, a new carburetor design was conceived, aimed specifically at providing a more uniform mixture to the cylinders. The carburetor design involves preheating the fuel, in combination with a novel flow-mixing section. Several production designs have been prepared for carburetors including these features. The X-1 prototype is the first of these proposals to be tested.

14.2.4.3 What Goes in the Objective?

The foreword has set the technical stage and shown why more work is required and, generally, what form it should take. The objectives section logically follows with a statement of exactly what the present program aims to accomplish. Freely draw upon the Motivating Questions.

The statement of objectives should not consist of a recipe for proceeding. It is not necessary to state exactly what will be done in terms of tests or data. What is desired here is the *objective* of the program, not the procedure.

An objective:

To determine the effect of the X-1 prototype carburetor on fuel economy and hydrocarbon emission of the 283 in.^{3} V-8 engine.

A procedure:

Install the X-1 prototype carburetor on a standard 283 in.^{3} engine and conduct standard tests for fuel economy and hydrocarbon emission (M.S. 364 Engine Test Code).

14.2.4.4 What Goes in the Results and Conclusions?

This section is frequently handled by a numerically coded list of statements: 1, 2, 3,..., etc. The first of these should be the most important conclusion. The last of these should be the least significant result.

Each statement may contain more than one sentence, but if it does, then the first sentence should be the most important statement – the conclusion – the rest contain the qualifications.

If you, the author, can't identify the most important conclusion, then you don't understand your objectives. You should sit back and reconsider what's going on. Sometimes you will conclude that the test you ran didn't really address the issue described in the objective section. Then you have to decide which section to rewrite. When the report is finally issued, the conclusions and the objectives have to make a sensible pair. If the objective commits you to answering a certain question, then the conclusion had better be the answer to that question!

Remember that whatever you, as author, place first in this list will be regarded by the readers as your impression of the most important thing you have to say.

An example, following on the same experiment:

Results and conclusions

i) Development of the X-1 prototype carburetor should be continued.
ii) Fuel consumption was reduced 3%, on average, over the speed and load range of M.S. 364 Engine Test Code.
iii) The test engine suffered a shaft seal failure, but this is not believed to be related to the carburetor design.

14.2.4.5 What Goes in the Discussion?

The discussion section starts off (in the first draft of the report) as the whole report that you could write if only your immediate supervisor, or someone equally familiar with your work, were going to read it. It is a compact, detailed description of the program from apparatus to results and conclusions in the most appropriate language. It would take a person "skilled in the art" and knowledgeable about your objectives and the problems you face to read it. That "skillfull" reader would not have to be told *why you did it* and *what your results mean*; only *what you did.*

The problem with that first draft is that very few people would understand it. The second draft must take into account the fact that your reports may be read by less knowledgeable persons. You, as author, have to describe the motivation (that material goes into the FOREWORD), extract the meat (that goes into the RESULTS AND CONCLUSIONS), and summarize the overall program (this becomes the ABSTRACT).

There are some editorial decisions that will need to be made along the line: should I put the explanation of why we did it this way into the FOREWORD, or right here, in the DISCUSSION? Those decisions are less important than recognizing that the explanation is needed and has to go somewhere.

The discussion, constituting the bulk of your report, requires organization to keep the reader oriented and to allow specific information to be found easily – i.e. more subheadings. One way to organize it is to follow the structure of the contributive parts of the report (there is no point in reviewing or amplifying material in the foreword): discuss the principal conclusions and results, because they form a good outline for the first section of the discussion. A line diagram of the apparatus will certainly appear in the report. The discussion of the apparatus can be structured around the line diagram: this will help organize your thoughts and remind you of each component. It also serves to review the procedure. The Uncertainty Analysis and sample calculations serve to outline this portion of the discussion. Each appendix provides a separate topic. Use subheadings freely to indicate the structure of the discussion, especially if it is long.

14.2.4.6 References

The purpose of citing references is twofold. First, to establish a traceable path back to the original source for information you quote but did not originate. Second, to give credit to the originators of work.

Engineers often fail to give credit where credit is due to colleagues whose insights lead to an understanding or to a particular form of a test or who provided a particularly useful interpretation of some mysterious results. I recommend that an "unofficial" source be acknowledged as a "personal communication." We are not bankers – we will not go broke by giving credit!

As to the form of the citation, each organization, each society, and each journal has a specific form for references. The citation should contain enough information to assure the reader's ability to obtain a copy of the desired work and to find the exact part which applies to your problem (to the nearest page, at least). When two or more sources are available, the most easily obtainable should be listed.

Don't use "second-party references." If Smith quotes Jones and you wish to use Jones's work, then reference Jones himself, not Smith's quotation of Jones. This is particularly important when using textbooks as sources for references. Very few of the citations within a textbook refer to work done by that author. If you feel guilty about citing a source you have not read yourself, adopt the form "Jones [19XX] as referenced by Smith [20YY]." Then you have preserved traceability and not implied that you, yourself, have read the original.

14.2.4.7 Figures

Figures are used to quickly convey an understanding of the relationships between the dependent and independent variables of a problem.

Certain information must appear on each figure:

- Title.
- Coordinate names.
- Scale markings.
- Author's initials (or name) and date.

Convention dictates that the independent variable be used as the abscissa, being placed at the bottom of the figure. If a figure is to be viewed from the edge of the sheet, then that is considered the "bottom" of the figure. Extra margin space must be allowed on the edge that will be used for binding.

The title must be sufficiently complete and descriptive to identify the content of the figure without ambiguity. You must take the view that this figure may come loose from the report and be found lying on the copy-room floor, all by itself. You want it still to be intelligible.

Don't use abbreviations in titles of figures. Include peripheral information wherever you, as author, feel that it is needed to prevent the reader from missing or misunderstanding the point of the figure.

Scales should begin with zero, at the lower left corner of the figure. Suppressed zero scales should be used with extreme caution and clearly "flagged" to warn the reader. The subjective importance of a change is based, first, on the appearance of the line. When several variables are to be plotted on the same sheet, the scale for each must be carefully chosen to avoid distorting the relationship between the variables. Choose scales for easy interpolation, in view of the grid chosen.

Don't use color to distinguish between different variables without also distinguishing the lines in some other way. Engineering reports are often reproduced in black and white. If all of your identification is by color code, then your reproduced figure will be a hopeless jumble.

Critically important data should be tabulated as well as shown in figures. This allows critical comparison of your work by readers without forcing them to introduce errors due to reading from the figures.

Lettering should be placed on the figure with care: a decimal point that occurs right at the intersection of two major grid lines may be concealed and lost. Use text boxes with clear white backgrounds.

Line diagrams and schematics of apparatus and instrumentation done on plain paper are very helpful in discussing the equipment you used. A "circuit diagram" approach is usually acceptable, using stylized symbols to represent the various pieces of apparatus, such as valves, pressure gauges, thermometers, etc. A photograph is still desirable, however, since it gives more detail than the schematic ever could.

14.2.4.8 Tables

With computerized data reduction becoming more and more common, the trend toward tabular data presentation is showing increasing strength. Such listings are inexpensive and convenient ways of getting information into the report. Use 12-point type, at least, in a standard font, and a simple tabular structure. That way the data will be intelligible to an optical character reader (OCR) program, even after it has been faxed a couple of times. Special characters should not be used in headings – they may not be properly interpreted.

When computer printouts are used as tables, include a translation index describing exactly what each printout label means. Don't attempt to conserve words in these descriptions. You get no star on your forehead by being so terse that you are ambiguous.

Each page of tables should contain a title reference, either partial or complete, so that it may be restored to its proper place when it is found on the floor of the copy room.

14.2.4.9 Appendices

Every engineering report contains certain points that could be amplified, derived more fully, or defended in more depth except for the fact that such a diversion would weaken the continuity of the main line of argument. Such material properly belongs in an appendix, if it is to be included at all.

Typically, the sample calculations, Uncertainty Analyses, lists of equipment and instruments, and derivations of formulae used will be relegated to appendices.

Each appendix must stand by itself, although it may refer to material in the body of the report or in other appendices.

Nomenclature used in the appendices should be consistent with that used in the body of the report.

14.2.5 The Mechanics of Report Writing

Technical reports can be written in a planned manner: you don't have to rely on inspiration to guide you. Here is a suggestion as to how to organize for the task of writing a report.

Let's assume that the experimental work has been completed and the data have been reduced to final form. It is now time to begin writing the report.

i) Decide on the length you want for the entire report and make a preliminary distribution among the different subheadings of the report.
ii) Prepare the equipment schematic and instrumentation line diagrams, and collect any photographs to be used.
iii) Plot the data in final form showing the most significant correlations. Arrange the figures in order of importance, most important first.

iv) Prepare any tabular summaries required.
v) Write the discussion in its most compact form, aiming for a skillful reader. Center the discussion on the figures. Look the discussions over and decide what must be addressed in the foreword.
vi) Write the foreword section.
vii) Write the objectives section.
viii) Write the results and conclusions.
ix) Check back:
  a) Do the conclusions follow from the results?
  b) Do the conclusions address the objective?
  c) Do the conclusions address the most important aspect of this work?
  d) Is everything in the report that you think needs to be preserved about this experiment?
  e) Is there material in the discussions that should be moved into the foreword?
  f) Is the report of acceptable length?
x) Write the abstract.

Reports are seldom written on a once-through basis: usually, they require iteration between what happened (the data), what it means (the conclusions), and how much you should say about it (the discussion).

14.2.6 Clear Language Versus "JARGON"

Each field of endeavor has its own peculiar vocabulary, developed to facilitate communication between coworkers. The common name for these vocabularies is "shop talk" or "jargon." The vocabulary of mechanical engineering would probably be entirely meaningless to a violin player, almost meaningless to an MD, and somewhat unclear to an electrical engineer. Within the field of mechanical engineering there are further subdivisions, each with its own collection of terms. A stress analyst and a thermodynamicist, for example, have different vocabularies. The author of a technical paper or report must be careful to adjust his or her vocabulary to coincide with that of probable readership. Thus, a highly compact, jargon-filled note might be left on a coworker's desk, but some "translation" to clear language (i.e. uncoded) might be desirable before the same information could be sent to a distant colleague. In another sense, jargon is the "slang" of technical writing: use it only in very informal communication.

A classic example of pure jargon is reproduced in the following pages: the Turbo-Encabulator (Panel 14.1). It is interesting to note that no reader can be certain that this is a hoax because no reader can guarantee that he (or she) understands the jargon used here! Much of the engineering literature looks like this to the nonspecialist.

Jargon is not always so clearly intentional as in the Turbo-Encabulator. A desire for compactness frequently leads to jargon-filled writing. The following paragraphs are quite intelligible to those persons working in the field of the turbulent boundary layer with mass addition. To nonspecialists, they would appear to be the second installment of the Turbo-Encabulator:

Panel 14.1 The Turbo-Encabulator

For a number of years now work has been proceeding in order to bring to perfection the crudely conceived idea of a machine that would not only supply inverse reactive current for use in unilateral phase detractors, but would also be capable of automatically synchronizing cardinal grammeters.[3] Such a machine is the TURBO-ENCABULATOR. Basically the only new principle involved is that instead of power being generated by the relative motion of conductors and fluxes, it is produced by the modial interaction of magneto reluctance and capacitive directance.

The original machine had a base plate of pre-fabulated amulite, surmounted by a malleable logarithmic casing in such a manner that the two spurving bearings were in direct line with the pentametric fan. The latter consisted simply of six hydro-coptic margelvances, so fitted to the ambigacient lunar vane shaft that side fumbling was effectively prevented. The main winding was of the normal lotus-odelta type placed in panendromic semi-beloid slots on the stator, every seventh conductor being connected by nonreversible tremic pipes to the differential girdle rings on the upper end of the grammeters. Forty-one manestically spaced grouting brushes were arranged to feed into the rotor slip-stream a mixture of high S-value phenylhydrobenzamine and 5% reminative tetryliodohexamine. Both of these lipids have specific percosities as given by the formula P = 2.5NO*7 where N is the diathetical evolute of retrogradial temperature phase disposition and C is, of course, Chomendulatcls annular refractive pilfrometer (for a description of this instrument see L.F. Rumpleverstein in "Zeitschrift-fur elektrotechnistatisshe – Donnerblitzen," Vol. VII). Up to date nothing has been found to equal the transcendental hopper dadoscope (see Peruvian Academy of Scatological Science Journal, June 1957). Electrical engineers will appreciate the difficulty of nubbing together a regurgative purvel and supremitive wannel sprocket. Indeed, this proved to be a stumbling block to further development until, in 1953, it was found that the use of anhydrous mangling pins enabled the dryptinastic boiling shim to be tankered. The early attempts to construct a sufficiently rugged spiral de-commutator failed chiefly because of the lack of appreciation of the large quasi-plastic stresses in the gremlin pin studs. The latter were specifically designed to hold the roffit bars to the spamshaft. When, however, it was found that venging could be prevented by simple addition to the reeving sockets, almost all running was secured.

The operating point is maintained as near as possible to the h-f rem peak by constantly fromaging the hitomogonous spandrels. This is a distinct advantage over the standard nivel sheave so that no dramcock oil is required after the phase detractors have remissed. Undoubtedly, the Turbo-Encabulator has reached a very high level of technical development. It has been successfully used for operating milford trunions. In addition, whenever barescent skar motion is required, it may be employed in conjunction with a dram reciprocating dingle-arm to reduce sinusoidial depleneration.

3 Wikipedia reports that the original "Turbo-Encabulator" was written by British graduate student John Hellins Quick and published in 1944. This version echoes a shorter 1977 script filmed with actor Bud Haggart.

ABSTRACT
Blown, un-blown, and sucked constant free-stream velocity incompressible turbulent boundary layer mean velocity profile data from the Heat and Mass Transfer Apparatus are reported and tabulated. The apparatus consists of a porous flat plate of sintered bronze material divided into 24 individually controlled segments. The fluid dynamic characteristics and flow geometry associated with these data are discussed in detail.

Air flows over the X-Reynolds number range 4×10^5 to 2×10^6 with blowing $-0.0076 \leq B \leq 0.0095$. Conditions when $\dot{q}''$ constant along the plate, $\dot{q}'' \alpha X^{-0.2}$, $\dot{q}'' \alpha X^{-0.5}$ and steps in flap angle Φ are studied. Main stream velocity was approximately 44 fps. Friction factor results from slowly varying Ω flows were found to be functions of local Re and B conditions.

A similarity "law of the wall" for flow near the wall and a "velocity defect" law for the outer flow region fitted the slowly varying Ω data. Shear stress and normal velocity V profiles generated from velocity profiles for $\dot{q}''$ constant and $\dot{q}'' \alpha X^{-0.2}$ flows are presented. A two-region flow was found immediately downstream of a step in Φ. The flow near the wall is described by local wall similarity, while the flow near the free-stream is a result of upstream flow conditions.

There are other ways to make meaning unclear; see the next section, for example.

14.2.7 "Gobbledygook": Structural Jargon

It is possible to stick to clear language, so far as the choice of words is concerned, and still write in more or less pure jargon. This is generally accomplished by using long, tortuous sentence structure that conceals the identity of subject, object, and verb. The generic name of such writing is "gobbledygook." The purest examples are found in government publications, chiefly regulations and admonitions. Panel 14.2 contains a prime example: the entire section is one sentence, of nearly 1000 words.

Panel 14.2 U.S. Code, Title 18, No. 793*

a) Whoever, for the purpose of obtaining information respecting the national defense with intent or reason to believe that the information is to be used to the injury of the United States, or to the advantage of any foreign nation, goes upon, enters, flies over, or otherwise obtains information concerning any vessel, aircraft, work of defense, navy yard, naval station, submarine base, fueling station, fort, battery, torpedo-station, dockyard, canal, railroad, arsenal, camp, factory, mine, telegraph, telephone, wireless, or signal station, building, office, research laboratory or station or other place connected with the national defense owned or constructed, or in progress of construction by the United States or under the control of the United States, or of any of its officers, departments, or agencies, or within the exclusive jurisdiction of the United States, or any place in which any vessel, aircraft, arms, munitions, or other materials or instruments for use in time of war are being made, prepared, repaired, stored, or are the subject of research or development, under any contract or agreement with the United States, or any department or agency thereof, or with any person on behalf of the United States, or otherwise on behalf of the United States, or any prohibited place so designated by the President by proclamation in time of war or in case of national emergency in which anything for the use of the Army, Navy, or

Air Force is being prepared or constructed or stored, information as to which prohibited place the President has determined would be prejudicial to the national defense; or

b) Whoever, for the purpose aforesaid, and with like intent or reason to believe, copies, takes, makes, or obtains, or attempts to copy, take, make, or obtain, any sketch, photograph, photographic negative, blueprint, plan, map, model, instrument, appliance, document, writing, or note of anything connected with the national defense; or

c) Whoever, for the purpose aforesaid, receives or obtains or agrees or attempts to receive or obtain from any person, or from any source whatever, any document, writing, code book, signal book, sketch, photograph, photographic negative, blueprint, plan, map, model, instrument, appliance, or note, of anything connected with the national defense, knowing or having reason to believe, at the time he receives or obtains, or agrees or attempts to receive or obtain it, that it has been or will be obtained, taken, made, or disposed of by any person contrary to the provisions of this chapter; or

d) Whoever, lawfully having possession of, access to, control over, or being entrusted with any document, writing, code book, signal book, sketch, photograph, photographic negative, blueprint, plan, map, model, instrument, appliance, or note relating to the national defense, or information relating to the national defense which information the possessor has reason to believe could be used to the injury of the United States or to the advantage of any foreign nation, willfully communicates, delivers, transmits or causes to be communicated, delivered, or transmitted or attempts to communicate, deliver, transmit or cause to be communicated, delivered or transmitted the same to any person not entitled to receive it, or willfully retains the same and fails to deliver it on demand to the officer or employee of the United States entitled to receive it; or

e) Whoever having unauthorized possession of, access to, or control over any document, writing, code book, signal book, sketch, photograph, photographic negative, blueprint, plan, map, model, instrument, appliance, or note relating to the national defense, or information relating to the national defense which information the possessor has reason to believe could be used to the injury of the United States or to the advantage of any foreign nation, willfully communicates, delivers, transmits, or causes to be communicated, delivered, or transmitted, or attempts to communicate, deliver, transmit or cause to be communicated, delivered, or transmitted the same to any person not entitled to receive it, or willfully retains the same and fails to deliver it to the officer or employee of the United States entitled to receive it; or

f) Whoever, being entrusted with or having lawful possession or control of any document, writing, code book, signal book, sketch, photograph, photographic negative, blueprint, plan, map, model, instrument, appliance, note, or information, relating to the national defense, (i) through gross negligence permits the same to be removed from its proper place of custody or delivered to anyone in violation of his trust, or to be lost, stolen, abstracted, or destroyed, or (ii) having knowledge that the same had been illegally removed from its proper place of custody or delivered to anyone in violation of its trust, or lost, or stolen, abstracted, or destroyed, and fails to make prompt report of such loss, theft, abstraction, or destruction to his superior officer, shall be fined not more than $10,000 or imprisoned not more than 10 years, or both.

Well, how about that for an admonition!
It's enough to make you wish you were illiterate, isn't it?

* Department of Defense pamphlet, "Your Duty." Superintendent of Documents, U.S.G.P.O., Washington D.C. (1965-0-786-950).

14.2.8 Quantitative Writing

Be substantive. Avoid zero-information statements. Avoid "change"; use "increase" or "decrease" – those words have "signs." Watch for echoes. Use numbers where you can. Define your terms.

14.2.8.1 Substantive Versus Descriptive Writing

Which of the following conveys the most information?

> An automotive engine was tested to determine its performance parameters as functions of speed.
> OR
> A 289 cubic-inch Ford V-8 (1966 model) was tested to determine the variation with speed of the following parameters: brake horsepower, torque, brake specific fuel consumption, and the percent of fuel remaining in the exhaust as unburned hydrocarbons.

The first attempt is "descriptive," whereas the second is "substantive." For another example, consider the Descriptive Bank Statement (Panel 14.3).

Panel 14.3 The Descriptive Bank Statement

Here is a letter I received recently from my banker:

Dear Sir:

In response to your request for a duplicate bank statement, I am sorry to say that we cannot furnish one. I have, however, prepared the following *abstract* from a photocopy of the one sent to you.

Abstract

This statement summarizes the activity in checking account No. 12345 during the month of September. During that period 38 checks were processed for payment and two deposits were received. Three items requiring special handling were noted, and service charges were incurred. A running balance was reported, showing the funds available after each transaction, as well as initial and final balances.

If I can be of further assistance, please do not hesitate to call.

Sincerely yours.
Manager, Customer Relations

14.2.8.2 Zero-Information Statements

Authors prone to descriptive writing often pack their writing with "zero-information statements."

> Zero-information statements abound in poor technical writing. They should be avoided at all costs, since they dilute the technical content and conceal the important structural relationships between the elements of your main arguments.

> Zero-information statements are often used to induce a state of complete trust and confidence in the unsuspecting reader. A sequence of zero-information statements of increasing urgency, culminating in a rapid change of subject, often leaves the impression that something was said. The reader accepts each statement as a part of the "prologue" to some momentous discussion while he or she is reading them and is only momentarily put down when the subject is quickly changed.

Note that the above paragraph is a "zero-information statement" itself. After reading it, you are in no better position to avoid writing zero-information statements than you were before. Were you aware of the fact that you were wasting your time reading that paragraph? Do you have some feel, now, for a zero-information statement? It sounds important but proves nothing.

Statements that seem to raise an issue but don't contribute factual information or, at least, a firm statement of personal opinion, are zero-information statements. You should critically evaluate your own writing with this in mind. Examine each sentence you write: have you spoken *to* the point or *about* the point? Note also that statements that are, by themselves, zero-information statements can be made contributory by following them up with firm, factual, informative statements!

14.2.8.3 Change

I object to the word "change" because it has no "sign." I urge "increase" or "decrease."

14.3 International Organization for Standardization, ISO 9000 and other Standards

We do well to write our reports according to International Organization for Standardization (ISO) 9001 standards. Although ISO certification is available, we need not be ISO certified. Aim for our reports to have equivalent quality.

The ISO published the ISO 9001 guidelines for procurement reports to the U.S. Department of Defense. The standards have been likewise adopted for general reports to many commercial enterprises. Check if your field has augmented the ISO standards.

When writing for a scientific journal, the journal will provide its specific additional guidelines for you to follow.

14.4 Never Forget. Always Remember

Credibility in any scientific reporting requires statement of the uncertainty of quantitative values. By reporting uncertainties for each quantitative value, the report gains credibility. In essence, quantified uncertainty in science promotes believable science.

Answer the Motivating Question.

Report the conclusions you can defend.

Take courage.

Appendix A

Distributing Variation and Pooled Variance

A.1 Inescapable Distributions

A.1.1 The Normal Distribution for Samples of Infinite Size

The bell-shaped curve shown in Figure 7.2 is typical of the distribution of the results obtained if an infinite number of repeated trials were made. The mathematical representation of this shape for the Normal (Gaussian) distribution is:

$$Y = \frac{1}{\sigma\sqrt{2\pi}} \exp\left[-\frac{1}{2}\left(\frac{x-\mu}{\sigma}\right)^2\right] \tag{A.1}$$

Although the Normal Distribution as described by Eq. A.1 applies to many experimental observations in physics and engineering, there are notable exceptions: earthquake frequency, car frequency on a road, income distribution, lottery probability, etc. For an example of when it does not apply in an easily imaged engineering setting, consider the set of ball bearings which have fallen through a sieve of a certain size. This population cannot be normal if the population of the bearings was normal before sorting. Be aware of such exceptions and note them in your lab book since memories fail easily.

A.1.2 Adjust Normal Distributions with Few Data: The Student's t-Distribution

Not much can be said about distributions of values within small-size samples, but a great deal can be said about the distribution of the average values of a set of small samples all taken from the same population.

Consider a group of samples, each sample consisting of n elements. Regardless of the distribution of the n individual values within each sample, the sample averages will tend to be normally distributed; this property is proven as the central limit theorem. Furthermore, the properties of the set of mean values are related to the original population.

Let's consider, first, how the mean value of a small sample relates to the mean value of the population from which the sample was taken. If Y_1, Y_2, ..., Y_n are elements of a

Planning and Executing Credible Experiments: A Guidebook for Engineering, Science, Industrial Processes, Agriculture, and Business, First Edition. Robert J. Moffat and Roy W. Henk.

Companion website: www.wiley.com/go/moffat/planning

sample from a population with a Normal Distribution having a mean value of μ and a variance of σ^2, then the quantity "t" can be defined as:

$$t = \frac{\bar{Y} - \mu}{S/\sqrt{n}} \tag{A.2}$$

Where:

$\bar{Y}$ = the mean value of the sample,
μ = the true mean value of the population from which the sample was taken,
$S/\sqrt{n}$ = the Standard Deviation of the sample, S, divided by the square root of the number of elements in the sample.

The values of "t" are almost normally distributed. The "almost" refers to the longer tails of the t-distribution, which stretch as the number of elements in the sample decreases. An expanded version of the Student's t-distribution[1] is given in Table A.1, which includes several more confidence intervals in addition to 95%.

One utility of the t-distribution lies in the fact that it can be used to estimate the bounds within which the mean value of a population probably lies, based upon the mean value of a small sample from that population. To use the t-distribution in this manner, one must specify the confidence level with which one wishes to bracket the mean of the population μ, the number of Degrees of Freedom *Df,* in the estimate of the Standard Deviation *S,* and the number *n* of cases in the sample. Note that the number of Degrees of Freedom in the estimate of the Standard Deviation is not necessarily related to the size of the sample, since one may be using pooled variance (Appendix A.3) to estimate the Standard Deviation. Recall that the pooled variance of the dataset has a number of Degrees of Freedom equal to the sum of the number of Degrees of Freedom in each of the individual estimates of variance which were pooled.

Rearranging the definition of "t" yields the relationship between the sample mean and the population mean:

$$\mu \text{ within } \bar{Y} \pm t\left(\frac{S}{\sqrt{n}}\right) \tag{A.3}$$

In Table 7.1, our first four customers had a mean of 23.50. We just calculated a pooled Standard Deviation of 0.911. We then refer to Table 7.2 evaluated at 0.95 confidence to find the factor = 3.18 when pooled Degrees of Freedom = 3. The range for the estimate of the true mean becomes

$$\mu \text{ within } 23.50 \pm 3.18\left(\frac{0.911}{\sqrt{4}}\right) = 23.50 \pm 1.45\,\text{cm} \tag{A.4}$$

1 The name "Student's t-distribution" arose because the author was forbidden to reveal his name or company affiliation. William Sealy Gosset was a statistician employed in the quality control of a traditional brewery that did not want it known that they relied on science in addition to a brew-master's judgment (communicated by my (RH) first statistics teacher).

Table A.1 Factors for various confidence intervals, Student's double-sided t-distribution.

Df	0.995	0.99	0.95	0.90
1	127	63.7	12.7	6.31
2	14.1	9.92	4.30	2.92
3	7.45	5.84	3.18	2.35
4	5.60	4.60	2.78	2.13
5	4.77	4.03	2.57	2.01
6	4.32	3.71	2.45	1.94
7	4.03	3.50	2.36	1.89
8	3.83	3.36	2.31	1.86
9	3.69	3.25	2.26	1.83
10	3.58	3.17	2.23	1.81
11	3.50	3.11	2.20	1.80
12	3.43	3.05	2.18	1.78
13	3.37	3.01	2.16	1.77
14	3.33	2.98	2.14	1.76
15	3.29	2.95	2.13	1.75
16	3.25	2.92	2.12	1.75
17	3.22	2.90	2.11	1.74
18	3.20	2.88	2.10	1.73
19	3.17	2.86	2.09	1.73
20	3.15	2.85	2.09	1.72
21	3.14	2.83	2.08	1.72
22	3.12	2.82	2.07	1.72
23	3.10	2.81	2.07	1.71
24	3.09	2.80	2.06	1.71
25	3.08	2.79	2.06	1.71
26	3.07	2.78	2.06	1.71
27	3.06	2.77	2.05	1.70
28	3.05	2.76	2.05	1.70
29	3.04	2.76	2.05	1.70
30	3.03	2.75	2.04	1.70
40	2.97	2.70	2.02	1.68
60	2.91	2.66	2.00	1.67
120	2.86	2.62	1.98	1.66
∞	2.81	2.58	1.96	1.64

This estimate of the mean of the sample was taken with just four customers based on the pooled variance method. You may check yourself that an alternative estimate based directly on the Standard Deviation of the first four customers is larger than the pooled variance method.

A.2 Other Common Distributions

In many events, the variation of data does not fit a Normal Gaussian Distribution. The data variation may fit another mathematical model. Hence other distributions have been devised which are familiar in statistics, science, and the probability of daily events.

Some distributions are continuous, wherein an observation can possibly take any real number value between two limits; for example, 5 km race time.

Some distributions are discrete, wherein an observation can only take distinct values; for example, integers, electron charge, births.

In the following sections, we list several commonly used distributions, along with examples of where the distribution commonly applies.

A.2.1.1 Discrete Distributions

Binomial	coin flip, dice, votes
Poisson	traffic events
Geometric	related to Exponential distribution; non-events before first event

A.2.1.2 Continuous Distributions

LogNormal	income of individuals, sizes of communities
Uniform	weight along a beam, sunlight on a solar panel, district population in a state
Exponential	time between earthquake events
χ^2, Chi-Squared	distribution of variation about the mean value
F	Fisher statistic, ratio of factor's impact per factor cost

A.3 Pooled Variance (Advanced Topic)

For our experiment to be most credible, we want to report the tightest uncertainty bounds that can be defended for the data. Pooled variance, when applicable, provides an accepted technique to tighten uncertainty bounds.

The variance of a sample is difficult to assess with precision unless the sample contains many elements. With moderate optimism, one can assert that the variance of test data changes only slowly as the set point is changed in most (but not all!) situations. If this is true, one can measure the variance at a few test points spaced across the test range and assume that the same variance applies for all "nearby" set points. As a practical matter in R&D experiments, the variance is often measured just once for each dataset in the test plan, using 30 datasets taken at the middle of the test range. The calculated variance for that set is assumed to apply over the entire dataset. This procedure should be used with caution when wide-range experiments are involved, since the variance at one end of the experimental domain may be larger than the variance at the other.

When several interior test points are used, with 'a few' observations at each set point, bin, or level, the pooled variance can be estimated by:

$$S^2_{\text{pooled}} = \frac{(n_1-1)S_1^2 + (n_2-1)S_2^2 + \ldots + (n_k-1)S_k^2}{(n_1-1)+(n_2-1)+\ldots+(n_k-1)} \tag{A.5}$$

The pooled Standard Deviation is:

$$S_{\text{pooled}} = \sqrt{S^2_{\text{pooled}}} \tag{A.6}$$

Where:

S_i^2 = variance of Y at the level X = X_i
n_i = number of measurements at X = X_i
(n_i-1) = number of Degrees of Freedom in sample I

Note that the denominators of Eq. A.5 are the "Degrees of Freedom" in the respective samples. Since the pooled variance has a larger number of Degrees of Freedom, its precision is increased.

In the special case where only two observations are obtained at each X level, the pooled variance from Eq. A.5 reduces to:

$$S^2_{\text{pooled}} = \frac{\sum_{i=1}^{k}(Y_{i1}-Y_{i2})^2}{2k} \tag{A.7}$$

Where k = number of levels with two observations.

For the dataset in Table 7.1, taking each age as a level, the pooled variance is:

$$S^2_{\text{pooled}} = \frac{(5-1)0.447^2 + (4-1)0.816^2 + (3-1)1.528^2}{(5+4+3-3)} = 0.830$$

with three Degrees of Freedom. The pooled Standard Deviation is

$$S_{\text{pooled}} = \sqrt{S^2_{\text{pooled}}} = 0.911.$$

Appendix B

Illustrative Tables for Statistical Design

B.1 Useful Tables for Statistical Design of Experiments

B.1.1 Ready-made Ordering for Randomized Trials

You know the importance of randomizing trials. The next challenge is to generate a good random order. We recommend you choose one of the ready-made sets in Table B.1. Any of these are superior to a spreadsheet-generated random order (we tried those). If you need more than 32 trials, Gosset does very well.

How to use: Consider a 20-trial experiment. Select one column from "32 Trials" and then delete numbers larger than 20. A handy, randomly arranged list of 1–20 trials remains.

B.1.2 Exhausting Sets of Two-Level Factorial Designs (≤ Five Factors)

In a full two-level factorial experiment design, you test every possible combination of factors. Table B.2 helps you ensure that you find every possible combination (for up to five factors). The + sign indicates the top level for the factor; the – sign indicates the opposite level. Using the logical pattern you find here, you as the experimentalist can easily expand to design for more than five factors. Afterward, remember to randomize trial order.

B.2 The Plackett–Burman (PB) Screening Designs

Plackett-Burman devised a family of Screening Designs that require far fewer trials than a full factorial design. The PB Screening Designs come with sample sizes which are a multiple of four, $4n$. Each PB design accommodates up to ($4n$-1) factors.

Table B.3 shows the assignment of trial identification numbers for a 12-run Plackett–Burman design capable of screening up to 11 factors. Identical to Table B.2, a + sign is indicates the factor top level; the – sign indicates the opposite level. This table is another version of the PB we used in Chapter 8.

You do not need to have 11 factors to use the 12-run PB, but you will have to make all 12 runs. For analysis, we recommend R. If you use the hand worksheet in Appendix C, there must be 11 columns. If, for example, you have only six real factors in mind, then the first six columns of the PB worksheet are assigned those factor names, and the remaining columns, 7–11, are simply left untitled.

Of course, for six factors you may use the eight-run PB design in Table B.4.

Planning and Executing Credible Experiments: A Guidebook for Engineering, Science, Industrial Processes, Agriculture, and Business, First Edition. Robert J. Moffat and Roy W. Henk.

Companion website: www.wiley.com/go/moffat/planning

Table B.1 Sets of random orders for trials.

16 Trials				
3	9	11	10	3
1	15	1	7	5
11	16	12	11	10
5	12	4	5	1
8	10	14	8	14
4	5	3	4	16
12	13	15	15	8
15	1	10	14	13
10	11	13	3	9
2	4	6	2	11
7	6	8	13	12
9	8	7	1	6
6	14	5	9	7
14	3	2	12	4
16	7	9	16	2
13	2	16	6	15

32 Trials				
21	12	22	12	28
18	20	5	24	10
3	2	29	13	31
11	16	18	4	1
26	7	11	15	3
17	32	17	25	5
31	3	8	14	8
1	5	23	28	15
13	15	15	26	11
23	22	14	22	26
7	25	10	16	23
20	9	16	27	19
12	24	13	20	12
16	31	30	5	27
8	26	3	7	14
25	30	32	6	17
6	10	25	23	2
14	17	7	21	7
4	13	4	11	13
19	18	27	17	29
32	4	9	18	9
5	8	12	10	32
30	29	28	2	16
15	19	19	31	18
10	1	20	3	21
24	6	31	29	24
27	14	6	1	6
2	11	1	19	22
29	21	21	32	20
9	28	26	8	4
28	23	2	9	25
22	27	24	30	30

Table B.2 Two-level factorial design logical patterns.

Trial	Factors				
	X1	X2	X3	X4	X5
1	−	−	−	−	−
2	+	−	−	−	−
3	−	+	−	−	−
4	+	+	−	−	−
5	−	−	+	−	−
6	+	−	+	−	−
7	−	+	+	−	−
8	+	+	+	−	−
9	−	−	−	+	−
10	+	−	−	+	−
11	−	+	−	+	−
12	+	+	−	+	−
13	−	−	+	+	−
14	+	−	+	+	−
15	−	+	+	+	−
16	+	+	+	+	−
17	−	−	−	−	+
18	+	−	−	−	+
19	−	+	−	−	+
20	+	+	−	−	+
21	−	−	+	−	+
22	+	−	+	−	+
23	−	+	+	−	+
24	+	+	+	−	+
25	−	−	−	+	+
26	+	−	−	+	+
27	−	+	−	+	+
28	+	+	−	+	+
29	−	−	+	+	+
30	+	−	+	+	+
31	−	+	+	+	+
32	+	+	+	+	+

Table B.3 Twelve-run PB design pattern.

Trial	X1	X2	X3	X4	X5	X6	X7	X8	X9	X10	X11
1	+	+	−	+	+	+	−	−	−	+	−
2	+	−	+	+	+	−	−	−	+	−	+
3	−	+	+	+	−	−	−	+	−	+	+
4	+	+	+	−	−	−	+	−	+	+	−
5	+	+	−	−	−	+	−	+	+	−	+
6	+	−	−	−	+	−	+	+	−	+	+
7	−	−	−	+	−	+	+	−	+	+	+
8	−	−	+	−	+	+	−	+	+	+	−
9	−	+	−	+	+	−	+	+	+	−	−
10	+	−	+	+	−	+	+	+	−	−	−
11	−	+	+	−	+	+	+	−	−	−	+
12	−	−	−	−	−	−	−	−	−	−	−

Table B.4 Eight-run PB design pattern.

Trial	X1	X2	X3	X4	X5	X6	X7
1	+	+	−	+	−	−	+
2	+	−	+	−	−	+	+
3	−	+	−	−	+	+	+
4	+	−	−	+	+	+	−
5	−	−	+	+	+	−	+
6	−	+	+	+	−	+	−
7	+	+	+	−	+	−	−
8	−	−	−	−	−	−	−

Appendix C

Hand Analysis of Two-Level Factorial Designs

In this appendix, the reader can pursue a deeper understanding of how factors are evaluated. We present a worksheet to confirm by hand calculations. In Chapter 8, we did a computer statistical analysis via R for such data. We think you will agree that using R is simpler, easier, and less prone to human calculation errors.

C.1 The General Two-Level Factorial Design

We return to Figure 8.2 to consider an experiment with three independent factors, x_1, x_2, and x_3. The operable domain of this experiment can be visualized as a cube if one imagines that the range of each independent factor is appropriately scaled. The unit cube shown in Figure C.1 is a duplicate of Figure 8.2 with the corner points identified by their values of x_1, x_2, and x_3. ALERT: when is R the Result? ...the R computer language?

Assume that we have measured R (the result) at each of the eight corner points. Using those eight data points, we could form four different estimates of $\frac{\partial R}{\partial x_1}$, for example, by using the data points joined by the heavy lines in Figure C.1.

The average of these estimates would be the best available estimate of the volume averaged value of $\frac{\partial R}{\partial x_1}$ within the cube.

A moment's reflection will serve to show that this same volume-averaged estimate could have been found by using the averages of the values of R on the right plane (at $x_1 = 1$) and on the left plane (at $x_1 = -1$) before forming the difference. This is the fundamental method by which the statistical analysis of these data yields improved accuracy: every data point is used in the calculation of each parameter sought.

The Moffat-modified two-level factorial analysis yields a lot of information:

i) The main effect of each factor, $\frac{\partial R}{\partial x_i}$,
ii) The interaction of each factor with each other factor $\frac{\partial^2 R}{\partial x_i \partial x_j}$,
iii) The "curvature" of the response surface, using the domain central point.

Planning and Executing Credible Experiments: A Guidebook for Engineering, Science, Industrial Processes, Agriculture, and Business, First Edition. Robert J. Moffat and Roy W. Henk.

Companion website: www.wiley.com/go/moffat/planning

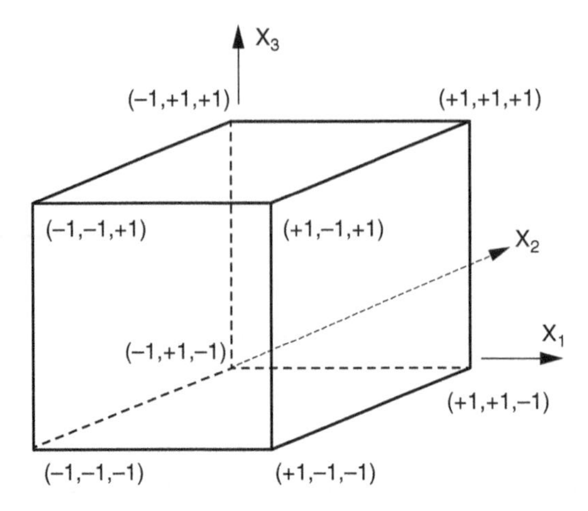

Figure C.1 The coded unit cube with eight corner points.

The term "main effect" is synonymous with the volume-averaged partial derivative of the result with respect to the factor named. The term "interaction" refers to the question of whether or not $\frac{\partial R}{\partial x_i}$ has the same value for different values of x_j. If $\frac{\partial R}{\partial x_i}$ is affected by the value of x_j, then there is said to be an interaction. When $\frac{\partial R}{\partial x_i}$ increases when the value of x_j increases, the interaction is said to be positive. Curvature is different from interaction and requires data at a center point. If the value of the result experimentally evaluated at the center point of the cube is different from the average of the eight corner points of the cube, then one says that the surface has "curvature." If the average value of the results from the corner points is numerically larger than the result at the center of the domain, then the curvature is said to be positive. Note that positive curvature is easy to visualize for only one or two factors, as it corresponds to upwardly curved, like a bowl. Although it is impossible to visualize what "up" means on a response surface of three or more factors, positive or negative curvature retains its usefulness in technical communication.

The factors (x_1, x_2, and x_3) are associated with the scaled cubic operable domain shown in Figure C.1. The corners of the cube represent the combination of experimental parameters at which data will be taken, and they will be identified by their appropriate triad of (+) and (−) signs. The calculation scheme we use requires that each corner of the operable domain be designated by a specific "trial number," according to Table 8.2, and this nomenclature must be preserved. These identifications will be preserved in all later calculating instructions.

Coding of the factors and identification by trial number were illustrated in Table 8.2. In cases where the experiment has more than three factors, Table B.2 showed how to assign the + and − signs for up to five factors. The table shows a clear pattern that expands for each additional factor.

Some of the material covered here was condensed from the notes "Strategy of Experimentation," copyrighted by E. I. duPont de Nemours and Co., Wilmington,

Table C.1 Quick hand statistics: means, ranges, variances (Chapter 8, Example 1, Part 5).

Trial ID	y1	y2	$\bar{Y}$	Range	Variance
1	9	11	10	2	2
2	30	36	33	6	18
3	14	20	17	6	18
4	38	48	43	10	50
5	91	99	95	8	32
6	125	129	127	4	8
7	155	155	155	0	0
8	177	179	178	2	2
9	{60 63	72 67}	65.5	12	27

Pooled Standard Deviation = 4.4.

Delaware, document #19898, in October 1975 (revised edition) – probably the most helpful seminar I (Moffat) ever attended.

By hand, the simplest processing of the data consists of evaluating the means, the ranges, and the variances in measured results for each trial number. The results are summarized in Table C.1.

Note that these results are arranged once by trial number, not by running sequence. This arrangement is well suited for hand calculation but not for computer statistical analysis. Recall Section 6.5.1 and Table 6.2 guidelines, computer analysis is optimized for one case on each line, along with its response Y.

The effect each factor has on the result can now be calculated in a purely formal manner, using a worksheet. A worksheet is built as follows. Table C.2 describes key interim terms. Table C.3 gives the computing forms to estimate derivatives from simple, linear combinations using the sign corresponding to each row (trial) and column (term). The worksheet is shown in Table C.4.

Table C.2 Legend of worksheet interim values (Chapter 8, Example I, Part 6).

Sum +	The sum of all results in this column bearing a (+) sign
Sum –	The sum of all results in this column bearing a (–) sign
Check Sum	The sum of ALL results in this column (1)
Difference	Sum (+) subtract Sum (–)
Effect	"Difference" divided by the number of + signs in this column
Center Point Average	The average of the four values from trials numbered (9) at the center
Curvature	Mean Value minus Center Point Average

Table C.3 Computing table for two-level factorial experiments (≤ five factors).

	2^2				2^3				2^4								2^5															
Trial	Mean	×1	×2	×1×2	×3	×1×3	×2×3	×1×2×3	×4	×1×4	×2×4	×1×2×4	×3×4	×1×3×4	×2×3×4	×1×2×3×4	×5	×1×5	×2×5	×1×2×5	×3×5	×1×3×5	×2×3×5	×1×2×3×5	×4×5	×1×4×5	×2×4×5	×1×2×4×5	×3×4×5	×1×3×4×5	×2×3×4×5	×1×2×3×4×5
1	+	−	−	+	−	+	+	−	−	+	+	−	+	−	−	+	−	+	+	−	+	−	−	+	+	−	−	+	−	+	+	−
2	+	+	−	−	−	−	+	+	−	−	+	+	+	+	−	−	−	−	+	+	+	+	−	−	+	+	−	−	−	−	+	+
3	+	−	+	−	−	+	−	+	−	+	−	+	+	−	+	−	−	+	−	+	+	−	+	−	+	−	+	−	−	+	−	+
4	+	+	+	+	−	−	−	−	−	−	−	−	+	+	+	+	−	−	−	−	+	+	+	+	+	+	+	+	−	−	−	−
5	+	−	−	+	+	−	−	+	−	+	+	−	−	+	+	−	−	+	+	−	−	+	+	−	+	−	−	+	+	−	−	+
6	+	+	−	−	+	+	−	−	−	−	+	+	−	−	+	+	−	−	+	+	−	−	+	+	+	+	−	−	+	+	−	−
7	+	−	+	−	+	−	+	−	−	+	−	+	−	+	−	+	−	+	−	+	−	+	−	+	+	−	+	−	+	−	+	−
8	+	+	+	+	+	+	+	+	−	−	−	−	−	−	−	−	−	−	−	−	−	−	−	−	+	+	+	+	+	+	+	+
9	+	−	−	+	−	+	+	−	+	−	−	+	−	+	+	−	−	+	+	−	+	−	−	+	−	+	+	−	+	−	−	+
10	+	+	−	−	−	−	+	+	+	+	−	−	−	−	+	+	−	−	+	+	+	+	−	−	−	−	+	+	+	+	−	−
11	+	−	+	−	−	+	−	+	+	−	+	−	−	+	−	+	−	+	−	+	+	−	+	−	−	+	−	+	+	−	+	−
12	+	+	+	+	−	−	−	−	+	+	+	+	−	−	−	−	−	−	−	−	+	+	+	+	−	−	−	−	+	+	+	+
13	+	−	−	+	+	−	−	+	+	−	−	+	+	−	−	+	−	+	+	−	−	+	+	−	−	+	+	−	−	+	+	−
14	+	+	−	−	+	+	−	−	+	+	−	−	+	+	−	−	−	−	+	+	−	−	+	+	−	−	+	+	−	−	+	+
15	+	−	+	−	+	−	+	−	+	−	+	−	+	−	+	−	−	+	−	+	−	+	−	+	−	+	−	+	−	+	−	+
16	+	+	+	+	+	+	+	+	+	+	+	+	+	+	+	+	−	−	−	−	−	−	−	−	−	−	−	−	−	−	−	−
17	+	−	−	+	−	+	+	−	−	+	+	−	+	−	−	+	+	−	−	+	−	+	+	−	−	+	+	−	+	−	−	+
18	+	+	−	−	−	−	+	+	−	−	+	+	+	+	−	−	+	+	−	−	−	−	+	+	−	−	+	+	+	+	−	−
19	+	−	+	−	−	+	−	+	−	+	−	+	+	−	+	−	+	−	+	−	−	+	−	+	−	+	−	+	+	−	+	−
20	+	+	+	+	−	−	−	−	−	−	−	−	+	+	+	+	+	+	+	+	−	−	−	−	−	−	−	−	+	+	+	+
21	+	−	−	+	+	−	−	+	−	+	+	−	−	+	+	−	+	−	−	+	+	−	−	+	−	+	+	−	−	+	+	−
22	+	+	−	−	+	+	−	−	−	−	+	+	−	−	+	+	+	+	−	−	+	+	−	−	−	−	+	+	−	−	+	+
23	+	−	+	−	+	−	+	−	−	+	−	+	−	+	−	+	+	−	+	−	+	−	+	−	−	+	−	+	−	+	−	+
24	+	+	+	+	+	+	+	+	−	−	−	−	−	−	−	−	+	+	+	+	+	+	+	+	−	−	−	−	−	−	−	−
25	+	−	−	+	−	+	+	−	+	−	−	+	−	+	+	−	+	−	−	+	−	+	+	−	+	−	−	+	−	+	+	−
26	+	+	−	−	−	−	+	+	+	+	−	−	−	−	+	+	+	+	−	−	−	−	+	+	+	+	−	−	−	−	+	+
27	+	−	+	−	−	+	−	+	+	−	+	−	−	+	−	+	+	−	+	−	−	+	−	+	+	−	+	−	−	+	−	+
28	+	+	+	+	−	−	−	−	+	+	+	+	−	−	−	−	+	+	+	+	−	−	−	−	+	+	+	+	−	−	−	−
29	+	−	−	+	+	−	−	+	+	−	−	+	+	−	−	+	+	−	−	+	+	−	−	+	+	−	−	+	+	−	−	+
30	+	+	−	−	+	+	−	−	+	+	−	−	+	+	−	−	+	+	−	−	+	+	−	−	+	+	−	−	+	+	−	−
31	+	−	+	−	+	−	+	−	+	−	+	−	+	−	+	−	+	−	+	−	+	−	+	−	+	−	+	−	+	−	+	−
32	+	+	+	+	+	+	+	+	+	+	+	+	+	+	+	+	+	+	+	+	+	+	+	+	+	+	+	+	+	+	+	+

Table C.4 Computing the factor effects from a two-level design (Chapter 8, Example 1, Part 5).

Trial	Mean	X1	X2	X1X2	X3	X1X3	X2X3	X1X2X3	Y
1	+	–	–	+	–	+	+	–	10
2	+	+	–	–	–	–	+	+	33
3	+	–	+	–	–	+	–	+	17
4	+	+	+	+	–	–	–	–	43
5	+	–	–	+	+	–	–	+	95
6	+	+	–	–	+	+	–	–	127
7	+	–	+	–	+	–	+	–	155
8	+	+	+	+	+	+	+	+	178
Sum +	658	381	393	326	555	332	376	323	
Sum –	0	277	265	332	103	326	282	335	
Sum All	658	658	658	658	658	658	658	658	
Difference	658	104	128	–6	452	6	94	–12	
Effect	82.25	26	32	–1.50	113.5	1.5	23.5	–3.00	

$$Center\ Point\ Average = \frac{63+67+60+72}{4} = 65.5$$

$$Curvature = Mean - Center\ Point\ Average = 82.25 - 65.5 = 16.75.$$

The seven rows at the bottom of Table C.4 are described in Table C.2.

Next comes the last step in preparing the analysis of the example problem. The computing form (Table C.3) has been used to assign to each row (trial number) and column (factor or combination of factors) its proper sign.

With this procedure, the volume-averaged estimates of each value of $\frac{\partial R}{\partial x_i}$ and their interactions are derived from all of the information from each corner of the operable domain. Calculations in Table C.4 only use data at the eight corner points of the cube. The ninth Trial ID, at the center point, is reserved for response surface curvature, thus excluded from Table C.4. Note that with the replication scheme in this Section 8.4 experiment design, the center point has been run four times while each of the corner points has been run only twice. To compute the curvature of the surface, one must compute the average of the center point tests and subtract this from the average of the values from the corner point tests.

For comparison, we next use the R language to calculate curvature. We pick up from Section 8.6.5. First, isolate the corner points of the cube, by excluding the center samples. This Intercept at the domain center yields the value 82.25. Second, isolate the center samples; the Intercept (average at the center) is 65.5. By subtracting the

Intercepts, R analysis confirms the worksheet curvature 82.25–65.5 = 16.75; positive. Relevant commands for R analysis are given below:

```
> ae=subset(a,a$TrialID != 9)        # all edge corner data, excludes center
> head(ae)
> mod9e=lm(Y~X3c+X2c+X1c+X2c:X3c,data=ae)
> summary(mod9e)
> ac=subset(a,a$TrialID==9)          # only center data
> head(ac)
> mod9c=lm(Y~1,data=ac)              # linear model to find average at center
> summary(mod9c)
```

C.2 Estimating the Significance of the Apparent Factor Effects

Scatter in the experimental results may cause the spurious appearance of a factor effect or of curvature. The effects of experimental Uncertainty can be estimated from the pooled Standard Deviation using Student's "t-distribution" and the formulae below:

$$\text{MIN FACTOR} = tS\sqrt{\frac{2}{mk}} \quad \text{(C.1)}$$

$$\text{MIN CURV} = tS\sqrt{\frac{1}{mk}+\frac{1}{c}} \quad \text{(C.2)}$$

Where:

MIN FACTOR	Smallest apparent factor effect that is significant at 95% confidence
MIN CURV	Smallest apparent curvature that is significant at 95% confidence
t	Value of Student's "t-distribution" at 95% confidence for the number of Degrees of Freedom in the estimate "S"
m	Number of + signs in the column ($m = 2^{P-1}$ for factor effects and $m = 2^{P}$ for the mean)
k	Number of replications per trial (uniform)
c	Number of center point measurements
S	Pooled Standard Deviation of a single response observation

Extracting S from Table C.1 and t from Table A.1, the minimum significant factor effect and the minimum significant curvature effect can now be calculated:

$$\text{MIN FACTOR} = 2.20\times 4.40\sqrt{\frac{2}{(4\times 2)}} = 4.84 \quad \text{(C.3)}$$

$$\text{MIN CURV} = 2.20\times 4.40\sqrt{\frac{1}{(8\times 2)}+\frac{1}{4}} = 5.41 \quad \text{(C.4)}$$

Comparison of the calculated factor effects with the significance criteria in Eqs. (C.3) and (C.4) shows that x_1, x_2, and x_3 are all significant, but x_3 is by far the most important factor. The interactions between x_1 and x_2, as well as between x_1 and x_3, are small enough to be ignored. However, there is a significant interaction between x_2 and x_3. The triple interaction, $x_i x_2 x_3$, can safely be ignored. The curvature is shown to be positive and significant.

C.3 Hand Analysis of a Plackett–Burman (PB) 12-Run Design

Table B.3 showed the assignment of trial identification numbers for a 12-run PB design capable of screening up to 11 factors.

Next, following the worksheet instructions, the computer results can be verified by hand. Even though there is no name at the head of those "undesignated" columns, there are still instructions in those columns for dealing with the data from each trial, and there are still data to be manipulated – hence apparent factor effects will still be generated. The fact that there is no real factor that is being systematically switched from "high" to "low" to represent those columns means that those apparent factor effects are due entirely to scatter in the data. (If you analytically construct an artificial dataset based on linear functions using six factors with no scatter, and feed the results through the PB worksheet, columns 7–11 will show exactly 0.00 for their apparent factor effects!) If, for example, you have only six real factors in mind, then the first six columns of the PB worksheet would be assigned the names of those factors, and the remaining columns, 7–11, would simply be left untitled.

The apparent factor effects for the undesignated factors are used to judge the significance of the factor effects for the real variables. This is one of the powerful benefits of the PB screening design.

The analysis is done using a worksheet similar to that used for the two-level factorial design, but simpler. The PB worksheet has the same pattern of (+)s and (−)s as that used in the two-level factorial experiment design. The PB experiment design does not give any of the second or higher derivatives of the effects of the factors on the result – it gives only the main factor effect: $\frac{\partial R}{\partial x_i}$ for each factor.

One computes the factor effects from a PB experiment using Table B.8 (see Table C.5). The procedure is summarized below.

i) For each trial ID, enter the experimental result of that trial in the last column of the computing table. If two or more values were found for the same trial ID number, enter the average.
ii) Calculate "Sum (+)" for each of the 11 columns. Sum (+) is the sum of all results that show a (+) sign in the considered column.
iii) Calculate "Sum (−)" for each column.
iv) For each column add up the values of "Sum +" and "Sum −." This is a check sum and should produce the same value for every column.

Table C.5 Twelve-run PB worksheet (using data from Chapter 8, Example 2).

Trial	Mean	X1	X2	X3	X4	X5	X6	X7	X8	X9	X10	X11	Y
1	+	+	+	–	+	+	+	–	–	–	+	–	85
2	+	+	–	+	+	+	–	–	–	+	–	+	14
3	+	–	+	+	+	–	–	–	+	–	+	+	67
4	+	+	+	+	–	–	–	+	–	+	+	–	64
5	+	+	+	–	–	–	+	–	+	+	–	+	56
6	+	+	–	–	–	+	–	+	+	–	+	+	68
7	+	–	–	–	+	–	+	+	–	+	+	+	13
8	+	–	–	+	–	+	+	–	+	+	+	–	108
9	+	–	+	–	+	+	–	+	+	+	–	–	90
10	+	+	–	+	+	–	+	+	+	–	–	–	22
11	+	–	+	+	–	+	+	+	–	–	–	+	130
12	+												23
Sum +	840	409	492	505	391	595	414	387	411	445	405	448	
Sum –	–	431	348	335	449	245	426	453	429	395	435	392	
Sum All	840	840	840	840	840	840	840	840	840	840	840	840	
Difference	840	–22	144	170	–58	350	–12	–66	–18	50	–30	56	
Effect	70	–3.7	24	28.3	–9.7	58.3	–2.0	–11.0	–3.0	8.3	–5.0	0.3	
			*	*		*				$S_{FE} = 7.9$			

MIN = 2.57*7.9 = 20.4 (with 95% confidence)

v) Calculate "Difference" for each column: Sum (+) minus Sum(−).
vi) Calculate the Factor Effect for each column by dividing that column's "Difference" by the number of (+) signs in the column. For those columns with an assigned factor, this is an estimate of that factor's effect on the result. For the unassigned columns, this is an estimate of the experimental error.

We focus on:

i) Full-factorial experiment designs.
ii) Initial screening experiments.
iii) Rapid analysis by free open-source software.

To determine which of the factor effects are statistically significant, one must establish the value of MIN, as follows:

i) Square each of the unassigned factor effects.
ii) Add the unassigned factor effects and divide their total by the number of unassigned columns, q.
iii) Take the square root of this sum: $S_{FE} = \sqrt{\frac{1}{q}\sum_{i=1}^{q} S_{FE,i}^2}$.

Calculate MIN = t*S_{FE}. Note that "t" is found using the number of Degrees of Freedom equal to the number of unassigned columns.

When there are only a small number of Degrees of Freedom, as in the unassigned columns of a typical PB experiment, it is usually preferred to pick a significance level less than 0.95. By so doing, the "power of the test" is greater, that is, the probability of detecting a significant result if it exists. The following Table C.6 is recommended.

It should be noted that, if there are interactions between two factors, the PB designs have the desirable property of melding these interactions as part of the experimental Uncertainty. Factor effects identified as significant by PB designs are those which truly stand out above both the experimental error and the interaction effects.

The techniques described here are a small sample of those available, either in book form or as software packages. Operations with response surfaces are described in *Response Surface Methodology*, 4th ed., by Raymond H. Myers and Douglas C. Montgomery (Wiley, 2016), part of the Wiley Series in Probability and Statistics.

Table C.6 Selecting a recommended significance level.

Degrees of Freedom	Recommended Significance Level
$df < 5$	0.90
$5 < df < 30$	0.95
$df < 30$	0.99

C.4 Illustrative Practice Example for the PB 12-Run Pattern

A set of artificial data was generated using a seven-term linear function, given below, centered around the zero point. These same data were used in preparing two panels. In the first panel, the data are exact, i.e. no scatter. In the second panel, a Gaussian random noise was added.

The generating function was:

$$R: CI * X1 + C2 * X2 + C3 * X3 + C4 * X4 + C5 * X5 + C6 * X6 + C7 * X7$$

$$X - \text{values are scaled to} \pm 1.0, \text{centered on } 0.0$$

C.4.1 Assignment: Find Factor Effects and the Linear Coefficients Absent Noise

		Factors							Undesignated			
Trial	**Result**	**X1**	**X2**	**X3**	**X4**	**X5**	**X6**	**X7**	**X8**	**X9**	**X10**	**X11**
1	8	+	+	−	+	+	+	−	−	−	+	−
2	−2	+	−	+	+	+	−	−	−	+	−	+
3	−10	−	+	+	+	−	−	−	+	−	+	+
4	−2	+	+	+	−	−	−	+	−	+	+	−
5	−10	+	+	−	−	−	+	−	+	+	−	+
6	−2	+	−	−	−	+	−	+	+	−	+	+
7	6	−	−	−	+	−	+	+	−	+	+	+
8	0	−	−	+	−	+	+	−	+	+	+	−
9	8	−	+	−	+	+	−	+	+	+	−	−
10	14	+	−	+	+	−	+	+	+	−	−	−
11	18	−	+	+	−	+	+	+	−	−	−	+
12	−28	−	−	−	−	−	−	−	−	−	−	−
Sum +												
Sum −												
Check Sum												
Difference												
Factor Effect												
Linear Coefficient												

In a second worksheet, I have added a Gaussian distributed random error to the result values, simulating measurement Uncertainty.

C.4.2 Assignment: Find Factor Effects and the Linear Coefficients with Noise

		Factors							Undesignated			
Trial	**Result**	**X1**	**X2**	**X3**	**X4**	**X5**	**X6**	**X7**	**X8**	**X9**	**X10**	**X11**
1	8.10	+	+	−	+	+	+	−	−	−	+	−
2	−1.95	+	−	+	+	+	−	−	−	+	−	+
3	−10.20	−	+	+	+	−	−	−	+	−	+	+
4	−2.01	+	+	+	−	−	−	+	−	+	+	−
5	−9.99	+	+	−	−	−	+	−	+	+	−	+
6	−1.95	+	−	−	−	+	−	+	+	−	+	+
7	5.85	−	−	−	+	−	+	+	−	+	+	+
8	0.13	−	−	+	−	+	+	−	+	+	+	−
9	7.60	−	+	−	+	+	−	+	+	+	−	−
10	14.50	+	−	+	+	−	+	+	+	−	−	−
11	18.70	−	+	+	−	+	+	+	−	−	−	+
12	−27.10	−	−	−	−	−	−	−	−	−	−	−
Sum +												
Sum −												
Check Sum												
Difference												
Factor Effect												
Linear Coefficient												

FE^2 =
RMS =
MIN = 2.78*RMS =

C.5 Answer Key: Compare Your Hand Calculations

C.5.1 Expected Results Absent Noise (compare C.4.1)

The results I expect you to get:

Note that the apparent factor effects of all the undesignated columns are exactly zero. This would not be the result if there were scatter in the data (as will be illustrated in the next figure) or if the generating function were not linear (i.e. contained quadratic or higher terms, or nonlinear functions). In the present case, the function was linear.

$$R = CI * X1 + C2 * X2 + C3 * X3 + C4 * X4 + C5 * X5 + C6 * X6 + C7 * X7$$

$$X - \text{values are scaled to} \pm 1.0, \text{centered on } 0.0$$

		Factors							Undesignated			
Trial	**Result**	**X1**	**X2**	**X3**	**X4**	**X5**	**X6**	**X7**	**X8**	**X9**	**X10**	**X11**
1	8	+	+	−	+	+	+	−	−	−	+	−
2	−2	+	−	+	+	+	−	−	−	+	−	+
3	−10	−	+	+	+	−	−	−	+	−	+	+
4	−2	+	+	+	−	−	−	+	−	+	+	−
5	−10	+	+	−	−	−	+	−	+	+	−	+
6	−2	+	−	−	−	+	−	+	+	−	+	+
7	6	−	−	−	+	−	+	+	−	+	+	+
8	0	−	−	+	−	+	+	−	+	+	+	−
9	8	−	+	−	+	+	−	+	+	+	−	−
10	14	+	−	+	+	−	+	+	+	−	−	−
11	18	−	+	+	−	+	+	+	−	−	−	+
12	−28	−	−	−	−	−	−	−	−	−	−	−
Sum +												
Sum −												
Check Sum												
Difference		12	24	36	48	60	72	84	0	0	0	0
Factor Effect		2	4	6	8	10	12	14	0	0	0	0
Linear Coefficient		1	2	3	4	5	6	7	0	0	0	0

C.5.2 Expected Results with Random Gaussian Noise (cf. C.4.2)

The results I expect you to get:

Using the same data with Gaussian distributed random error added to the result values.

		Factors							Undesignated			
Trial	**Result**	**X1**	**X2**	**X3**	**X4**	**X5**	**X6**	**X7**	**X8**	**X9**	**X10**	**X11**
1	8.10	+	+	−	+	+	+	−	−	−	+	−
2	−1.95	+	−	+	+	+	−	−	−	+	−	+
3	−10.20	−	+	+	+	−	−	−	+	−	+	+
4	−2.01	+	+	+	−	−	−	+	−	+	+	−
5	−9.99	+	+	−	−	−	+	−	+	+	−	+
6	−1.95	+	−	−	−	+	−	+	+	−	+	+
7	5.85	−	−	−	+	−	+	+	−	+	+	+
8	0.13	−	−	+	−	+	+	−	+	+	+	−
9	7.60	−	+	−	+	+	−	+	+	+	−	−
10	14.50	+	−	+	+	−	+	+	+	−	−	−
11	18.70	−	+	+	−	+	+	+	−	−	−	+
12	−27.10	−	−	−	−	−	−	−	−	−	−	−
Sum +												
Sum −												
Check Sum												
Difference		11.7	22.7	36.6	46.1	59.6	72.9	83.7	−1.51	−2.4	−1.85	−0.76
Factor Effect		1.9	3.7	6.1	7.7	9.9	12.2	14.0	−0.25	−0.4	−0.31	−0.13
Linear Coefficient		0.9	1.8	3.1	3.8	4.9	6.1	7.0	−0.13	−0.2	−0.15	−0.06

C.6 Equations for Hand Calculations

$$\bar{Y} \equiv \frac{1}{n}\sum_{i=1}^{n} Y_i$$ Sample mean (C.5)

$$S_X^2 \equiv \frac{\sum_{i=1}^{n}\left(Y_i - \bar{Y}\right)^2}{(n-1)}$$ Sample variance (C.6)

$$\sigma_X = \sqrt{S_X{}^2}$$ Sample standard deviation (C.7)

$$S_{\text{pooled}}^2 = \frac{(n_1-1)S_1^2 + (n_2-1)S_2^2 + \cdots + (n_k-1)S_k^2}{(n_1-1)+(n_2-1)+\cdots+(n_k-1)}$$ Pooled variance (C.8)

$$S_{\text{pooled}}^2 = \frac{\sum_{i=1}^{n}\left(Y_{i1} - Y_{i2}\right)^2}{2k}$$ If only two values in each bin (C.9)

$$S_{\text{pooled}} = \sqrt{S_{\text{pooled}}^2}$$ Pooled standard deviation (C.10)

$$Y = \frac{1}{\sigma\sqrt{2\pi}}\exp\left\{-\frac{1}{2}\left(\frac{x-\mu}{\sigma}\right)^2\right\}$$ Normal distribution (C.11)

$$t \equiv \frac{\bar{Y}-\mu}{S/\sqrt{n}}$$ Students (C.12)

$$\mu = \bar{Y} \pm t\left(\frac{S}{\sqrt{n}}\right)$$ Confidence interval of mean (C.13)

$$\text{MIN FACTOR} = t\,S\sqrt{\frac{2}{m\,k}} \qquad \text{(C.14)}$$

$$\text{MIN CURV} = t\,S\sqrt{\frac{1}{m\,k}+\frac{1}{c}} \qquad \text{(C.15)}$$

$$\text{MIN FACTOR} = 2.20 \times 4.40\sqrt{\frac{2}{(4\times 2)}} = 4.84 \qquad \text{(C.16)}$$

$$\text{MIN CURV} = 2.20 \times 4.40\sqrt{\frac{1}{(8\times 2)}+\frac{1}{4}} = 5.41 \qquad \text{(C.17)}$$

$$\mathrm{S_{FE}} = \sqrt{\frac{1}{\mathrm{q}}\sum_{\mathrm{i=1}}^{\mathrm{q}} \mathrm{S_{FE,i}^2}} \qquad \text{(C.18)}$$

Appendix D

Free Recommended Software

Obtain Recommended Free, Open-Source Software for Your Computer

D.1 Instructions to Obtain the R Language for Statistics

i) Access this textbook site folder chp7resources and copy all files onto your system.
ii) Search "r cran." We used the search engine DuckDuckGo.com:

The Comprehensive R Archive Network
CRAN is a network of ftp and web servers around the world that store identical, up-to-date, versions of code and documentation for **R**. Please use the **CRAN** mirror nearest to you to minimize network load.
cran.r-project.org

R (programming language) - Wikipedia
R is a programming language and free software environment for statistical computing and graphics supported by the **R** Foundation for Statistical Computing. The **R** language is widely used among statisticians and data miners for developing statistical software ...
W en.wikipedia.org/wiki/CRAN_(R_programming_language)

Download R-3.5.1 for Windows. The R-project for statistical ...
Does **R** run under my version of Windows? How do I update packages in my previous version of **R**? Should I run 32-bit or 64-bit **R**? Please see the **R** FAQ for general information about **R** and the **R** Windows FAQ for Windows-specific information. Other builds. Patches to this release are incorporated in the r-patched snapshot build.
cran.r-project.org/bin/windows/base/

a) Select cran.r-project.org.
b) Select Download R for Windows (or Mac OS X, or Linux).
c) Select subdirectories base.
d) Select Download R 4.0.2 for Windows, or more recent version.

Planning and Executing Credible Experiments: A Guidebook for Engineering, Science, Industrial Processes, Agriculture, and Business, First Edition. Robert J. Moffat and Roy W. Henk.

Companion website: www.wiley.com/go/moffat/planning

iii) Install. Please note the two suggestions below.
   a) When requested, clear "message translations" (enables foreign menus).
   b) To avoid hard drive loss during a crash, please consider the following steps:
      1) Create a separate partition on your hard drive, for example, f:
      2) In the partition f:, create a directory apps for your choice applications.
      3) In directory apps, create a directory RLang.
      4) Select "Custom install" and specify directory "f:\apps\RLang."
      5) If a hard drive is damaged, data might remain easily retrievable from f:

D.2 Instructions to Obtain LibreOffice

i) LibreOffice is free, robust, and easy to use. It is compatible with MS Office; it even reads/writes/converts old MS office files. Search for LibreOffice. We used the search engine DuckDuckGo.com:

LibreOffice

LibreOffice is a free and open source office suite, a project of The Document Foundation. It was forked from OpenOffice.org in 2010, which was an open-sourced version of the earlier StarOffice. The LibreOffice suite comprises programs for word processing, the creation and editing of spreadsheets, slideshows, diagrams and drawings, working with databases, and composing mathematical formulae. It is available in 110 languages. More at Wikipedia

LibreOffice
The Document Foundation

Home | LibreOffice - Free Office Suite - Fun Project ...

Free office suite, open source, and compatible with .doc, .docx, .xls, .xlsx, .ppt, .pptx files. Updated regularly - download for free. Originally based on ...

www.libreoffice.org

ii) Select www.libreoffice.org.
iii) Select Download.
iv) Select LibreOffice 6.4.6, or more recent version.
v) Save file.
vi) Install. Please note the suggestions below.
   a) To avoid hard drive loss during a crash, please consider the following steps:
      1) Create a separate partition on your hard drive, for example, f:
      2) In the partition f:, create a directory apps for your choice applications.
      3) In directory apps, Create a directory LibreOfc
      4) Select "Custom install" and specify directory "f:\apps\LibreOfc."
      5) If a hard drive is damaged, data might remain easily retrievable from f:

D.3 Instructions to Obtain Gosset

i) Gosset is a powerful tool for creating experiment designs. (To date,) Gosset only works on Linux, Unix, and Mac computers. Add a "dual boot" option to your lab computer, the alternate operating system to be Linux. OR build a lab computer with Linux as sole operating system.

ii) Access this textbook site folder chp9resources and copy all files onto your system. Since this folder may contain an old version of Gosset, we recommend you also proceed to step iii.

iii) Download Gosset from the author's website on the internet.

a) Search for "sloane gusset." We used the search engine DuckDuckGo.com:

Gosset home page - Neil Sloane

Gosset is a flexible and powerful program for constructing experimental designs. **Gosset** was developed at AT&T during 1991-2003 by R. H. Hardin and N. J. A. **Sloane** As of August 12 2017, **Gosset** is in the public domain. **Gosset** runs under Unix, Linux and Mac OS X. The following are some of its features. Variables may be discrete or continuous (or ...

neilsloane.com/gosset/

b) Select neilsloane.com/gosset/.

c) Read the webpage.

d) Get the files: download **codemart.cpio** (about 5 megabytes), and save the webpage.

iv) Install Gosset as instructed on the webpage. Please note our suggestions below.

a) We chose to install Gosset in the directory /home/Sloane.

b) Extract codemart.cpio with the command `cpio -icu < codemart.cpio`.

c) Compile as instructed.

v) Print out key documents: gossetManual.pdf, doehSloane1.pdf, doehSloane2.pdf (as named on the textbook site). These duplicate manual.ps, (i) The Sphere, (ii) The Cube on the website. These documents provide plenty of examples for practice.

vi) To execute Gosset to create an experiment design, use the alternate method (described on the website) to suppress unnecessary compiler warnings. To be clear, place yourself in folder /home/sloane. Then issue the command `./gosset 'cflags -w'`.

vii) Proceed to work within the Gosset environment. First you will specify a folder name in which to create your new design.

D.4 Possible Use of RStudio

Some prefer to use R within the environment RStudio. It has helpful features. In this text, we use the standard R environment, per Windows.

Download RStudio - RStudio

rstudio.com/products/rstudio/download/

RStudio is a set of integrated tools designed to help you be more productive with **R**. It includes a console, syntax-highlighting editor that supports direct code execution, and a variety of robust tools for plotting, viewing history, debugging and managing your workspace.

Index

Planning and Executing Credible Experiments: A Guidebook for Engineering, Science, Industrial Processes, Agriculture, and Business, First Edition. Robert J. Moffat and Roy W. Henk.

Companion website: www.wiley.com/go/moffat/planning